Engineering Creative Design in Robotics and Mechatronics

Maki K. Habib
The American University in Cairo, Egypt

J. Paulo Davim
University of Aveiro, Portugal

A volume in the Advances in
Mechatronics and Mechanical Engineering
(AMME) Book Series

ENGINEERING
SCIENCE REFERENCE
An Imprint of IGI Global

Managing Director:	Lindsay Johnston
Editorial Director:	Joel Gamon
Production Manager:	Jennifer Yoder
Publishing Systems Analyst:	Adrienne Freeland
Development Editor:	Christine Smith
Assistant Acquisitions Editor:	Kayla Wolfe
Typesetter:	Christina Henning
Cover Design:	Jason Mull

Published in the United States of America by
Engineering Science Reference (an imprint of IGI Global)
701 E. Chocolate Avenue
Hershey PA 17033
Tel: 717-533-8845
Fax: 717-533-8661
E-mail: cust@igi-global.com
Web site: http://www.igi-global.com

Library of Congress Cataloging-in-Publication Data

Engineering creative design in robotics and mechatronics / Maki K. Habib and J. Paulo Davim, editors.
      pages cm
  Includes bibliographical references and index.
  Summary: "This book captures the latest research developments in the subject field of robotics and mechatronics and provides relevant theoretical knowledge in this field"-- Provided by publisher.
    ISBN 978-1-4666-4225-6 (hardcover) -- ISBN 978-1-4666-4227-0 (print & perpetual access) -- ISBN 978-1-4666-4226-3 (ebook) 1. Robotics. 2. Mechatronics. I. Habib, Maki K., 1955- editor of compilation. II. Davim, J. Paulo, editor of compilation.
  TJ211.E56 2013
  629.8'92--dc23
                                          2013014069

This book is published in the IGI Global book series Advances in Mechatronics and Mechanical Engineering (AMME) Book Series (ISSN: Pending; eISSN: pending)

British Cataloguing in Publication Data
A Cataloguing in Publication record for this book is available from the British Library.

Advances in Mechatronics and Mechanical Engineering (AMME) Book Series

J. Paulo Davim
University of Aveiro, Portugal

ISSN: Pending
EISSN: Pending

MISSION

With its aid in the creation of smartphones, cars, medical imaging devices, and manufacturing tools, the mechatronics engineering field is in high demand. Mechatronics aims to combine the principles of mechanical, computer, and electrical engineering together to bridge the gap of communication between the different disciplines.

The **Advances in Mechatronics and Mechanical Engineering (AMME) Book Series** provides innovative research and practical developments in the field of mechatronics and mechanical engineering. This series covers a wide variety of application areas in electrical engineering, mechanical engineering, computer and software engineering; essential for academics, practitioners, researchers, and industry leaders.

COVERAGE

- Bioengineering Materials
- Biologically Inspired Robotics
- Computational Mechanics
- Computer-Based Manufacturing
- Control Systems Modelling and Analysis
- Intelligent Navigation
- Manufacturing Methodologies
- Mechanisms and Machines
- Nanomaterials and Nanomanufacturing
- Tribology and Surface Engineering

IGI Global is currently accepting manuscripts for publication within this series. To submit a proposal for a volume in this series, please contact our Acquisition Editors at Acquisitions@igi-global.com or visit: http://www.igi-global.com/publish/.

Titles in this Series

For a list of additional titles in this series, please visit: www.igi-global.com

Engineering Creative Design in Robotics and Mechatronics
Maki K. Habib (The American University in Cairo, Egypt) and J. Paulo Davim (University of Aveiro, Portugal)
Engineering Science Reference • copyright 2013 • 366pp • H/C (ISBN: 9781466642256) • US $195.00 (our price)

Computational Methods for Optimizing Manufacturing Technology Models and Techniques
J. Paulo Davim (University of Aveiro, Portugal)
Engineering Science Reference • copyright 2012 • 395pp • H/C (ISBN: 9781466601284) • US $195.00 (our price)

DISSEMINATOR of KNOWLEDGE
www.igi-global.com
701 E. Chocolate Ave., Hershey, PA 17033
Order online at www.igi-global.com or call 717-533-8845 x100
To place a standing order for titles released in this series, contact: cust@igi-global.com
Mon-Fri 8:00 am - 5:00 pm (est) or fax 24 hours a day 717-533-8661

Editorial Advisory Board

Table of Contents

Section 3

Section 4

Detailed Table of Contents

Section 1

Chapter 1

Ken Saito, Nihon University, Japan
Minami Takato, Nihon University, Japan
Yoshifumi Sekine, Nihon University, Japan
Fumio Uchikoba, Nihon University, Japan

Insect type 4.0, 2.7, 2.5 mm. width, length, height size silicon micro-robot system with active hardware neural networks locomotion controlling system is presented in this chapter. The micro-robot system was made from a silicon wafer fabricated by Micro-Electro Mechanical Systems (MEMS) technology. The mechanical system of the robot equipped with millimeter-size rotary type actuators, link mechanisms, and six legs to realize the insect-like switching behavior. In addition, the authors constructed the active hardware neural networks by analog CMOS circuits as a locomotion controlling system. Hardware neural networks consisted of pulse-type hardware neuron models as basic components. Pulse-type hardware neuron model has same basic features of biological neurons such as threshold, refractory period, spatio-temporal summation characteristics, and enables the generation of continuous action potentials. The hardware neural networks output the driving pulses using synchronization phenomena such as biological neural networks. Four output signal ports are extracted from hardware neural networks, and they are connected to the actuators. The driving pulses can operate the actuators of silicon micro-robot directly. Therefore, the hardware neural networks realize the robot control without using any software programs or A/D converters. The micro-robot emulates the locomotion method and the neural networks of an insect with rotary type actuators, link mechanisms, and hardware neural networks. The micro-robot performs forward and backward locomotion, and also changes direction by inputting an external trigger pulse. The locomotion speed was 26.4 mm/min when the step width was 0.88 mm.

Chapter 2

Shinya Aoi, Kyoto University, Japan

Recently, interest in the study of legged robots has increased, and various gait patterns of the robots have been established. However, unlike humans and animals, these robots still have difficulties in achieving adaptive locomotion, and a huge gap remains between them. This chapter deals with the gait transition of a biped robot from quadrupedal to bipedal locomotion. This gait transition requires drastic changes in the robot posture and the reduction of the number of supporting limbs, so the stability greatly changes during the transition. A locomotion control system is designed to achieve the gait transition based on the physiological concepts of central pattern generator, phase resetting, and kinematic synergy, and the usefulness of this control system is verified by the robot experiment.

Chapter 3
Suguru N. Kudoh, Kwansei Gakuin University, Japan

A neurorobot is a model system for biological information processing with vital components and the artificial peripheral system. As a central processing unit of the neurorobot, a dissociated culture system possesses a simple and functional network comparing to a whole brain; thus, it is suitable for exploration of spatiotemporal dynamics of electrical activity of a neuronal circuit. The behavior of the neurorobot is determined by the response pattern of neuronal electrical activity evoked by a current stimulation from outer world. "Certain premise rules" should be embedded in the relationship between spatiotemporal activity of neurons and intended behavior. As a strategy for embedding premise rules, two ideas are proposed. The first is "shaping," by which a neuronal circuit is trained to deliver a desired output. Shaping strategy presumes that meaningful behavior requires manipulation of the living neuronal network. The second strategy is "coordinating." A living neuronal circuit is regarded as the central processing unit of the neurorobot. Instinctive behavior is provided as premise control rules, which are embedded into the relationship between the living neuronal network and robot. The direction of self-tuning process of neurons is not always suitable for desired behavior of the neurorobot, so the interface between neurons and robot should be designed so as to make the direction of self-tuning process of the neuronal network correspond with desired behavior of the robot. Details of these strategies and concrete designs of the interface between neurons and robot are be introduced and discussed in this chapter.

Chapter 4
Sayyed Farideddin Masoomi, University of Canterbury, New Zealand
XiaoQi Chen, University of Canterbury, New Zealand
Stefanie Gutschmidt, University of Canterbury, New Zealand
Mathieu Sellier, University of Canterbury, New Zealand

Efficient cruising, maneuverability, and noiseless performance are the key factors that differentiate fish robots from other types of underwater robots. Accordingly, various types of fish-like robots have been developed such as RoboTuna and Boxybot. However, the existing fish robots are only capable of a specific swimming mode like cruising inspired by tuna or maneuvering inspired by labriforms. However, for accomplishing marine tasks, an underwater robot needs to be able to have different swimming modes. To address this problem, the Mechatronics Group at University of Canterbury is developing a fish robot with novel mechanical design. The novelty of the robot roots in its actuation system, which causes its efficient cruising and its high capabilities for unsteady motion like fast start and fast turning. In this chapter, the existing fish robots are introduced with respect to their mechanical design. Then the proposed design of the fish robot at University of Canterbury is described and compared with the existing fish robots.

Chapter 5

Tüze Kuyucu, Doshisha University, Japan

Ivan Tanev, Doshisha University, Japan

Katsunori Shimohara, Doshisha University, Japan

In Genetic Programming (GP), most often the search space grows in a greater than linear fashion as the number of tasks required to be accomplished increases. This is a cause for one of the greatest problems in Evolutionary Computation (EC): scalability. The aim of the work presented here is to facilitate the evolution of control systems for complex robotic systems. The authors use a combination of mechanisms specifically designed to facilitate the fast evolution of systems with multiple objectives. These mechanisms are: a genetic transposition inspired seeding, a strongly-typed crossover, and a multiobjective optimization. The authors demonstrate that, when used together, these mechanisms not only improve the performance of GP but also the reliability of the final designs. They investigate the effect of the aforementioned mechanisms on the efficiency of GP employed for the coevolution of locomotion gaits and sensing of a simulated snake-like robot (Snakebot). Experimental results show that the mechanisms set forth contribute to significant increase in the efficiency of the evolution of fast moving and sensing Snakebots as well as the robustness of the final designs.

Chapter 6

Tomohiro Yamaguchi, Nara National College of Technology, Japan

Takuma Nishimura, Nara National College of Technology, Japan & NTT WEST, Japan

Keiki Takadama, The University of Electro-Communications, Japan

This chapter describes the interactive learning system to assist positive change in the preference of a human toward the true preference. First, an introduction to interactive reinforcement learning with human in robot learning is given; then, the need to estimate the human's preference and to consider its changes by interactive learning system is described. Second, requirements for interactive system as being human adaptive and friendly are discussed. Then, the passive interaction design of the system to assist the awareness for a human is proposed. The system behaves passively to reflect the human intelligence by visualizing the traces of his/her behaviors. Experimental results show that subjects are divided into two groups, heavy users and light users, and that there are different effects between them under the same visualizing condition. They also show that the system improves the efficiency for deciding the most preferred plan for both heavy users and light users.

Section 2

Chapter 7

Huei Ee Yap, Waseda University, Japan

Shuji Hashimoto, Waseda University, Japan

In this chapter, the authors present the design and implementation of a step-traversing two-wheeled robot. Their proposed approach aims to extend the traversable workspace of a conventional two-wheeled robot. The nature of the balance problem changes as the robot is in different phases of motion. Maintaining balance with a falling two-wheeled robot is a different problem than balancing on flat ground.

Active control of the drive wheels during flight is used to alter the flight of the robot to ensure a safe landing. State dependent feedback controllers are used to control the dynamics of the robot on ground and in air. Relationships between forward velocity, height of step, and landing angle are investigated. A physical prototype has been constructed and used to verify the viability of the authors' control scheme. This chapter discusses the design attributes and hardware specifications of the developed prototype. The effectiveness of the proposed control scheme has been confirmed through experiments on single- and continuous-stepped terrains.

Chapter 8

Kai Liu, Tsinghua University, China

Hongbo Li, Tsinghua University, China

Zengqi Sun, Tsinghua University, China

In this chapter, the authors tackle the task of picking parts from a bin (bin-picking task), employing a 6-DOF manipulator on which a single hand-eye camera is mounted. The parts are some cylinders randomly stacked in the bin. A Quasi-Random Sample Consensus (Quasi-RANSAC) ellipse detection algorithm is developed to recognize the target objects. Then the detected targets' position and posture are estimated utilizing camera's pin-hole model in conjunction with target's geometric model. After that, the target, which is the easiest one to pick for the manipulator, is selected from multi-detected results and tracked while the manipulator approaches it along a collision-free path, which is calculated in work space. At last, the detection accuracy and run-time performance of the Quasi-RANSAC algorithm is presented, and the final position of the end-effecter is measured to describe the accuracy of the proposed bin-picking visual servoing system.

Chapter 9

Craig Schlenoff, NIST, USA & University of Burgundy, France

Anthony Pietromartire, NIST, USA & University of Burgundy, France

Zeid Kootbally, NIST, USA

Stephen Balakirsky, NIST, USA

Sebti Foufou, University of Burgundy, France & Qatar University, Qatar

In this chapter, the authors describe a novel approach for inferring intention during cooperative human-robot activities through the representation and ordering of state information. State relationships are represented by a combination of spatial relationships in a Cartesian frame along with cardinal direction information. The combination of all relevant state relationships at a given point in time constitutes a state. A template matching approach is used to match state relations to known intentions. This approach is applied to a manufacturing kitting operation, where humans and robots are working together to develop kits. Based upon the sequences of a set of predefined high-level state relationships that must be true for future actions to occur, a robot can use the detailed state information presented in this chapter to infer the probability of subsequent actions. This would enable the robot to better help the human with the operation or, at a minimum, better stay out of his or her way.

This chapter addresses smart sensor systems. In recent years, goods identification technology using a soft magnetic barcode, radio frequency identification, and automated wheelchair guidance technology using a magnetic field usable in dirty environments as part of Robotics and Mechatronics are becoming important in many areas, such as factories, physical distribution, office, security, etc. These identification and guidance technologies are based on sensing of magnetic field. Therefore, smart magnetic sensing technologies suitable for these identification and guidance techniques are described in this chapter.

Section 3

This chapter presents the design and calculation procedure for a teleoperation and remote control of a medical robot that can help a doctor to use his hands/fingers to examine patients in remote areas. This teleoperation system is simple and low cost, connected to the global Internet system, and through the interaction with the master device, the medical doctor is able to communicate control signals for the slave device. This controller is robust to the time-variant delays and the environment uncertainties while assuring the stability and the high transparent performance. A novel theoretical framework and algorithms are developed with time forward observer-based adaptive controller and neural network-based multiple model. The system allows the medical doctor to feel the real sense of the remote environments.

Teleoperation of forestry machinery is a difficult problem. The difficulties arise because forestry machines are primarily used in unstructured and uncontrolled environments. However, improvements in technology are making implementation of teleoperation for forestry machines feasible with off-the-shelf computing and networking hardware. The state-of-the-art in teleoperation of forestry machinery is reviewed as well as teleoperation in similarly unstructured and uncontrolled environments such as mining and underwater. Haptic feedback in a general sense is also reviewed, as while haptic feedback has been implemented on some types of heavy machinery it has not yet been implemented on forestry machinery.

This chapter presents the state-of-art of the bilateral teleoperation field. It starts with a discusion of the early class of techniques, which are based on passivity and scattering theory. The main issue in bilateral telerobotic systems is the communication delay between the operator and the remote site (environment), which (if not treated) can lead the system to instability. The chapter continues by presenting the evolution of modern control techniques for stabilization and compensation of the time delay consequences. These techniques include predictive control, adaptive control, sliding-mode robust control, neural learning control, fuzzy control, and neurofuzzy control. Four case studies are reviewed that show what kind of results can be obtained.

Section 4

This chapter deals with research activities that have been carried out so far in the field of modelling and simulation of gas turbines for system optimization purposes. It covers major white-box and black-box gas turbine models and their applications to control systems.

This chapter describes the development of a robotic CAM system for an articulated industrial robot from the viewpoint of robotic servo controller. It is defined here that the CAM system includes an important function that allows an industrial robot to move along not only numerical control data (NC data) but also cutter location data (CL data) consisting of position and orientation components. A reverse post-processor is proposed for the robotic CAM system to online generate CL data from the NC data generated for a five-axis NC machine tool with a tilting head, and the transformation accuracy about orientation components in CL data is briefly evaluated. The developed CAM system has a high applicability to other industrial robots with an open architecture controller whose servo system is technically opened to end-users, and also works as a straightforward interface between a general CAD/CAM system and an industrial robot. The basic design of the robotic CAM system and the experimental result are presented, in which an industrial robot can move based on not only CL data but also NC data without any teaching.

Utilizing robotic hands for manipulating objects and assembly requires one to deal with problems like immobility, grasp planning, and regrasp planning. This chapter integrates some essential subjects on robotic grasping: the first section presents a concise taxonomy of robotic grippers and hands. Then the basic concepts of grasping are provided, including immobility, form-closure, and force-closure, 2D and 3D grasping, and Coulomb friction. Next, the principles of grasp planning, measures of grasping quality, pre-grasp, stable grasps, and regrasp planning are presented. The chapter presents comparisons for robotic grippers, a new classification of measures of grasp quality, and a new categorization of regrasp planning approaches.

The Knowledge-Intensive Sustainable Evolution Dynamics (KISBED) (patent pending), a platform the authors use in their "use-cases," shows that it works. Cyber, infrastructure, and product are integrated in the Cyberinfra Product "function." The perception properties are not long tagged or have no carriers, and the signal travels a short distance before it collides. The authors prove the KISBED through some examples.

Preface

Robotics and Mechatronics successfully fuse (but are not limited to) mechanics, electrical, electronics, sensors and perception, informatics and intelligent systems, control systems and advanced modeling, optics, smart materials, actuators, systems engineering, artificial intelligence, intelligent computer control, precision engineering, virtual modeling, etc. into a unified framework that enhances the design of products and manufacturing processes.

The synergy in engineering creative design and development enables a higher level of interdisciplinary research that leads to high quality performance, smart and high functionality, precision, robustness, power efficiency, application flexibility and modularity, improved quality and reliability, enhanced adaptability, intelligence, maintainability, better spatial integration of subsystems (embodied systems), miniaturization, embedded lifecycle design, sustainable development, and cost effective approach. The adoption of such a synergized inter- or trans-disciplinary approach to engineering design implies a greater understanding of the design process.

While the technologies are advancing in different directions and there continues to be progressive evolution of interdisciplinary development in terms of research, education, and product development, there is continuous and growing interest in the fields of robotics and mechatronics.

This book aims to capture the state-of-art research developments in the subject area of engineering creative design in robotics and mechatronics, and to provide relevant theoretical knowledge in the field, technological evolution, and new findings. This book includes 17 chapters, divided into four sections.

The first section covers chapters 1-6, which present robotics-mechatronics and biomimetics as an interdisciplinary engineering science. It covers topics on: "Silicon Micro-Robot with Neural Networks; "Gait Transition Control of a Biped Robot from Quadrupedal to Bipedal Locomotion Based on Central Pattern Generator, Phase Resetting, and Kinematic Synergy"; "Design for Information Processing in Living Neuronal Networks"; "Novel Swimming Mechanism for a Robotic Fish"; "Efficient Evolution of Modular Robot Control via Genetic Programming"; and "Awareness-Based Recommendation: Toward the Human Adaptive and Friendly Interactive Learning System."

The second section includes chapters 7-10. It introduces research topics related to advancement in robotics with main focus on control and stability, visual servoing, inferring intention, and sensors.

The third section covers chapters 11-13. It focuses on teleoperation and associated research issues in different applications, such as: "Development and Simulation of an Adaptive Control System for the Teleoperation of Medical Robots"; "Design and Development of Teleoperation for Forest Machines: An Overview"; "Time Delay and Uncertainty Compensation: State-of-Art and with Case Studies."

The fourth section consists of chapters 14-17. These chapters discuss different topics related to "Modeling and Simulation Approaches for Gas Turbine System Optimization"; "Robotic CAM System Available for Both CL and NC Data"; "Robotic Grippers, Grasping, and Grasp Planning"; and "Cyber Infra Product Concept and its Prototyping Strategies."

Creative engineering design in robotics and mechatronics helps to prepare graduate students, engineers, and scientists who are looking to develop innovative, intelligent, and bioinspired ideas for autonomous and smart interdisciplinary products and systems to meet today's most pressing challenges.

This book is aimed for senior students in mechatronics and relevant fields, engineers, and scientists, and also for graduate students, robotics engineers and researchers, and practicing engineers who wish to enhance and broaden their knowledge and expertise on the fundamentals, practices, technologies, applications, and the evolution of robotics and mechatronics.

Maki K. Habib
The American University in Cairo, Egypt

J. Paulo Davim
University of Aveiro, Portugal

Section 1

Chapter 1
Silicon Micro–Robot with Neural Networks

Ken Saito
Nihon University, Japan

Yoshifumi Sekine
Nihon University, Japan

Minami Takato
Nihon University, Japan

Fumio Uchikoba
Nihon University, Japan

ABSTRACT

Insect type 4.0, 2.7, 2.5 mm. width, length, height size silicon micro-robot system with active hardware neural networks locomotion controlling system is presented in this chapter. The micro-robot system was made from a silicon wafer fabricated by Micro-Electro Mechanical Systems (MEMS) technology. The mechanical system of the robot equipped with millimeter-size rotary type actuators, link mechanisms, and six legs to realize the insect-like switching behavior. In addition, the authors constructed the active hardware neural networks by analog CMOS circuits as a locomotion controlling system. Hardware neural networks consisted of pulse-type hardware neuron models as basic components. Pulse-type hardware neuron model has same basic features of biological neurons such as threshold, refractory period, spatio-temporal summation characteristics, and enables the generation of continuous action potentials. The hardware neural networks output the driving pulses using synchronization phenomena such as biological neural networks. Four output signal ports are extracted from hardware neural networks, and they are connected to the actuators. The driving pulses can operate the actuators of silicon micro-robot directly. Therefore, the hardware neural networks realize the robot control without using any software programs or A/D converters. The micro-robot emulates the locomotion method and the neural networks of an insect with rotary type actuators, link mechanisms, and hardware neural networks. The micro-robot performs forward and backward locomotion, and also changes direction by inputting an external trigger pulse. The locomotion speed was 26.4 mm/min when the step width was 0.88 mm.

DOI: 10.4018/978-1-4666-4225-6.ch001

INTRODUCTION

Many studies have been done on micro robot for several applications such as precise manipulation, medical field, and so on (e.g., Shibata, Aoki, Otsuka, Idogaki, & Hattori, 1997; Takeda, 2001; Habib, Watanabe, & Izumi, 2007; Habib, 2011; Baisch, Sreetharan, & Wood, 2010). However, further miniaturizations and higher functionalization on the micro-robot system are required to play an important role in these fields. Although the miniaturization of the robot has conventionally been progressed by mechanical machining and assembling, some difficulty has appeared in order to achieve further miniaturizations. In particular, frame parts, actuators, motion controllers, power sources and sensors (e.g., Tsuruta, Mikuriya, & Ishikawa, 1999). Instead of the conventional mechanical machining, Micro-Electro Mechanical Systems (MEMS) technology based on the IC production lines has been studied for making the simple components of the micro-robot (e.g., Donald, Levey, McGray, Paprotny, & Rus, 2006; Edqvist, et al., 2009; Suematsu, Kobayashi, Ishii, Matsuda, Sekine, & Uchikoba, 2009). In addition, the development of the actuator is important subjects. The type of the micro-actuator by MEMS technology is categorized into two groups. For example, uses the field forces. Otherwise uses the property of the material itself (e.g., Tang, Nguyen, & Howe, 1989; Sniegowski, & Garcia, 1996; Asada, Matsuki, Minami, & Esashi, 1994; Suzuki, Tani, & Sakuhara, 1999; Surbled, Clerc, Pioufle, Ataka, & Fujita, 2001). In particular, shape memory alloy actuator shows a large displacement such as 50% of the total length in millimeter size. However, micro-actuators using field forces or piezoelectric elements to the micro-robot had a weakness for moving on the uneven surface. Therefore, micro-robot which could locomote by step pattern was desired.

Programmed control by a digital systems based on microcontroller has been the dominant system among the robot control. On the other hand, insects realize the autonomous operation using excellent structure and active neural networks control by compact advanced systems. Therefore, some advanced studies of artificial neural networks have been paid attention for applying to the robot. A lot of studies have reported both on software models and hardware models (e.g., Matsuoka, 1987; Ikemoto, Nagashino, Kinouchi, & Yoshinaga, 1997; Nakada, Asai, & Amemiya, 2003). However, using the mathematical neuron models in large scale neural network is difficult to process in continuous time because the computer simulation is limited by the computer performance, such as the processing speed and memory capacity. In contrast, using the hardware neuron model is advantageous because even if a circuit scale becomes large, the nonlinear operation can perform at high speed and process in continuous time. Therefore, the construction of a hardware model that can generate oscillatory patterns is desired. For this reason, we are studying about millimeter size micro-robot system which can control the locomotion by active hardware neural networks.

In this chapter, we will propose the active hardware neural networks controlled 4.0, 2.7, 2.5 mm. width, length, height size silicon micro-robot system from silicon wafer fabricated by MEMS technology.

SILICON MICRO-ROBOT

We constructed the miniaturized robot by MEMS technology. In this section, the basic components of the fabricated silicon micro-robot were shown. The number of the legs of the silicon micro-robot was six. The structure and the step pattern of the robot was emulated those of an ant. The micro-robot consisted of frame parts, rotary type actuators and link mechanisms.

Figure 1 (a) shows the shape and dimensions of components of silicon micro-robot. The micro-fabrication of the silicon wafer was done by the MEMS technology. The shapes were machined

Figure 1. Mechanical parts of silicon micro-robot

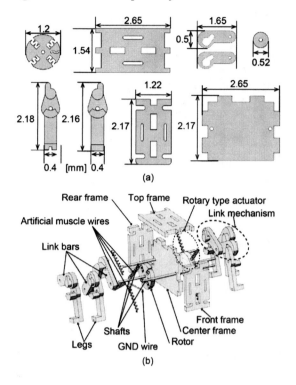

(a)

(b)

by photolithography based Inductively Coupled Plasma (ICP) dry etching (Bhardwaj, & Ashraf, 1995). Figure 1 (b) shows the developed figure of the silicon micro-robot. The frame components, the rotary type actuators and the link mechanism were made from silicon wafer. We used the 100, 200, 385, 500 μm thickness silicon wafer depends on each mechanical part. The frame parts consisted of a top frame, a rear frame, a front frame and a center frame. The rotary type actuators consisted of a rotor, GND wire, shaft and 4 pieces of helical artificial muscle wires. The frame components and the rotary type actuators were connected by the helical artificial muscle wire which was the shape memory alloy (see Homma, 2003). We used the BioMetal® Helix BMX50 to the helical artificial muscle wire (http://www.toki.co.jp). The basic characteristics of the helical artificial muscle wire were as follows. The standard coil diameter of the artificial muscle wire was 0.2 mm

where wire diameter was 0.05 mm. The practical maximum force produced 3 to 5 gf where kinetic displacement was 50%. The standard drive current was 50 to 100 mA where standard electric resistance was 3600 ohm/m. The rotary type actuator generated the locomotion of the robot by supplying the electrical current to the helical artificial muscle wires. The wire shrank at high temperature and extended at low temperature. In this study, the wire was heated by electrical current flowing, and cooled by stopping the flowing. The rotational movement of the each actuator was obtained by changing the flowing sequence. The link mechanisms consisted of a link bars, shafts and legs. The front leg and the rear leg were connected to the middle leg by link bars, respectively. The middle leg is connected to the rotor by the shaft. Therefore, the rotational phase was same as the rotor. On the contrary, the other two legs are connected by the link bar that generates 90 degree phase shift. Also, backward step was obtained by the counter rotation of the actuator.

Figure 2 (a) shows the schematic diagram of the silicon micro-robot. The design size was 4.0, 2.7, 2.5 mm, width, length, height, respectively. The link mechanism and rotary type actuator were connected by shafts such as shown in Figure 2 (a). Figure 2 (b) shows the picture of the rotary type actuator. The one side of the artificial muscle wires connected to the copper wires and the other side connected to the rotor by using solder paste. The GND wire was connected to the rotor directly. Figure 2 (c) shows the picture of the fabricated silicon micro-robot. The copper wires above the silicon micro-robot were two GND wires and eight signal wires (power supply wires of the artificial muscle wires). In the case of connecting the signal wires to the power supply sources (hardware neural networks, microcontroller, waveform generator), the silicon micro-robot could locomote.

Figure 2. Schematic diagram and picture of silicon micro-robot

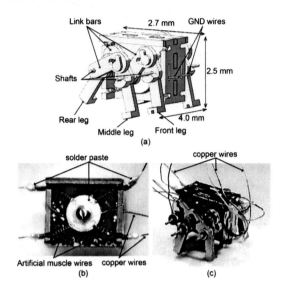

(a)

(b)

(c)

Figure 3. Schematic diagram of locomotion of the silicon micro-robot

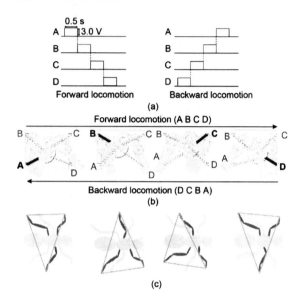

(a)

(b)

(c)

LOCOMOTION MECHANISMS OF THE SILICON MICRO-ROBOT

Insects can locomote smoothly on the uneven surface by using step pattern. On the other hand, many micro-robots were weak in moving on the uneven surface. Therefore, micro-robot which could locomote by step pattern was desired. We replicate the locomotion of an ant by using rotary type actuator and link mechanism. In this section, we will discuss about locomotion mechanisms of the silicon micro-robot.

Figure 3 (a) shows the schematic diagrams of the waveform to actuate the silicon micro-robot. Figure 3 (b) shows the schematic diagrams of the locomotion. The forward locomotion and backward locomotion was realized by changing the waveform. Figure 3 (c) shows the schematic diagrams of example of locomotion of ant. The silicon micro-robot imitated the locomotion method of ants. The touch with ground timing of the legs are same in Figure 3 (b) and (c).

To heat the helical artificial muscle wires, we required to input the pulse width 0.5 s, pulse period 2 s and pulse amplitude 3 V (See Figure

3 (a)). Therefore, the micro-robot required 2 s to finish the 1 cycle locomotion. Figure 3 (b) shows the schematic diagram of locomotion of silicon micro-robot. The silicon micro-robot could move by the rotational actuator. The helical artificial muscle wire had a characteristic of changing length according to temperature. In the case of heating the wire shrunk and in the case of cooling the wire extended. In particular, when heating the helical artificial muscle wires from A to D, the silicon micro-robot would move forward. In contrast, heating the helical artificial muscle wires from D to A, the robot moved backward. The locomotion pattern is 180-degree phase shift on each side to represent the locomotion of insect. The silicon micro-robot imitated the locomotion method of ants (See Figure 3 (c) The touch with ground timing of the legs are same.). In the case of inputting shorter than pulse width 0.5 s, the artificial muscle wire could not shrunk because the joule heat by drive current was not enough. In contrast, in the case of inputting longer than pulse width 2 s, the artificial muscle wire could not extended because the joule heat by drive current was too much and cooling must be difficult.

HARDWARE NEURAL NETWORKS

It is well known that locomotion rhythms of living organisms are generated by Central Pattern Generator (CPG). Previously, we proposed the CPG model using pulse-type hardware neuron model (see Okazaki, Ogiwara, Yang, Sakata, Saito, Sekine, & Uchikoba. 2011; Saito, Matsuda, Saeki, Uchikoba, & Sekine. 2011; Saito, Takato, Sekine, & Uchikoba. 2012). CPG model was board level circuit using surface-mounted components. The board level circuit was 10 cm square size. Therefore, it was impossible to integrate on the silicon micro-robot system. Proposing hardware neural networks is a simplified model of CPG model for the purpose of further miniaturizations.

The pulse-type hardware neuron model has the same basic features of biological neurons such as threshold, refractory period, spatio-temporal summation characteristics, and enables the generation of continuous action potentials. The pulse-type hardware neuron model consists of a synaptic model and cell body mode. Figure 4 (a) shows the circuit diagram of synaptic model by CMOS. The deference of excitatory synaptic model and inhibitory synaptic model is only direction of electrical current. The synaptic model has the spatio-temporal summation characteristics similar to those of living organisms, spatio-temporal summate the output of cell body models. The circuit parameters of synaptic model were as follows: $C_{ES}=C_{IS2}=1$ pF, M_{ES1-3}, M_{IS1-5}: W/L=1. The voltage source $V_{DD}=5$ V. Figure 4 (b) shows the circuit diagram cell body model by CMOS. The circuit parameters of the cell body model were as follows: $C_G=39$ μF, $C_M=270$ nF, M_{C1}, M_{C2}: W/L=10, M_{C3}: W/L=0.1, M_{C4}: W/L=0.3. The voltage source $V_A=3.3$ V. The input voltage v_{ESin}, v_{ISin} of synaptic model were output voltage v_{Cout} of cell body model. The input current i_{Cin} of cell body model was output current i_{ESout} and i_{ISout} of synaptic model. Figure 4 (c) shows the schematic diagrams of cell body model, excitatory synaptic model and inhibitory synaptic model.

Figure 4. Pulse-type hardware neuron model

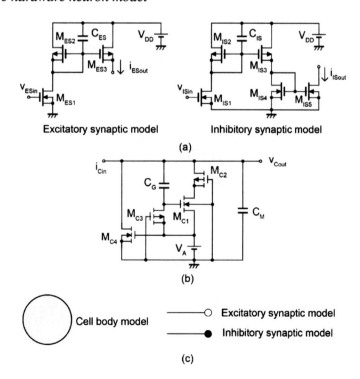

E and I indicate cell body model with excitatory synaptic model and cell body model with inhibitory synaptic model, respectively. Figure 5 (a) shows the excitatory mutual coupling. The cell body model connected by excitatory synaptic model cause of in-phase synchronization. Figure 5 (b) shows the inhibitory mutual coupling. The cell body model connected by inhibitory synaptic model cause of anti-phase synchronization. Therefore, we use the inhibitory mutual coupling to generate the driving pulses which can operate the actuators of silicon micro-robot.

LOCOMOTION CONTROL USING HARDWARE NEURAL NETWORKS

Using the anti-phase synchronization phenomena of Figure 5 (b), we constructed the hardware neural networks. The constructed hardware neural networks were shown in Figure 6.

Figure 6 (a) shows the connection diagram and layout pattern of hardware neural networks. The four cell body models are mutually coupled by twelve inhibitory synaptic models. Four output

ports were extracted from hardware neural networks and they were connected to the artificial muscle wires. In addition, four trigger pulse input port were extracted to hardware neural networks. According to the input timing of external trigger pulse, hardware neural networks could change the sequence of output waveform. This IC chip was made by On-Semiconductor, HOYA Corporation, and KYOCERA Corporation. The design rule of the IC chip was double metal double poly CMOS 1.2 μm rule. The chip was a square 2.3 mm x 2.3 mm in size. The sizes of capacitors of cell body model are too large to implement to the CMOS IC chip. Therefore, the capacitors C_G and C_M were mounted externally of bare chip. Figure 6 (b) shows the example of output waveform of neural networks. Hardware neural networks are coupled neural networks system, which can generate the locomotion rhythms such as living organisms. It is shown that hardware neural networks could output the waveform of forward locomotion and backward locomotion such as Figure 3 (a). Thus, our hardware neural networks are effective to generate the locomotion of the silicon micro-robot.

Figure 5. Synchronization phenomena of pulse-type hardware neuron models

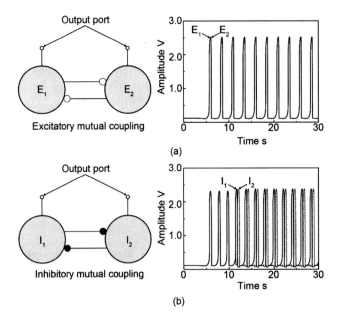

Figure 6. Hardware neural networks

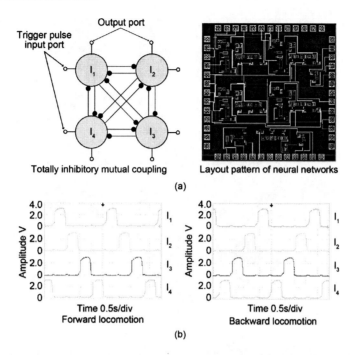

Four signal ports were extracted from hardware neural networks and they were connected to the artificial muscle wires.

Hardware neural networks can output the waveform of forward locomotion and backward locomotion which is necessary to actuate the silicon micro-robot. In particular, our constructed hardware neural networks can control the movements of silicon micro-robot. The locomotion speed was 26.4 mm/min where the step width was 0.88 mm. (See Figure 7).

CONCLUSION AND DISCUSSIONS

In this chapter, we proposed the hardware neural networks controlled silicon micro-robot. As a result, we developed the following conclusions.

1. 4.0, 2.7, 2.5 mm. width, length, height size micro-robot could fabricated by MEMS technology.
2. Hardware neural networks can output the waveform of forward locomotion and

backward locomotion which is necessary to actuate the silicon micro-robot. In particular our constructed hardware neural networks can control the movements of silicon micro-robot. The locomotion speed of silicon micro-robot was 26.4 mm/min where the step width was 0.88 mm.

Our silicon micro-robot could locomote by step pattern. Therefore, our robot could locomote more smoothly on the uneven surface compared with the other micro-robot (For example, micro-robot using field forces actuator or piezoelectric actuator.).

Our silicon micro-robot system targeting the integration of mechanical system, CMOS neural networks controlling system, sensory system, and power source system. The mechanical systems consist of frame parts, small size rotary type actuators and link mechanisms. The locomotion of silicon micro-robot was controlled by CMOS neural networks controlling system. The sensory system and power source system are under development. Therefore, this chapter discuss about integration

Figure 7. Locomotion of the silicon micro-robot

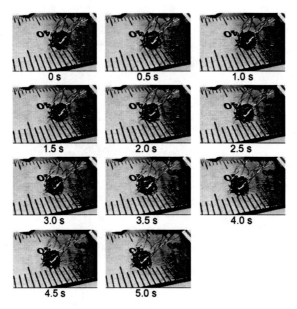

of mechanical system and CMOS neural networks controlling system. Our target is advanced package integration of controllers, actuators, sensors, and power sources on the silicon frame of silicon micro-robot. Now, we are studying about sensory system and power source system of the silicon micro-robot. In near future, we will achieve the whole silicon micro-robot system (See Figure 8).

Figure 8. Conceptual diagram of the silicon micro-robot

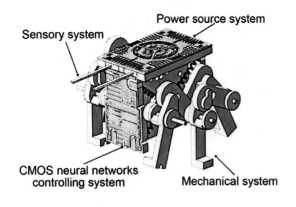

ACKNOWLEDGMENT

This research was supported by JSPS KAKENHI (23760243), Nihon University College of Science and Technology Project Research, Nihon University Academic Research Grant (Total research, "11-002"). The fabrication of the silicon micro-robot was supported by Research Center for Micro Functional Devices, Nihon University. The VLSI chip in this study has been fabricated in the chip fabrication program of VLSI Design and Education Center (VDEC), the University of Tokyo in collaboration with On-Semiconductor, HOYA Corporation, KYOCERA Corporation, and Cadence Design Systems, Inc.

REFERENCES

Asada, N., Matsuki, H., Minami, K., & Esashi, M. (1994). Silicone micromachined two-dimensional galvano optical scanner. *IEEE Transactions on Magnetics, 30,* 4647–4649. doi:10.1109/20.334177.

Baisch, A. T., Sreetharan, P. S., & Wood, R. J. (2010). Biologically-inspired locomotion of a 2g hexapod robot. In *Proceedings of EEE IROS 2010,* (pp. 5360-5365). IROS. doi:10.1109/IROS.2010.5651789

Bhardwaj, J. K., & Ashraf, H. (1995). Advanced silicon etching using high-density plasmas. In *Proceedings of SPIE Micromachining and Micro Fabrication Process Technology,* (Vol. 2639, pp. 224-233). SPIE. doi:10.1117/12.221279

Donald, B. R., Levey, C. G., McGray, C. D., Paprotny, I., & Rus, D. (2006). An untethered, electrostatic, globally controllable MEMS micro-robot. *Journal of Microelectromechanical Systems, 15,* 1–15. doi:10.1109/JMEMS.2005.863697.

Edqvist, E., Snis, N., Mohr, R. C., Scholz, O., Corradi, P., & Gao, J. et al. (2009). Evaluation of building technology for mass producible millimeter-sized robots using flexible printed circuit boards. *Journal of Micromechanics and Microengineering,* 1–11. doi: doi:10.1088/0960-1317/19/7/075011.

Habib, M. K. (2011). Biomimetcs: Innovations and Robotics. *International Journal of Mechatronics and Manufacturing Systems, 4*(2), 113–134. doi:10.1504/IJMMS.2011.039263.

Habib, M. K., Watanabe, K., & Izumi, K. (2007). Biomimetcs robots: From bio-inspiration to implementation. In *Proceedings of the 33rd Annual Conference of the IEEE Industrial Electronics Society,* (pp. 143-148). IEEE. doi:10.1109/IECON.2007.4460382

Homma, D. (2003). Metal artificial muscle bio metal fiber. *RSJ, 21,* 22–24.

Ikemoto, T., Nagashino, H., Kinouchi, Y., & Yoshinaga, T. (1997). oscillatory mode transitions in a four coupled neural oscillator model. In *Proceedings of the International Symposium on Nonlinear Theory and its Applications,* (pp. 561-564). Retrieved from http://ci.nii.ac.jp/naid/110003291511/en/

Matsuoka, K. (1987). Mechanism of frequency and pattern control in the neural rhythm generators. *Biological Cybernetics, 56,* 345–353. doi:10.1007/BF00319514 PMID:3620533.

Nakada, K., Asai, T., & Amemiya, Y. (2003). An analog CMOS central pattern generator for inter-limb coordination in quadruped locomotion. *IEEE Transactions on Neural Networks, 14,* 1356–1365. doi:10.1109/TNN.2003.816381 PMID:18244582.

Okazaki, K., Ogiwara, T., Yang, D., Sakata, K., Saito, K., Sekine, Y., & Uchikoba, F. (2011). Development of pulse control type MEMS micro robot with hardware neural network. *Artificial Life and Robotics, 16*(2), 229–233. doi:10.1007/s10015-011-0925-9.

Saito, K., Matsuda, A., Saeki, K., Uchikoba, F., & Sekine, Y. (2011). Synchronization of coupled pulse-type hardware neuron models for CPG model. In *The relevance of the time domain to neural network models* (pp. 117–133). New York: Springer.

Saito, K., Takato, M., Sekine, Y., & Uchikoba, F. (2012). Biomimetics micro robot with active hardware neural networks locomotion control and insect-like switching behaviour. *International Journal of Advanced Robotic Systems,* 1–6. doi:10.5772/54129.

Shibata, T., Aoki, Y., Otsuka, M., Idogaki, T., & Hattori, T. (1997). Microwave energy transmission system for microrobot. *IEICE Transactions on Electronics*, *80*(2), 303–308. Retrieved from http://search.ieice.org/bin/summary.php?id=e80-c_2_303.

Sniegowski, J. J., & Garcia, E. J. (1996). Surface-micromachined gear trains driven by an on-chip electrostatic microengine. *IEEE Electron Device Letters*, *17*, 366–368. doi:10.1109/55.506369.

Suematsu, H., Kobayashi, K., Ishii, R., Matsuda, A., Sekine, Y., & Uchikoba, F. (2009). MEMS type micro robot with artificial intelligence system. In *Proceedings of International Conference on Electronics Packaging*, (pp. 975-978). ICEP.

Surbled, P., Clerc, C., Pioufle, B. L., Ataka, M., & Fujita, H. (2001). Effect of the composition and thermal annealing on the transformation temperature sputtered TiNi shape memory alloy thin films. *Thin Solid Films*, *401*, 52–59. doi:10.1016/S0040-6090(01)01634-0.

Suzuki, Y., Tani, K., & Sakuhara, T. (1999). Development of a new type piezo electric micromotor. In *Proceedings of Transduceres '99* (pp. 1748–1751). Transduceres.

Takeda, M. (2001). Applications of MEMS to industrial inspection. In *Proceedings of IEEE MEMS 2001*, (pp. 182-191). IEEE. doi:10.1109/MEMSYS.2001.906510

Tang, W. C., Nguyen, T. H., & Howe, R. T. (1989). Laterally driven poly silicon resonant microstructure. In *Proceedings of IEEE Micro Electro Mechanical Systems: An Investigation of Micro Structures, Sensors, Actuators, Machines and Robots*, (pp. 53-59). IEEE. doi:10.1016/0250-6874(89)87098-2

Tsuruta, K., Mikuriya, Y., & Ishikawa, Y. (1999). Microsensor developments in Japan. *Sensor Review*, *19*(1), 7–42. doi:10.1108/02602289910255568.

Chapter 2

Gait Transition Control of a Biped Robot from Quadrupedal to Bipedal Locomotion based on Central Pattern Generator, Phase Resetting, and Kinematic Synergy

Shinya Aoi
Kyoto University, Japan

ABSTRACT

Recently, interest in the study of legged robots has increased, and various gait patterns of the robots have been established. However, unlike humans and animals, these robots still have difficulties in achieving adaptive locomotion, and a huge gap remains between them. This chapter deals with the gait transition of a biped robot from quadrupedal to bipedal locomotion. This gait transition requires drastic changes in the robot posture and the reduction of the number of supporting limbs, so the stability greatly changes during the transition. A locomotion control system is designed to achieve the gait transition based on the physiological concepts of central pattern generator, phase resetting, and kinematic synergy, and the usefulness of this control system is verified by the robot experiment.

1. INTRODUCTION

Humans and animals create various locomotor behaviors, such as walk, run, turn, crawl, skip, and jump, depending on the situation. One objective in robotics is to reproduce such locomotor behaviors using legged robots from the engineering view point. So far, many studies have investigated methods to achieve stable locomotor behaviors for various gait patterns of legged robots. However, their transitions have not been thoroughly examined. In particular, the gait transition from

DOI: 10.4018/978-1-4666-4225-6.ch002

quadrupedal to bipedal locomotion needs drastic changes in the robot posture and the reduction of the number of supporting limbs. Therefore, the stability greatly changes during the transition and this transition poses a challenging task in robotic studies.

Recently, interest in the study of legged robots has been growing. However, unlike humans and animals, these robots still have difficulties in achieving adaptive locomotion in various situations, and a huge gap remains between them. To create new control strategies, it is natural to use ideas inspired from biological systems. In this study, we construct a locomotion control system of a biped robot to achieve the gait transition from quadrupedal to bipedal locomotion based on the physiological concepts of Central Pattern Generator (CPG), phase resetting, and kinematic synergy, and verify the usefulness of this control system by the robot experiment.

2. GAIT TRANSITION STRATEGY

2.1. Problems in the Gait Transition

Because locomotor behavior is rhythmic motion, a steady gait of a legged robot indicates a stable limit cycle in the state space. Therefore, we need to control the robot to establish a limit cycle to produce a stable gait. Because different steady gaits have different stable limit cycles, a change of the gait pattern implies that the robot state moves from one limit cycle to another. Even if a robot obtains steady gaits, many difficulties remain for the gait transition.

For the gait transition, the following two issues are crucial: (1) because a robot has many Degrees Of Freedom (DOFs), it is difficult to determine how to produce robot movements to connect one gait pattern to another, in other words, how to construct adequate constraint conditions in motion planning; (2) even if the robot establishes

stable gait patterns, it may fall over during the gait transition and it is difficult to establish stable gait transition without falling over.

2.2. Solving the Redundancy Problem Based on Kinematic Synergy

In Aoi and Tsuchiya (2007), we investigated the turning behavior of a biped robot, where we produced the turning by changing the gait pattern between straight and curved walking. The turning consisted of two types of limit cycles (straight and curved walking) and the transition between them. Although we dealt with a gait transition, we did not need to solve the redundancy problem. This was because there was a phase at which the robot joint configurations were identical between the robot movements to produce straight and curved walking, as shown in Figure 1a. Therefore, we did not need to create additional robot movements to connect these two gait patterns; we just needed to change the gait pattern at the phase. However, in this study we have to solve the redundancy problem to produce the gait transition of a biped robot from quadrupedal to bipedal locomotion, because no such phase exists for robot movements to produce quadrupedal and bipedal walking (Figure 1b).

Physiological findings suggest the importance of muscle synergies for controlling movements in humans and animals (d'Avella, et al., 2003; d'Avella & Bizzi, 2005; Dominici, et al., 2011; Drew, et al., 2008; Danna-dos-Santos, et al., 2007; Ivanenko, et al., 2005; Ting & Macpherson, 2005; Todorov & Jordan, 2002); these concepts have been analyzed using principal component analysis and factor analysis and are viewed as one solution to handle the redundancy problem in biological systems. Muscle synergy is related to the co-variation of muscle activities. Ivanenko *et al.* (2004, 2006) reported that, although recorded electromyographic data during human bipedal locomotion are complicated, the data can be ac-

Figure 1. Schematic diagrams of desired trajectories in joint space for (a) straight and curved walking and (b) quadrupedal and bipedal walking. (a) has a phase at which desired joints are identical between two gait patterns unlike (b). (b) needs additional connection to change gait pattern.

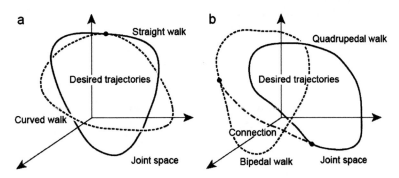

counted for by a combination of only five basic patterns. Furthermore, similar movement patterns such as obstacle avoidance and kick motion share the same basic patterns and can be explained by adding another pattern.

In addition to muscle synergies, kinematic synergies have been investigated during human whole-body movements (Alexandrov, et al., 1998, 2001). Most studies have associated kinematic synergies with simultaneous movements in several joints. Freitas *et al.* (2006) investigated joint angle co-variation patterns during a rhythmic whole-body motion and suggested that two kinematic synergies account for the joint co-variation and contribute to the stabilization of two performance variables: the Center Of Mass (COM) location and trunk orientation. This means that some DOFs are functionally connected by objects, depending on the task, which reduces the number of DOFs and solves the problem of motor redundancy. Such angle co-variation behavior is also observed during the locomotion of humans and animals (Bianchi, et al., 1998; Funato, et al., 2010; Poppele & Bosco, 2003). When elevation angles at the thigh, shank, and foot during locomotion are plotted, the angles describe regular loops constrained on a plane, and hint that such a kinematically coordinative structure is embedded in the motion generation mechanism.

In this study, based on these physiological findings, we produce robot movements during gait transition to connect two gait patterns by constructing coordinative structures. To change the gait pattern from quadrupedal to bipedal locomotion, the robot raises its trunk and its arms leave the ground. Because the number of limbs that support the body is reduced, an adequate location between the supporting limb locations and the COM location is crucial when raising the trunk; otherwise, the robot easily falls over. In addition, because raising the trunk induces a large change in the robot posture and the COM location, the appropriate change in trunk orientation is crucial. Therefore, we use the COM location relative to the supporting positions of the limbs and the trunk orientation as performance variables and produce robot movements by constructing two kinematically coordinative structures that control these two performance variables.

2.3. Improving Robustness Based on CPG and Phase Resetting

Physiological studies have suggested that the CPG in the spinal cord greatly contributes to the generation of rhythmic limb movement, such as locomotion (Grillner, 1975; Orlovsky, et al., 1999; Shik & Orlovsky, 1976). Based on the physiological concept of the CPG, locomotion control systems of legged robots have been developed to create adaptive walking of the robots in various environments (Aoi & Tsuchiya, 2005, 2007; Ijspeert, 2008; Kimura, et al., 2007; Liu, et al., 2008; Nakanishi, et al., 2004; Nakanishi, et al., 2006; Steingrube, et al., 2010).

The CPG can produce oscillatory behaviors even without rhythmic input and proprioceptive feedback. However, it must use sensory feedback to produce effective locomotor behavior. Physiological studies have shown that locomotor rhythm and its phase are modulated by producing phase shift and rhythm resetting based on sensory afferents and perturbations (phase resetting) (Conway, et al., 1987; Duysens, 1977; Lafreniere-Roula & McCrea, 2005; Schomburg, et al., 1998). Such rhythm and phase modulations by phase resetting have for the most part been investigated during fictive locomotion in cats, and their functional roles during actual locomotion remain largely unclear. However, simulation studies of human bipedal locomotion have demonstrated that phase resetting plays important roles in generating adaptive locomotor behavior (Aoi, et al., 2010; Yamasaki, et al., 2003). In addition, biped robots with phase resetting have established adaptive locomotion to force disturbances and environmental variations (Aoi & Tsuchiya, 2005, 2007; Nakanishi, et al., 2004; Nakanishi, et al., 2006). Therefore, to address the second issue regarding walking stabil-

ity during gait transition, we use the locomotion control system constructed based on the CPG and phase resetting.

3. BIPED ROBOT

Figure 2a shows a biped robot used in this study that consists of a trunk composed of two parts (upper and lower trunk), a pair of arms composed of two links, and a pair of legs composed of five links (Aoi, et al., 2012). Figure 2b shows the schematic model of the robot. Each link is connected to the others through a rotational joint with a single DOF. Each joint has a motor and an encoder to manipulate the angle. Four touch sensors are attached to the corners of the sole of each foot and one touch sensor is attached to the tip of the hand of each arm. Table 1 shows the physical parameters of the robot.

The left and right legs are numbered Legs 1 and 2, respectively. The joints of the legs are also numbered: Joints 1…5 from the side of the trunk, where Joints 1 and 2 are the roll and pitch hip joints, respectively, Joint 3 is the pitch knee joint, and Joints 4 and 5 are the pitch and roll ankle joints, respectively. The arms are numbered in a similar manner: Joint 1 is the pitch shoulder joint and Joint 2 is the pitch elbow joint. The trunk consists of the upper and lower parts connected by the pitch waist joint, named Waist Joint. To

Figure 2. Biped robot: (a) robot and (b) schematic model

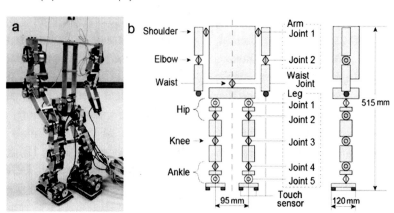

Table 1. Physical parameters of the biped robot

Link	Mass [kg]	Length [cm]
Trunk	1.42	27.2
Arm	0.53	22.2
Leg	1.40	24.3
Total	5.28	51.5

describe the configuration of the robot, we introduce angles θ_{W}, $\theta_{\mathrm{A}j}^{i}$, and $\theta_{\mathrm{L}k}^{i}$ (i=1,2, j=1,2, k=1,...,5), which are the rotation angles of Waist Joint, Joint j of Arm i, and Joint k of Leg i.

The robot walks on a flat floor with no elevation. The electric power is externally supplied and the robot is controlled by an external host computer (Intel Pentium 4 2.8 GHz, RT-Linux), which calculates the desired joint movements and solves the oscillator phase dynamics in the locomotion control system. It receives the command signals at intervals of 1 ms. The robot is connected with the electric power unit and the host computer by cables that are held up during the experiment to avoid influencing the walking behavior.

4. LOCOMOTION CONTROL SYSTEM

4.1. CPG-Based Locomotion Control System

Although the organization of the CPG in biological systems remains largely unclear, physiological findings suggest that the CPG consists of hierarchical networks composed of a Rhythm Generator (RG) and Pattern Formation (PF) networks (Burke, et al., 2001; Lafreniere-Roula & McCrea, 2005; Rybak, et al., 2006). The RG network generates the basic rhythm and modulates it in response to sensory afferents and perturbations. The PF network shapes the rhythm into spatiotemporal patterns of motor commands. That is, the CPG separately controls the locomotor rhythm and the motor commands in the RG and PF networks, respectively.

In this study, we construct a locomotion control system based on this two-layer hierarchical network model. For the RG network model, we develop a rhythm generator using six simple phase oscillators (Leg 1, Leg 2, Arm 1, Arm 2, Trunk, and Inter oscillators) to produce the rhythm information for locomotor behavior. The rhythm information is regulated by phase resetting in response to touch sensor signals. For the PF network model, we construct a trajectory generator, gait generator, and motor controller to create motor torques based on the rhythm information from the rhythm generator to produce the joint movements. Our locomotion control system consists of a motion generator and motion controller (Figure 3). The motion generator has the rhythm generator, gait generator, and trajectory generator. The motion controller has the motor controller.

4.2. Trajectory Generator

The trajectory generator produces the desired movements for all joints using the oscillator phases in the rhythm generator (see Section 4.3) and kinematic parameters determined in the gait generator (see Section 4.4). First, we introduce

Figure 3. Locomotion control system

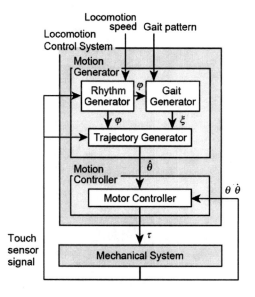

φ_{L}^{i}, φ_{A}^{i}, φ_{T}, and φ_{I} ($i = 1, 2$) for the phases of Leg i, Arm i, Trunk, and Inter oscillators, respectively.

We design the desired limb kinematics consisting of the swing and stance phases in the pitch plane (Figure 4). For the leg movement (Figure 4a), Joint 4 (ankle pitch joint) follows a simple closed curve relative to the trunk during the swing phase, which includes an Anterior Extreme Position (AEP) and a Posterior Extreme Position (PEP). Joint 4 starts from PEP and continues until the foot touches the ground. During the stance phase, Joint 4 traces out a straight line from the Landing Position (LP) to PEP. Therefore, this leg movement depends on the timing of foot contact in each step cycle. In both the swing and stance phases, the angular movement of Joint 4 is designed so that the foot is parallel to the line that connects points AEP and PEP. We use D to denote the distance between AEP and PEP. We denote the

swing and stance phase durations by T_{sw} and T_{st}, respectively. The duty factor β, the stride length S, and the locomotion speed v are then given by

$$\beta = \frac{T_{\mathrm{st}}}{T_{\mathrm{sw}} + T_{\mathrm{st}}}, \; S = \frac{T_{\mathrm{sw}} + T_{\mathrm{st}}}{T_{\mathrm{st}}} D, \; v = \frac{D}{T_{\mathrm{st}}}$$

These values are satisfied regardless of the gait pattern. The lower trunk is at an angle of ψ_{H} to the line perpendicular to the line connecting points AEP and PEP. We denote the height and forward bias from the center of points AEP and PEP to Joint 2 (hip pitch joint) by Δ_{L} and H_{L}, respectively. These two trajectories for the swing and stance phases provide the desired movements $\hat{\theta}_{\mathrm{L}j}^{i}$ ($i=1,2$, $j=2,3,4$) of Joint j (hip, knee, and ankle pitch joints) of Leg i by the function of phase φ_{L}^{i} of Leg i oscillator, where we use $\varphi_{\mathrm{L}}^{i} = 0$ at point PEP and $\varphi_{\mathrm{L}}^{i} = \varphi_{\mathrm{AEP}}$ at point AEP.

To increase the stability of bipedal locomotion in three-dimensional space, we use the roll joints in the legs. We design the desired movements $\hat{\theta}_{\mathrm{L}1}^{i}$ and $\hat{\theta}_{\mathrm{L}5}^{i}$ ($i=1,2$) of Joints 1 and 5 (hip and ankle roll joints) of Leg i by the functions of phase φ_{T} of Trunk oscillator by

$$\hat{\theta}_{\mathrm{L}1}^{i} = R\cos(\varphi_{\mathrm{T}} + \phi),$$
$$\hat{\theta}_{\mathrm{L}5}^{i} = -R\cos(\varphi_{\mathrm{T}} + \phi) \qquad i = 1,2$$

where R is the amplitude of the roll motion and ϕ determines the phase relationship between the leg movements in the pitch and roll planes.

For the arm movement, the hand follows a simple closed curve during the swing phase and a straight line during the stance phase, as is similar to the legs except for the bend direction between Joint 2 of the arm (elbow pitch joint) and Joint 3 of the leg (knee pitch joint) (Figure 4b). The upper trunk is at an angle of $\psi_{\mathrm{H}} + \psi_{\mathrm{W}}$ to the

Figure 4. Desired kinematics composed of swing and stance phases for (a) leg and (b) arm

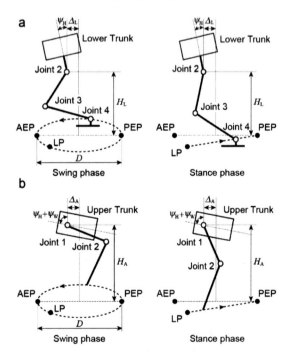

line perpendicular to the line that connects points AEP and PEP, where angle ψ_W is the pitch angle of Waist joint. We denote the height and forward bias from the center of points AEP and PEP to Joint 1 (shoulder pitch joint) of the arm by Δ_A and H_A, respectively. For the distance D, the swing phase durations T_{sw}, and the stance phase duration T_{st}, we use the same values as those used for the legs. These two trajectories for the swing and stance phases give the desired movements $\hat{\theta}_{Aj}^i$ (i,j=1,2) of Joint j (shoulder and elbow pitch joints) of Arm i by the function of phase φ_A^i of Arm i oscillator, where we use $\varphi_A^i = 0$ at point PEP and $\varphi_A^i = \varphi_{AEP}$ at point AEP.

We use these desired joint movements for both quadrupedal and bipedal locomotion and change the seven kinematic parameters Δ_A, Δ_L, H_A, H_L, ψ_H, ψ_W, and R depending on the gait pattern (see Section 4.4). To achieve the desired joint movements, the motor controller manipulates the joint angles based on PD feedback control using high-gain feedback gains by

$$\tau_W = -\kappa_W \left(\theta_W - \hat{\theta}_W\right) - \sigma_W \dot{\theta}_W$$

$$\tau_{Aj}^i = -\kappa_A \left(\theta_{Aj}^i - \hat{\theta}_{Aj}^i\right) - \sigma_A \dot{\theta}_{Aj}^i \qquad i = 1,2, j = 1,2$$

$$\tau_{Lk}^i = -\kappa_L \left(\theta_{Lk}^i - \hat{\theta}_{Lk}^i\right) - \sigma_L \dot{\theta}_{Lk}^i \qquad i = 1,2, k = 1,...,5$$

where τ_W, τ_{Aj}^i, and τ_{Lk}^i (i=1,2,j=1,2,k=1,...,5) are joint torques of Waist Joint, Joint j of Arm i, and Joint k of Leg i and κ_W, κ_A, κ_L, σ_W, σ_A, and σ_L are gain constants.

4.3. Rhythm Generator

The rhythm generator creates rhythm information for the locomotor behavior through interactions of the robot mechanical system, the oscillator control system, and the environment. It has six phase oscillators (Leg 1, Leg 2, Arm 1, Arm 2,

Trunk, and Inter oscillators), which produce the basic rhythm for locomotion based on commands related to the locomotion speed and also receive touch sensor signals to modulate the rhythm by phase resetting. The oscillator phases follow the dynamics:

$$\dot{\varphi}_I = \omega + g_{II}$$

$$\dot{\varphi}_T = \omega + g_{IT}$$

$$\dot{\varphi}_A^i = \omega + g_{IA}^i + g_{2A}^i \qquad i = 1,2$$

$$\dot{\varphi}_L^i = \omega + g_{IL}^i + g_{2L}^i \qquad i = 1,2$$

where $\omega \left(= 2\pi / (T_{sw} + T_{st})\right)$ is the basic oscillator frequency that uses the same value for all the oscillators; g_{II}, g_{IT}, g_{IA}^i, and g_{IL}^i (i=1,2) are functions related to the interlimb coordination (see Section 4.3.1); and g_{2A}^i and g_{2L}^i (i=1,2) are functions related to the phase and rhythm modulation by phase resetting in response to the touch sensor signals (see Section 4.3.2).

4.3.1. Interlimb Coordination

To establish stable locomotion, interlimb coordination is an essential factor. For example, both legs generally move in antiphase to prevent toppling over, both arms also move in antiphase, and one arm and the contralateral leg move in phase. Because the desired limb kinematics is designed by the corresponding oscillator phase, the interlimb coordination pattern is given by the phase difference between the oscillators. We determine the desired phase differences by $\varphi_A^1 - \varphi_A^2 = \pi$, $\varphi_L^1 - \varphi_L^2 = \pi$, and $\varphi_A^1 - \varphi_L^2 = 0 \left(\varphi_A^2 - \varphi_L^1 = 0\right)$ so that both legs and arms generally move in antiphase and one arm and the contralateral leg move in phase. We use the phase differences between the oscillators based on Inter oscillator and determine the functions g_{II}, g_{IT}, g_{IA}^i, and g_{IL}^i (i=1,2) by

$$g_{11} = -\sum_{i=1}^{2} K_A \sin\left(\varphi_1 - \varphi_A^i - (-1)^i \pi / 2\right)$$

$$-\sum_{i=1}^{2} K_L \sin\left(\varphi_1 - \varphi_L^i + (-1)^i \pi / 2\right)$$

$$g_{1T} = -K_T \sin(\varphi_T - \varphi_1)$$

$$g_{1A}^i = -K_A \sin\left(\varphi_A^i - \varphi_1 + (-1)^i \pi / 2\right) \qquad i = 1, 2$$

$$g_{1L}^i = -K_L \sin\left(\varphi_L^i - \varphi_1 - (-1)^i \pi / 2\right) \qquad i = 1, 2$$

where K_L, K_A, and K_T are gain constants.

4.3.2. Phase Resetting

Phase resetting regulates the locomotor rhythm and its phase by producing phase shift and rhythm resetting based on sensory information. We incorporate this phase resetting mechanism by using the functions g_{2A}^i and g_{2L}^i (i=1,2). In particular, when the hand of Arm i (the foot of Leg i) lands on the ground, phase φ_A^i of Arm i oscillator (phase φ_L^i of Leg i oscillator) is reset to φ_{AEP}. Therefore, the functions g_{2A}^i and g_{2L}^i are expressed as

$$g_{2A}^i = (\varphi_{AEP} - \varphi_A^i)\delta(t - t_{\text{Aland}}^i),$$

$$g_{2L}^i = (\varphi_{AEP} - \varphi_L^i)\delta(t - t_{\text{Lland}}^i) \qquad i = 1, 2$$

where $\varphi_{AEP} = 2\pi(1 - \beta)$, t_{Aland}^i $\left(t_{\text{Lland}}^i\right)$ is the time when the hand of Arm i (the foot of Leg i) lands on the ground (i=1,2), and $\delta(\cdot)$ denotes Dirac's delta function. Note that the touch sensor signals not only modulate the locomotor rhythm and its phase but also switch the arm and leg movements from the swing to the stance phase, as described in Section 4.2.

4.4. Gait Generator

By using the desired limb kinematics in Figure 4, we show the desired robot kinematics for quadrupedal and bipedal locomotion in Figure 5,

where COM indicates the center of mass of the upper trunk, l_U and l_L are the lengths from COM to Joint 1 of the arm (shoulder pitch joint) and Waist Joint in the pitch plane, respectively, l_W is the length from Waist Joint to Joint 1 of the leg (hip pitch joint), L_A and L_L are the forward biases from COM to the centers of the desired foot and hand trajectories, respectively, and $()^Q$ and $()^B$ indicate the parameters for quadrupedal and bipedal walking, respectively. Note that the parameters L_A and L_L are determined by the parameters Δ_A, Δ_L, ψ_H, and ψ_W, as follows:

$$L_A = l_U \sin(\psi_H + \psi_W) + \Delta_A$$

$$L_L = l_L \sin(\psi_H + \psi_W) + l_W \sin \psi_H + \Delta_L$$

Also, note that L_A^B and L_L^B are both set to 0, as shown in Figure 5b. Since the hands do not touch the ground during bipedal locomotion, the tip of the hand follows the closed curve.

Because the gait pattern is determined by the kinematic parameters Δ_A, Δ_L, H_A, H_L, ψ_H, and ψ_W for the pitch motion and the kinematic parameter R that constructs the robot movements in the roll plane for bipedal locomotion, the gait generator determines these kinematic parameters based on the upper command for the gait pattern.

4.5. Gait Transition Control

To change the gait pattern, robot motions must connect one pattern to another. However, thousands of methods can create such robot motions, due to the redundancy of DOFs of the robot. Because our control system designs robot motions for quadrupedal and bipedal locomotion using the same kinematic parameters, robot motions during the transition can be generated by changing these parameters from Δ_A^Q, Δ_L^Q, H_A^Q, H_L^Q, ψ_H^Q, ψ_W^Q, and R^Q to Δ_A^B, Δ_L^B, H_A^B, H_L^B, ψ_H^B, ψ_W^B, and R^B based on the upper command while the robot is walking. However, many parameters still exist, and we must determine how to change them.

Figure 5. Schematic diagrams and kinematic parameters for (a) quadrupedal and (b) bipedal locomotion

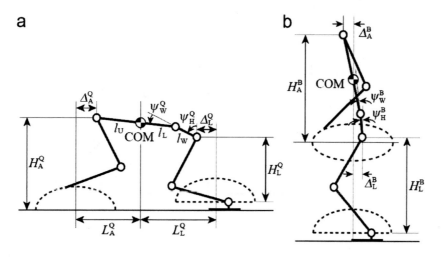

To handle the redundancy problem in the motion planning for gait transition, we construct the synergistic structures that depend on this transition task. To change the gait pattern from quadrupedal to bipedal locomotion, the robot raises its trunk and its arms leave the ground. Because the number of limbs that support the body is reduced, an adequate location between the supporting limb locations and the COM location is crucial when raising the trunk; otherwise, the robot easily falls over. In addition, because raising the trunk induces a large change in the robot posture and the COM location, the appropriate change in trunk orientation is crucial. Therefore, we use two performance variables, ξ_1 and ξ_2, that control the COM location relative to the supporting positions of the limbs and the trunk orientation, respectively, and produce robot motions by constructing a kinematically coordinative structure using these two performance variables. In particular, we encode these seven kinematic parameters Δ_A, Δ_L, H_A, H_L, ψ_H, ψ_W, and R using the performance variables ξ_1 and ξ_2 by

$$\Delta_A(\xi_1,\xi_2) = \Delta_A^Q - \{l_U \sin(\psi_H(\xi_1,\xi_2)$$
$$+\psi_W(\xi_1,\xi_2)) + \Delta_A^Q\}\xi_1$$
$$\Delta_L(\xi_1,\xi_2) = \Delta_L^Q - \{l_L \sin(\psi_H(\xi_1,\xi_2)$$
$$+\psi_W(\xi_1,\xi_2) + l_W \sin \psi_H(\xi_1,\xi_2)) + \Delta_L^Q\}\xi_1$$
$$H_A(\xi_1,\xi_2) = H_A^Q + (H_A^B - H_A^Q)\xi_2$$
$$H_L(\xi_1,\xi_2) = H_L^Q + (H_L^B - H_L^Q)\xi_2$$
$$\psi_H(\xi_1,\xi_2) = \psi_H^Q + (\psi_H^B - \psi_H^Q)\xi_2$$
$$\psi_W(\xi_1,\xi_2) = \psi_W^Q + (\psi_W^B - \psi_W^Q)\xi_2$$
$$R(\xi_1,\xi_2) = R^Q + (R^B - R^Q)(1 - e^{-\kappa\xi_2})$$

These functions mean that the performance variables ξ_1 and ξ_2 are used to horizontally move the foot and hand trajectories closer to the COM and to raise the trunk, respectively. By using this controller, gait transition is achieved by simply changing the performance variables (ξ_1, ξ_2) from $(0, 0)$ to $(1, 1)$, as shown in Figure 6. Because the roll motion is crucial during bipedal locomotion, we determine the kinematic parameter R by using an exponential function of the performance variable ξ_2, where κ is a parameter.

Figure 6. Performance variables ξ_1 and ξ_2 for gait transition from quadrupedal to bipedal locomotion

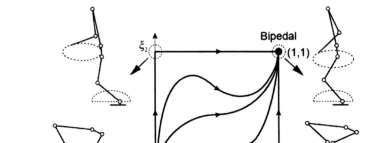

To prepare adequate kinematic coordination during the transition, we design the change of the performance variables ξ_1 and ξ_2 by the following two steps:

Step 1: When the control system receives a command to change the gait pattern, it increases the performance variable ξ_1 from 0 to $\bar{\xi}_1 \left(0 \leq \bar{\xi}_1 \leq 1\right)$ during time interval T_1.

Step 2: After Step 1 and when phase φ_1 of Inter oscillator is at φ_{raise}, the performance variables ξ_1 and ξ_2 are increased from $\bar{\xi}_1$ to 1 and from 0 to 1, respectively, during time interval T_2.

This strategy moves the foot and hand trajectories closer to the COM during Step 1 while it walks quadrupedally. During Step 2, the robot raises its trunk to start walking bipedally.

5. RESULTS

We perform the robot experiment to change the gait pattern from quadrupedal to bipedal locomotion using the proposed locomotion control system, where we use the following parameters for the robot experiment: $D = 3.0$ cm, $T_{\text{sw}} = 0.35$ s, $T_{\text{st}} =$ 0.35 s, $\phi = -150°$, $K_{\text{L}} = 2.0$, $K_{\text{A}} = 2.0$, $K_{\text{T}} = 10.0$, $\kappa = 6.0$, $\bar{\xi}_1 = 0.7$, $\varphi_{\text{raise}} = 0.83\,\pi$, $T_1 = 5.0$ s, and $T_2 = 2.0$ s; the remaining parameters are shown in Table 2.

The experimental result shows that the robot establishes stable quadruped locomotion and successfully changes the gait pattern to bipedal locomotion through Steps 1 and 2 (Figure 7). After the gait transition, the robot continues stable bipedal locomotion.

Table 2. Kinematic parameters for quadrupedal and bipedal locomotion

Parameter	Quadrupedal	Bipedal
Δ_{A} [cm]	0	0.7
Δ_{L} [cm]	0	-2.4
H_{A} [cm]	19	18
H_{L} [cm]	14	15
ψ_{H} [°]	31.4	0
ψ_{W} [°]	64.7	5.0
R [°]	0	4.0

Figure 7. Snapshots of the gait transition from quadrupedal to bipedal locomotion

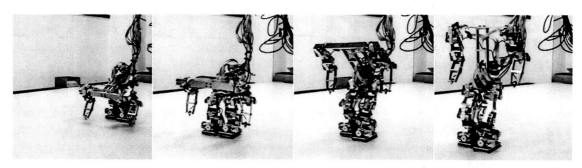

When we do not use phase resetting where the phase and rhythm are not modulated and the swing and stance phases of the desired leg and arm trajectories change at points AEP and PEP (Figure 4) independently of the touch sensor signals, the robot can not establish the gait transition and easily falls over. This shows the important contribution of phase resetting for a successful gait transition.

Since the robot raises its trunk to start walking bipedally during Step 2, the time interval T_2 is crucial for successful gait transition. In our experiment, it took almost three steps ($T_2 = 2.0$ s, $T_{sw} = 0.35$ s, $T_{st} = 0.35$ s). The Japanese macaque (*Macaca fuscata*) establishes the gait transition from quadrupedal to bipedal almost within one gait cycle (Ogihara *et al.*, 2009). To establish such an abrupt gait transition in the robot experiment, we may need more precise motion planning by considering the conservation of momentum and the zero moment point criteria (Vukobratović, et al., 1990), as well as the COM location and the trunk orientation.

planning for the gait transition and to improve the robustness of locomotion during the gait transition, we used nonlinear oscillators for the control system and designed the robot movements inspired from the physiological concepts of the CPG, phase resetting, and kinematic synergy. By incorporating the proposed control system, the robot successfully established the gait transition.

So far, many studies have investigated the gait transition from quadrupedal to bipedal walking using monkeys and examined the biomechanical and physiological differences in the control systems (Mori, et al., 2001; Mori, et al., 2006; Nakajima, et al., 2004; Ogihara, et al., 2009, 2011). Animals generate highly coordinated and skillful motions by integrating nervous, sensory, and musculoskeletal systems. By synthesizing biomechanical and physiological knowledge, robotics research is expected to contribute to the elucidation of the mechanisms in motion generation and the control strategies in the future, as well as to create new design principles for the control systems of legged robots.

6. CONCLUSION

In this study, we developed a locomotion control system for a biped robot to establish the gait transition from quadrupedal to bipedal locomotion. To tackle the redundancy problem in the motion

ACKNOWLEDGMENT

This study is supported in part by a Grant-in-Aid for Scientific Research (B) No. 23360111 from the Ministry of Education, Culture, Sports, Science, and Technology of Japan, and by JST, CREST.

REFERENCES

Alexandrov, Frolov, & Massion. (1998). Axial synergies during human upper trunk bending. *Experimental Brain Research, 118*, 210–220. doi:10.1007/s002210050274 PMID:9547090.

Alexandrov, Frolov, & Massion. (2001). Biomechanical analysis of movement strategies in human forward trunk bending. *Biological Cybernetics, 84*, 425–434. doi:10.1007/PL00007986 PMID:11417054.

Aoi, Egi, & Sugimoto, Yamashita, Fujiki, & Tsuchiya. (2012). Functional roles of phase resetting in the gait transition of a biped robot from quadrupedal to bipedal locomotion. *IEEE Transactions on Robotics, 28*(6), 1244–1259. doi:10.1109/TRO.2012.2205489.

Aoi, Ogihara, & Funato, Sugimoto, & Tsuchiya. (2010). Evaluating functional roles of phase resetting in generation of adaptive human bipedal walking with a physiologically based model of the spinal pattern generator. *Biological Cybernetics, 102*(5), 373–387. doi:10.1007/s00422-010-0373-y PMID:20217427.

Aoi & Tsuchiya. (2005). Locomotion control of a biped robot using nonlinear oscillators. *Autonomous Robots, 19*(3), 219–232. doi:10.1007/s10514-005-4051-1.

Aoi & Tsuchiya. (2007). Adaptive behavior in turning of an oscillator-driven biped robot. *Autonomous Robots, 23*(1), 37–57. doi:10.1007/s10514-007-9029-8.

Bianchi, Angelini, Orani, & Lacquaniti. (1998). Kinematic coordination in human gait: Relation to mechanical energy cost. *Journal of Neurophysiology, 79*, 2155–2170. PMID:9535975.

Burke, Degtyarenko, & Simon. (2001). Patterns of locomotor drive to motoneurons and last-order interneurons: Clues to the structure of the CPG. *Journal of Neurophysiology, 86*, 447–462. PMID:11431524.

Conway, Hultborn, & Kiehn. (1987). Proprioceptive input resets central locomotor rhythm in the spinal cat. *Experimental Brain Research, 68*, 643–656. doi:10.1007/BF00249807 PMID:3691733.

d'Avella, Saltiel, & Bizzi. (2003). Combinations of muscle synergies in the construction of a natural motor behavior. *Nature Neuroscience, 6*, 300–308. doi:10.1038/nn1010 PMID:12563264.

d'Avella & Bizzi. (2005). Shared and specific muscle synergies in natural motor behaviors. *Proceedings of the National Academy of Science USA, 102*(8), 3076-3081.

Danna-dos-Santos, Slomka, Zatsiorsky, & Latash. (2007). Muscle modes and synergies during voluntary body sway. *Experimental Brain Research, 179*, 533–550. doi:10.1007/s00221-006-0812-0 PMID:17221222.

Dominici, Ivanenko, Cappellini, d'Avella, Mondì, Cicchese, Lacquaniti. (2011). Locomotor primitives in newborn babies and their development. *Science, 334*, 997–999. doi:10.1126/science.1210617 PMID:22096202.

Drew, Kalaska, & Krouchev. (2008). Muscle synergies during locomotion in the cat: A model for motor cortex control. *The Journal of Physiology, 586*(5), 1239–1245. doi:10.1113/jphysiol.2007.146605 PMID:18202098.

Duysens. (1977). Fluctuations in sensitivity to rhythm resetting effects during the cat's step cycle. *Brain Research, 133*(1), 190-195.

Freitas, Duarte, & Latash. (2006). Two kinematic synergies in voluntary whole-body movements during standing. *Journal of Neurophysiology*, *95*, 636–645. doi:10.1152/jn.00482.2005 PMID:16267118.

Funato, Aoi, Oshima, & Tsuchiya. (2010). Variant and invariant patterns embedded in human locomotion through whole body kinematic coordination. *Experimental Brain Research*, *205*(4), 497–511. doi:10.1007/s00221-010-2385-1 PMID:20700732.

Grillner. (1975). Locomotion in vertebrates: Central mechanisms and reflex interaction. *Physiology Review, 55*(2), 247-304.

Ijspeert. (2008). Central pattern generators for locomotion control in animals and robots: a review. *Neural Networking, 21*(4), 642-653.

Ivanenko, Cappellini, & Dominici, Poppele, & Lacquaniti. (2005). Coordination of locomotion with voluntary movements in humans. *The Journal of Neuroscience*, *25*(31), 7238–7253. doi:10.1523/JNEUROSCI.1327-05.2005 PMID:16079406.

Ivanenko, Poppele, & Lacquaniti. (2004). Five basic muscle activation patterns account for muscle activity during human locomotion. *The Journal of Physiology*, *556*, 267–282. doi:10.1113/jphysiol.2003.057174 PMID:14724214.

Ivanenko, Poppele, & Lacquaniti. (2006). Motor control programs and walking. *The Neuroscientist*, *12*(4), 339–348. doi:10.1177/1073858406287987 PMID:16840710.

Kimura, Fukuoka, & Cohen. (2007). Adaptive dynamic walking of a quadruped robot on natural ground based on biological concepts. *The International Journal of Robotics Research, 26*(5), 475–490. doi:10.1177/0278364907078089.

Lafreniere-Roula & McCrea. (2005). Deletions of rhythmic motoneuron activity during fictive locomotion and scratch provide clues to the organization of the mammalian central pattern generator. *Journal of Neurophysiology*, *94*, 1120–1132. doi:10.1152/jn.00216.2005 PMID:15872066.

Liu, Habib, Watanabe, & Izumi. (2008). Central pattern generators based on Matsuoka oscillators for the locomotion of biped robots. *Artificial Life and Robotics*, *12*(1), 264–269. doi:10.1007/s10015-007-0479-z.

Mori, Mori, & Nakajima. (2006). Higher nervous control of quadrupedal vs bipedal locomotion in non-human primates: Common and specific properties. In H. Kimura, K. Tsuchiya, A. Ishiguro, & H. Witte (Eds.), *Adaptive Motion of Animals and Machines*, (pp. 53-65). New York: Springer.

Mori, Tachibana, & Takasu, Nakajima, & Mori. (2001). Bipedal locomotion by the normally quadrupedal Japanese monkey. *Acta Physiologica et Pharmacologica Bulgarica*, *26*(3), 147–150. PMID:11695527.

Nakajima, Mori, & Takasu, Mori, Matsuyama, & Mori. (2004). Biomechanical constraints in hindlimb joints during the quadrupedal versus bipedal locomotion of M. fuscata. *Progress in Brain Research*, *143*, 183–190. doi:10.1016/S0079-6123(03)43018-5 PMID:14653163.

Nakanishi, Morimoto, & Endo, Cheng, Schaal, & Kawato. (2004). Learning from demonstration and adaptation of biped locomotion. *Robotics and Autonomous Systems*, *47*(2-3), 79–91. doi:10.1016/j.robot.2004.03.003.

Nakanishi, Nomura, & Sato. (2006). Stumbling with optimal phase reset during gait can prevent a humanoid from falling. *Biological Cybernetics*, *95*, 503–515. doi:10.1007/s00422-006-0102-8 PMID:16969676.

Ogihara, Aoi, & Sugimoto, Tsuchiya, & Nakatsukasa. (2011). Forward dynamic simulation of bipedalwalking in the Japanese macaque: Investigation of causal relationships among limb kinematics, speed, and energetics of bipedal locomotion in a non-human primate. *American Journal of Physical Anthropology*, *145*(4), 568–580. doi:10.1002/ajpa.21537 PMID:21590751.

Ogihara, Makishima, & Aoi, Sugimoto, Tsuchiya, & Nakatsukasa. (2009). Development of an anatomically based whole-body musculoskeletal model of the Japanese macaque. *American Journal of Physical Anthropology*, *139*(3), 323–338. doi:10.1002/ajpa.20986 PMID:19115360.

Orlovsky, Deliagina, & Grillner. (1999). *Neuronal control of locomotion: From mollusk to man.* Oxford, UK: Oxford University Press.

Poppele & Bosco. (2003). Sophisticated spinal contributions to motor control. *Trends in Neurosciences*, *26*, 269–276. doi:10.1016/S0166-2236(03)00073-0 PMID:12744844.

Rybak, Shevtsova, Lafreniere-Roula, & McCrea. (2006). Modelling spinal circuitry involved in locomotor pattern generation: Insights from deletions during fictive locomotion. *The Journal of Physiology*, *577*(2), 617–639. doi:10.1113/jphysiol.2006.118703 PMID:17008376.

Schomburg, Petersen, Barajon, & Hultborn. (1998). Flexor reflex afferents reset the step cycle during fictive locomotion in the cat. *Experimental Brain Research*, *122*(3), 339–350. doi:10.1007/s002210050522 PMID:9808307.

Shik & Orlovsky. (1976). Neurophysiology of locomotor automatism. *Physiological Reviews*, *56*(3), 465–501. PMID:778867.

Steingrube, Timme, Wörgötter, & Manoonpong. (2010). Self-organized adaptation of a simple neural circuit enables complex robot behaviour. *Nature Physics*, *6*, 224–230. doi:10.1038/nphys1508.

Ting & Macpherson. (2005). A limited set of muscle synergies for force control during a postural task. *Journal of Neurophysiology*, *93*, 609–613. PMID:15342720.

Todorov & Jordan. (2002). Optimal feedback control as a theory of motor coordination. *Nature Neuroscience*, *5*, 1226–1235. doi:10.1038/nn963 PMID:12404008.

Vukobratović, B. Surla, & Stokić. (1990). Biped locomotion-dynamics, stability, control and application. New York: Springer-Verlag.

Yamasaki, Nomura, & Sato. (2003). Possible functional roles of phase resetting during walking. *Biological Cybernetics*, *88*, 468–496. PMID:12789495.

Chapter 3
Design for Information Processing in Living Neuronal Networks

Suguru N. Kudoh
Kwansei Gakuin University, Japan

ABSTRACT

A neurorobot is a model system for biological information processing with vital components and the artificial peripheral system. As a central processing unit of the neurorobot, a dissociated culture system possesses a simple and functional network comparing to a whole brain; thus, it is suitable for exploration of spatiotemporal dynamics of electrical activity of a neuronal circuit. The behavior of the neurorobot is determined by the response pattern of neuronal electrical activity evoked by a current stimulation from outer world. "Certain premise rules" should be embedded in the relationship between spatiotemporal activity of neurons and intended behavior. As a strategy for embedding premise rules, two ideas are proposed. The first is "shaping," by which a neuronal circuit is trained to deliver a desired output. Shaping strategy presumes that meaningful behavior requires manipulation of the living neuronal network. The second strategy is "coordinating." A living neuronal circuit is regarded as the central processing unit of the neurorobot. Instinctive behavior is provided as premise control rules, which are embedded into the relationship between the living neuronal network and robot. The direction of self-tuning process of neurons is not always suitable for desired behavior of the neurorobot, so the interface between neurons and robot should be designed so as to make the direction of self-tuning process of the neuronal network correspond with desired behavior of the robot. Details of these strategies and concrete designs of the interface between neurons and robot are be introduced and discussed in this chapter.

1. INTRODUCTION

A neurorobot is a model system in which a living neuronal circuit is electrically connected to a robot body (Bakkum, 2004; Kudoh, 2007, 2011; Novellino, 2007). The neurorobot serves as a simple model for reconnection of brain and peripherals and reconstruction of biological intelligence. The neurorobot concept was pioneered by Potter and colleagues in 2003, who developed "Hybrot," a contraction of hybrid robot (Bakkum, 2004). Hybrot is so to speak a kind of extension

DOI: 10.4018/978-1-4666-4225-6.ch003

of "Animat" reported by the same group in 2001 (DeMarse, 2001), in which a living neuronal circuit interacts with a computer-simulated environment. Although Hybrot replaced the simulated environment of Animat with a real robot body, both designs shared a common purpose: to elucidate mechanisms of biological intelligence, especially self-organization process of the intelligence during interactions between a neuronal network and the environment. Because the robot body is a mediator between neurons and environment, Hybrot exemplifies the field of "Embodied Cognitive Science," which emphasizes the critical roles of a body on biological intelligence (Brooks, 1986; Pfeifer, 1999). Although the concept was foresighted, Animat, which simply connects neurons and the environment, could not deliver its intended behavior. It appears that "certain premise rules" should be embedded in the relationship between spatiotemporal activity of neurons and intended behavior—premise rules do not autonomously emerge from random interactions.

As a strategy for embedding premise rules, two ideas are proposed. The first is "shaping" (Chao, 2008), by which a neuronal circuit is trained to deliver a desired output. This strategy requires a supervisor to examine the output. The biological analog of the supervisor is the reward system.

The second strategy is "coordinating" (Kudoh, 2007, 2011). A living neuronal circuit is regarded as the central processing unit of the neurorobot. No supervisor is assigned, and the neuronal circuit is not manipulated by other components of the neurorobot. The desired behavior is generated by simple rules embedded in the connections between the neurons and robot body. The differences between the two strategies will be discussed later in this chapter.

To date, the characteristic features of biological intelligence have been scarcely replicated by artificial intelligence. The gap between artificial intelligence and real life has been often attributed to the frame problem and symbol grounding problem. However, embodied cognitive science offers appropriate solutions to these opt-expressed problems. In the context of embodied cognitive science, the premise rules for desired behavior are embedded in the relationships between sensors and actuators. The simple subsumption architecture of this approach can be remarkably adaptable and reliable in complex environments (Brooks, 1986). A neurorobot is realization of embodiment cognitive science by incorporating a living neuronal circuit into robot system. Originally, a neurorobot and its "small brain" constituted an effective model of biological intelligence, with the robot itself performing no useful work. Currently, it is recognized as a model of biological intelligence, especially in terms of embodied cognitive science. In addition, the system can be used for testing tube for such as regenerative neuromedicine and Brain-Machine Interface (BMI). The neurorobot system is simple and versatile enough to probe related neuroengineering technologies such as neuroelectrodes and BMI decoding algorithms.

This chapter introduces the concept of the neurorobot system and its use in various technologies. Predicted future works are also discussed.

2. OUTLINE OF A NEUROROBOT

A neurorobot consists of three principal components: a Living Neuronal Circuit (LNC), neuro-device interface, and robot body.

A LNC uses a reconstructed semi-artificial neuronal network in a dissociated culture system, or a slice preparation of a brain (Bakkum, 2007). In a broad sense, a local circuit in a whole living brain can constitute the LNC of a neurorobot (Kawato, 2008); however, such configuration is generally regarded as a BMI rather than a neurorobot model.

The interface includes hardware such as electrodes, an amplifier, A/D converters, and a computer to implement software such as programs for decoding neuronal activity and stimulating the LNC. Interface software that connects the neurons to the outer world may be divided into several components, for example, an interfacing unit for the I/O of LNC and a manipulating unit for the robot. Most importantly, the robot body requires sensors that receive environmental input and actuators for manipulating the environment.

3. COMPONENTS OF A NEUROROBOT

3.1. Living Neuronal Network

As mentioned before, several LNC options are available. In this section, I focus on dissociated neuron cultures (Banker, 1977), which were adopted in our own neurorobot system (Kudoh, 2001; Kiyohara, 2010) and in the first neurorobot (DeMarse, 2001) proposed by the Potter group.

Brain tissues that are frequently used are those of rat cortex and hippocampus. Large quantities of cortical neurons are easily prepared because the cortex covers a wide area of the brain, whereas hippocampal culture is amongst the most widely used and understood techniques. In living brain tissue, neurons are connected in complex ways. Prior to culturing, a rat brain is dissected and sliced, and the neurons and surrounding supportable glial cells are mildly dissociated by a digestive enzyme such as papain or trypsin. Dissociated neuronal cells are then seeded on a culture dish precoated with adhesive materials such as electrically charged polymers or adhesive proteins. We adopted a modified conventional Banker's method, which is well known and extensively researched (Banker, 1977).

When a Multielectrodes Array (MEA) is used as an interface electrode, neuronal cells are directly cultured on the MEA dish, enabling neurons to contact to the electrodes (Gross, 1977; Pine, 1980; Robinson, 1993; Oka, 1999).

Figure 1 shows rat hippocampal neurons and glial cells cultured on an MEA dish. The number of neurons is stable after several weeks of culturing (Wolters, 2004; Kiyohara, 2010).

3.2. Stimulation and Recording Signals

In the neurorobot system, LNC electrical activity and stimulation are preferably achieved in a noninvasive manner. Typical methods for detecting neurorobot signals are extracellular potential recordings or bioimaging (Jimbo, 1993).

These methods guarantee noninvasive multisite measurements of network activity over long periods. MEA recording allows us to stimulate a LNC as well as record its activity. This feature is critical for completing closed-loop interactions. An actual neuronal circuit cultured on MEA dish with an extracellular potential multi-site recording system is popular in the area of neurorobotics (DeMarse, 2001; Bakkum, 2008; Kudoh, 2007; Novellino, 2007). Major recording and stimulating systems are available in the market, such as MEA-system and MED64 system (Oka, 1999; Kudoh, 2003). In

Figure 1. An example of a LNC on an MEA dish (E18DIV20). The black squares indicated by an arrowhead in the microphotograph are planer microelectrodes. An arrow indicates a representative neuron.

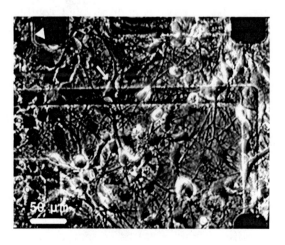

the MED64 system, the electrical signals of LNC action potentials are measured by an electrode array on an MED probe, amplified 1000-fold, converted to digital signals, and stored on the hard disk of a PC/AT compatible computer. The A/D conversion is performed at 20-kHz sampling frequency and 16 quantum bits.

3.3. Robot Body

Numerous robot body choices are available. Previously, small moving robots such as Khepera or e-puck have been adopted as the neurorobot body. An originally constructed robot, for example, a robot arm with camera vision, can also be used, provided that it is equipped with sensors and actuators. Because the neurorobot functions via closed-loop interactions between the neurons and environment, it should be able to detect environmental changes. Other commercially available robot constructs include Mindstorm NXT kit, a programmable robotics kit equipped with a 32-bit ARM7 microcontroller. Using Mindstorm NXT kit, robots of various types can be constructed, thereby allowing researchers to assess the influence of embodiment on neuron/environment interactions.

4. PRINCIPLES OF A NEUROROBOT

In this section, I focus on a neurorobot system whose LNC comprises a dissociated neuronal culture on an MEA dish. An almost reproducible pattern of neuronal activity is evoked by a particular input to this network. In addition, living neurons in the network retain synaptic plasticity (Jimbo, 1999; Van Pelt, 2005; Murata, 2011).

4.1. Autonomous Activity

The LNC is electrically stimulated and its electrical activity is recorded by the MEA-system (our group employed the MED64; Alpha MED Scientific, Japan). The LNC responses are im-

perfectly reproduced, i.e., the same input does not evoke precisely the same response (Figure 2 (a)). Depending on culture conditions, autonomous (spontaneous) electrical activity is often observed in the LNC, which influences the evoked response. Consequently, the response comprises both evoked and autonomous activity. Figure 2 (a) and (b) indicate some responses evoked by the stimulation of electrode #51 and electrode #8, respectively. The spatiotemporal response patterns evoked by the same input (a single shot of electrical stimulation to the same electrode) are very similar, unlike the responses evoked by different inputs (indicated in Figures 2 (a) and (b)). These results suggest that the LNC can express several patterns independently (Kudoh, 2003, 2011; Wagenaar, 2006; Rolston, 2007).

Although the responses evoked by the same inputs are more similar than those evoked by distinct inputs, they are nonetheless subject to fluctuations. Moreover, the responses of the cultured LNC to the same stimulation protocol are nonidentical. These fluctuations should be considered in the neurorobot decoder design.

Figure 2. Almost reproducible responses evoked by the same inputs. Left and right panels indicate the spatiotemporal patterns of responses evoked by #52 and #8 electrode, respectively. Arrows indicates electrodes for stimulation.

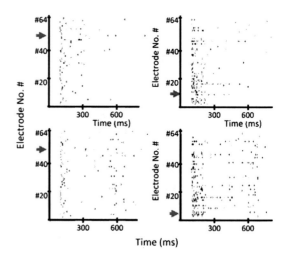

4.2. Shaping of Living Neurons

I previously mentioned "shaping" as a strategy for embedding premise rules. Essentially, shaping trains the LNC to generate a meaningful behavior of the neurorobot. Potter and colleagues, who pioneered the concept, applied appropriate training stimuli to modify the functional connections in a neuronal circuit until desired outputs were obtained. Their neurorobot was controlled by a novel statistic, the Center of neural activity of responses (CA) within 100 ms after each "proving" electrical stimulus (Bakkum, 2008; Chao, 2007, 2008) (where a "proving stimulus" is a mimicked sensory input to the brain). These stimuli were applied to the neural network every 5 s. LNC training was applied whenever the supervisor detected an unsuitable or undesired response. Training was achieved by applying Patterned Training Stimulation (PTS) to induce synaptic strength changes. If the behavior was corrected, training was terminated. Thus, Potter and colleagues succeeded in performing goal-directed behavior. They used well-designed stimulation protocols to tame the LNC (Chao, 2008). Another group elicited desired neurorobot behavior by constructing physical LNC structures (Cozzi, 2005; Novellino, 2007).

Both strategies presume that meaningful behavior requires manipulation of the LNC. Our group has proposed a different concept, in which the LNC is the sole decision maker of the neurorobot system and behavior is generated in the absence of an additional supervisor.

4.3. Living Neurons as Central Processor

We propose that the LNC is the highest information processing unit of the system (Kudoh, 2007, 2011), and should not be influenced by a supervisor. In other words, LNC is never explicitly manipulated. Instead of employing a supervisor, we adapt the relationships between neuronal activity and LNC sensor inputs. The strategy coordinates the surrounding components of LNC. Instinctive behavior is provided as premise control rules, which are embedded into the relationship between the LNC and robot. The direction of self-tuning process (such as synaptic modification) of LNC is not always suitable for desired behavior of the neurorobot, so we have to design the interface between neurons and robot so as to make the direction of self-tuning process of LNC correspond with desired behavior of the robot. Because the machine and biological component lacks a supervisor and reward system, respectively, nonteacher learning of the LNC is necessarily selected for this scheme.

In addition, the LNC network activity is detected in the absence of proving stimuli. Autonomous activity (with or without stimulation), rather than evoked responses, is evaluated continuously. The LNC is stimulated only when sensors are activated above a certain threshold. Thus, our neurorobot performs stimulation and measures activity independently and simultaneously. Consequently, the system does not discriminate autonomous activity from evoked activity. In fact, brain neurons cannot distinguish between signals originating from a sensor input and autonomous signals from other neurons in brain. Because autonomous activity can influence responses to external inputs, it can be regarded as an internal state of the LNC, whereas response fluctuations result from certain information processing. The internal states themselves can be regarded as representations of behavior or internal representations of an external object. In conclusion, a neuronal network response is not tightly coupled to environmental input, and each input recalls one of the several LNC states. Autonomous activity has been generally treated as noise (Potter, 2008). In contrast, we consider such fluctuations not as noise but as a certain type of LNC information processing (Kiyohara, 2011).

Defining the LNC as the master of the neurorobot is equivalent to creating a semi-artificial intelligence equipped with a body. The neurorobot is expected to develop cognitive functions in response to the real environment, analogous to infant development (See Figure 3). A baby constructs cognitive perception by multimodal sensing such as simultaneous looking and grasping at an object, and intelligence emerges from the relationships between corresponding perceptions circuits in the brain. The local circuits are regarded as neuron assemblies, proposed by Hebb (1949). In addition, neuron assembly and synaptic plasticity are preserved in the LNC. The remainder of biological intelligence develops from the interface between neurons and the external environment. The LNC-based neurorobot is complemented by the mobile robot body.

The embodiment of brain within body is perfected in animals. The features of external object are divided into sub classes of features detected by sensors. Evolution provides suitable (reasonable) rules to be embedded in the relationships between brain/peripheral interface (the sensors and actuators of the animal body).

In the dissociated culture of the neurorobot, no genetic constitution of suitable connections exists to guide purposive behavior. In addition, nonteacher learning must be applied to a system lacking a reward system. The connection between the LNC and robot body should favor ideal performance of the nonteacher learning (See Figure 4). We refer to a neurorobot exhibiting the previous

characteristics as "Vitroid," signifying a test tube for developing semi-living artificial intelligence (Kudoh, 2011).

5. INTERPRETERS

5.1. Overview

Complex algorithms are not required to generate simple purposive behavior. For example, in a simple collision avoidance task, the robot requires to only recognize obstacles blocking its left and right side. Thus, the LNC must discriminate just two distinct spatiotemporal patterns of activity, corresponding to the inputs from different stimulation electrodes.

A simple method of interfacing neurons and the outer environment to the control robot is weighted averaging of the spike frequencies of selected electrodes. Prior to our collision avoidance experiment, three highly active electrodes were selected for electrical input responses to obstacles located at the left and right of the robot. Collision avoidance is activated by assigning suitable weights to six spike numbers. This procedure embeds rules for reasonable behavior into the coupling algorithm.

Figure 3. Development of infant intelligence

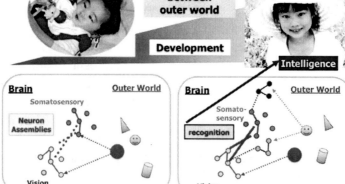

Figure 4. Embodiment schematics of an animal and a neurorobot

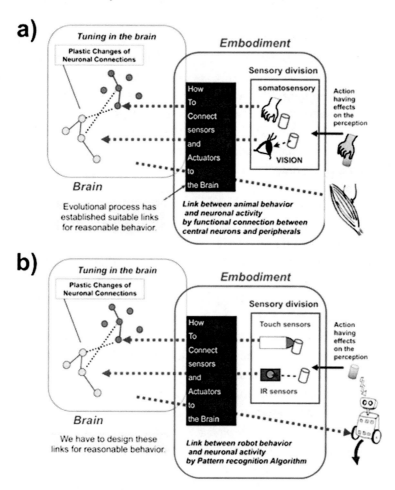

When response patterns are very discrete, simple weighted averaging of responses is sufficient. In such cases, neuronal activity is expected to settle into a stable pattern. In fact, the standard deviation in the number of evoked responses decreases during robot running. However, this result is ambiguous because the coupling rule for the neurorobot also decreases the standard deviation in the number of spontaneous activities. Because closed-loop interaction develops self-organized changes in LNC activity for stabilizing robot behavior, a more suitable linking algorithm should be sought. Such research is essential to understanding how robot behavior is linked to the neuronal network and to establishing suitable interaction schemes between the neurons and environment.

5.2. Strategy for Pattern Recognition of LNC Activity

As already mentioned, evoked LNC activity is influenced by the internal state of the network, which generates response fluctuations. In addition, the LNC activity gradually changes as the network develops. Accounting for this fluctuation and gradual change of network activity is essential. To evaluate the performance of different interfacing algorithms, the neurorobot should contain modular structures for exchanging multiple algorithms. Our Vitroid neurorobot is equipped with two interfacing programs, Brain Interface and Client. Brain Interface detects spike activity from noisy electrical signals and controls electrical stimulation. Two decoders, "output interpreter" and "input interpreter," are implemented in Client (See Figure 5). The output interpreter translates neuronal electrical activity patterns into robot behavior, whereas the input interpreter determines the stimulation pattern input to the LNN on the basis of the sensory inputs (sensors on the robot body).

Roughly two strategies are considered for pattern recognition of imperfectly reproducible LNC activity. The first is a pattern-matching approach with "fixed" pattern templates and flexible output modification (Figure 6 (a)). In this method, the templates of the output interpreter are priorly generated from the (almost) reproducible response patterns, and are thereafter fixed. The templates of the interpreter correspond to several situations; for example, the robot is located near the right wall (situation A) or midway between the left and right walls (situation B). If the input from the real world evokes a specific pattern that is 86% similar to situation A and 66% similar to situation B, the output interpreter selects a value intermediate between the ideal control values of each situation.

This strategy can be implemented using soft computing technology. In particular, fuzzy reasoning (Mamdani, 1974) is useful for such ambiguous

changes. An output interpreter is realized with a Fuzzy Pattern Template (FPT) matching method (Kudoh, 2007, 2011). The pattern templates are implemented as patterns of fuzzy "high" and "low" labels in antecedent clauses. The input patterns are compared to these templates, and the compatibility degrees are calculated on the basis of the similarities between the input pattern and each template. The actuator speed is the weighted average of the consequent-clause (then-part) values of each fuzzy rule. In this method, templates are totally fixed, whereas the actuator speed is flexibly calculated by simplified fuzzy reasoning.

The second approach uses dynamic pattern templates (Figure 6 (b)), which are the pattern templates are not fixed but updated according to the input response pattern of the neuronal network. In this strategy, the output interpreter has several classes, which are linked to particular robot behaviors. These classes include several similar patterns and are generated by a clustering process. The response pattern of the LNN is compared to class prototypes by nearest neighbor method and is allocated to a certain class. The classes are continuously updated during robot movement. If the uniformity within each class changes to meet certain criteria imposed by incoming new

Figure 5. Schematic of the proposed "vitroid" system

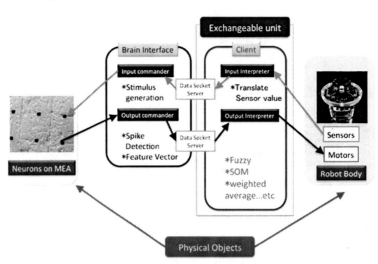

Figure 6. Strategies for pattern recognition of imperfectly reproducible LNC activity

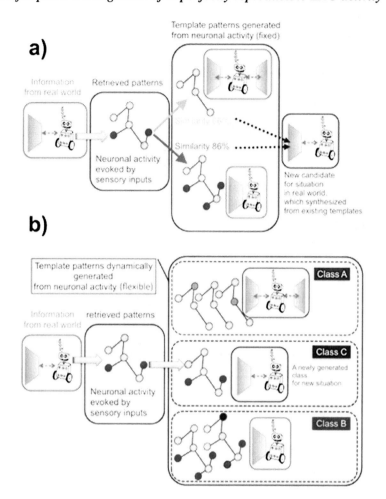

data, the output interpreter generates a new class for the new situation. In other words, the output interpreter only updates the range of classes (or the mapping between the input data and its representative position in the clustering space). Thus, the output interpreter generates a new candidate class for the real-world situation from its existing templates.

This type of output interpreter is implemented using a Self-Organizing Map (SOM; Kohonen, 2001). In our method, instead of performing clustering, a SOM gathers similar data onto an output layer map. If the distance between the center of the class and a dividing line is defined, classes are also defined. By simultaneous pattern recognition and learning, the SOM updates the mapping between the input data and their representative positions in the clustering space.

5.3. Fuzzy Template Matching

We previously the Vitroid implemented by Fuzzy Pattern Template (FPT) matching method. A set of 256 fuzzy rules are applied to eight inputs of the output interpreter. The fuzzy rules share a common set of 256 antecedent clauses (if-part) and two independent sets of 256 consequent clauses. The consequent clause of the same antecedent clause maintains the premise relationship such as symmetrical values.

This large number (256) of fuzzy rules describes all possible patterns invoked by combinations of eight inputs (2^8). The eight inputs from eight electrodes in the LNC represent the number of detected action potentials within a 50-ms time window. Each input of the fuzzy reasoning unit has two fuzzy labels, high-frequency and low-frequency. The maximum number of input spike activity of all electrodes defines the maximum of the horizontal axis in a membership function. The maximum of the membership function assigned to the high- and low-frequency labels corresponds to three-quarters and one-quarter of the points of maximum frequency, respectively. These antecedent clauses are used as pattern templates for the eight LNC inputs. The input patterns are compared to these templates, and the compatibility degrees are calculated on the basis of the similarities between the input pattern and each template. The speed of an actuator is the weighted average of consequent clause (then-part) values of each fuzzy rule. To determine the quantitative relationships between the input patterns and actuator speeds, we adjust the consequent clauses of each fuzzy rule by teacher learning.

Teacher signal is a set of speed values of both sides of the actuators, derived from the values of the IR sensors on the robot body. If the sum of the IR sensor values on the left side of the robot body exceeds that of the right side, the teacher signal is set to the desired values of actuator speeds; 10 and 0 for the left actuator and right actuator, respectively. This process links the spatiotemporal pattern of electrical neuron activity (evoked by an electrical input determined by IR sensor signals) to the desired behavior (generated by balancing the speeds of actuators on both sides of the robot body) (See Figure 7).

5.4. Self-Organization Map

Electrical activity patterns within a certain time window (in our case, 50 ms) constitute a feature vector of 64 spike numbers detected at each electrode. Basic SOM algorithms are employed for

Figure 7. Fuzzy reasoning used for pattern recognition

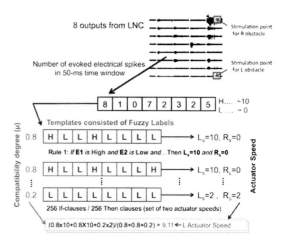

the output interpreter. The input layer containing 64 nodes receives the feature vector. The SOM features a two-dimensional output layer containing 10×10 nodes for dimensional reduction of the feature vector. The SOM determines a winner node in the output layer encoded in an input vector, as shown in Figure 8. Because the winner node is located in a 2D map, a 64-dimensional vector is mapped to 2D winner node. The node with the shortest Euclidean distance between an input feature vector and a corresponding reference vector is designated the winner for the input vector. The reference vector of the winner is then updated to more closely match the input vector. At the same time, the reference vectors of the neighboring

Figure 8. Schematic diagram of a SOM for the neurorobot

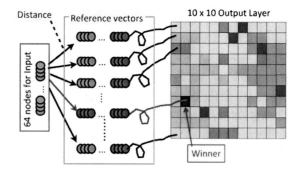

nodes of the winner node are updated toward the input vector. This procedure places nodes with similar reference vectors within neighborhoods. Consequently, spatial relationships between the nodes in the output layer correspond to the relationships between the input vectors coupled to those nodes. These steps are implemented in an orthodox SOM algorithm. We aim to define a reproducible pattern that represents information in the living neuronal network. The coupling of gathered nodes to similar feature vectors ensures that similar spatiotemporal patterns of network activity are gathered together.

The spatial location of the winner depends on the initial state of the reference vectors. If a certain electrode (say E1) is stimulated by an input from the robot sensor, a winner node should be located at a certain position in the output layer of the SOM. If a winner activated by E2 stimulation is located near that activated by E1, the outputs of the living neuronal network responding to the E1 and E2 inputs will be similar. In generating the robot behavior, instinctive behaviors such as collision avoidance should be incorporated as premise behaviors. In this case, ideal winner distribution will be predetermined. If the map location of a representative winner for the E1 input is specified, a suitable behavioral response to the E1

input can be defined. For example, if the E1 input signals an obstacle at the left side of the robot, the robot should turn to the right; thus, the left motor speed should be set higher than that of the right motor. To constrain the winner corresponding response to a particular input, the winner for that input should be fixed at a seed node and the E1 input should be repeatedly applied during the initial learning process (Figure 9). Thereafter, winners responding to the E1 input are expected to be gathered near the seed node. The robot is controlled by the position of the winner for a specified input activity. For example, the left and right motor speeds are calculated by the weighted average of the total network activity, where the weights are the Euclidian distances between the winner and E1 or E2 seed node.

6. EXPERIMENTS USING VITROID

6.1. Vitroid with FPT Matching Algorithm

Vitroid operating with the FPT matching algorithm navigated a course between two parallel walls without collision (Figure 10). The length of the course was 1200 mm, and the distance between the

Figure 9. Generation of robot behavior with SOM

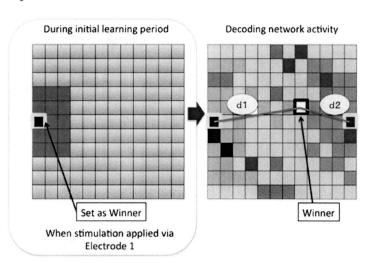

Figure 10. Test course for robot running and robot body (Khepera II)

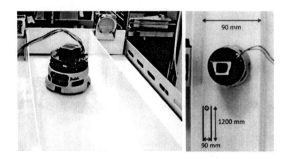

walls was 90 mm. The robot was initially placed midway between the walls at one end of the course, and connected to a set of control programs, Brain Interface and Client.

At each of the eight recording sites, the number of spikes evoked within a 50-ms time window and the fuzzy grades of the 256 rules were analyzed during robot running. In the latter part of the running, collision avoidance became gradually delayed (Figure 11(a)). While the variance in the number of electrical spikes evoked by the inputs tends to decrease during robot movement, the standard deviation in the number of electrical spikes evoked by spontaneous activity clearly increases. This suggests that the network dynamics are modified by the robot behavior, especially in response to the inputs (Figure 11 (b)).

Figure 11. (a) An example of the trajectory, classified inputs (L, R, SPT), and number of inputs corresponding to the activities in eight electrodes, and the fuzzy grade (compatibility degrees) of the top five rules during an experiment with 1 mM Mg²⁺ recording solution. (b) An example of the variance in the number of electrical spikes evoked by the inputs at the middle and latter stages of the experiment.

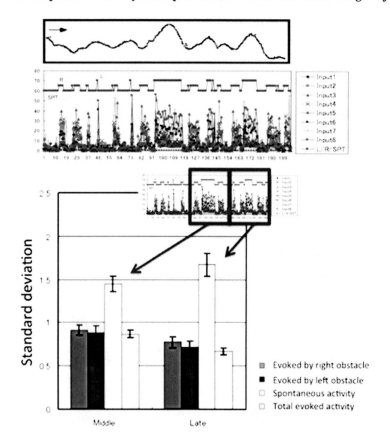

6.2. SOM Identifier for Spatiotemporal Pattern of the Spike Activity

The recognition resolution of electrical spike patterns in a neuronal network depends on the number of nodes in the SOM output layer. Adequate node number and nodal arrangement can only be determined empirically. To this end, the number of nodes was initially set as 10×10. At this grid setting, spike patterns evoked by the inputs from two distinct electrodes were separated. After 40 seedings, the winner nodes corresponding to the input spike patterns had been almost separated, with most winners correctly assigned to the left or right section of the output layer. Only a small minority of nodes was assigned the wrong location (Figure 12).

Figure 12. (a) An example of the spatial distribution of winner nodes for 20 input feature vectors. The grey squares indicate the winner nodes for the input feature vector of spike pattern evoked by the stimulation labeled "L." The black squares indicate the winner nodes for the stimulation labeled "R." (b) Plot of the number of occurrences of a selected number, as a function of the number of selection as winner nodes for L stimulation (solid line) and R stimulation (dotted line).

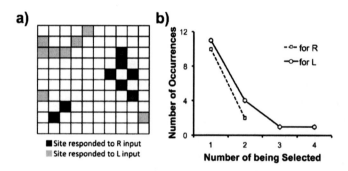

Figure 13. An example of the trajectory of an e-puck robot controlled by the proposed method with a SOM. This image was synthesized by merging frames extracted every 1-s from a 40-s movie. Electrical stimuli were applied to a living neuronal circuit via two electrodes linked to the left and right turns of the e-puck robot. One of these electrodes was randomly selected and electrical stimulation was applied every 1-s. The e-puck robot successfully behaved according to the spike pattern of a living neuronal circuit. The solid circle and arrow indicate starting position and initial moving direction of the robot.

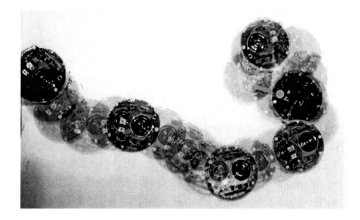

The number of selection as winner node was rarely more than one or two out of 40 selections. The spike pattern is expected to be periodic and such patterns should appear repeatedly. The classification of spike patterns was possibly too sensitive for a small number of representative patterns of electrical spikes evoked by just two types of stimulation. At this stage, we have confirmed that electrical spike patterns from 64 sites in a living neuronal network can be recognized by our method. In addition, the robot body (e-puck, implemented using EPFL) can be controlled by SOM-generated behaviors (Figure 13). Currently, we are examining the optimal number and arrangement of nodes in the output layer of the SOM.

7. DISCUSSION

Neurorobots have been proposed by several groups as a model system for investigating brain and peripheral connections and reconstructing biological intelligence. In a neurorobot system, a LNC is connected to a moving sensor-equipped robot via interface hardware and decoding software. Two main strategies exist for the connecting scheme; shaping and coordinating of the LNC. In addition, it is critical that whether autonomous activity and response fluctuations are regarded as mere noise or results of information processing. Handling such fluctuations is especially critical in the coordinating strategy, in which the LNC is regarded as the highest central processor of a neurorobot.

Decoding algorithms have been proposed, such as CA, simple weighted averaging, FTP matching, and SOM. These methods incorporate embedded rules and specific summations of input signals.

Placing excessive faith in fluctuated activity defeats the purposive behavior of the neurorobot. Furthermore, LNC activity is developing during culturing. Therefore, to maintain the desired robot behavior, buffering or coordinating is required. On the other hand, over-buffering the fluctuations may destroy the effective closed-loop interaction between the LNC and its external environment. The LNC neurons receive input from the environment in response to their prior activity. If the interpreter absorbs the fluctuations too effectively, the neurons cannot correctly interpret their external environment. Besides the behavioral performance of the neurorobot, linking algorithms should also be evaluated from this perspective.

One fundamental problem is the innate behavior of the LNC. The reproducibility of spike pattern evoked by a certain input and the number of separable patterns depend on the network structure and functional connectivity between living neurons. This issue is pertinent to the stimulation method. Biological conditions and LNC treatment methods are critical in realizing a stable control of the neurorobot.

These discussions and investigations of the neurorobot will contribute to the development of new methods for regenerative neuromedicine and the creation of novel algorithms for information processing.

REFERENCES

Bakkum, D. J., Gamblen, P. M., Ben-Ary, G., Chao, Z. C., & Potter, S. M. (2007). MEART: The semi-living artist. *Frontiers in Neurorobotics*, *1*(5), 1–10. doi: doi:10.3389/neuro.12.005.2007 PMID:18958272.

Bakkum, D. J., Shkolnik, A. C., Ben-Ary, G., Gamblen, P., DeMarse, T. B., & Potter, S. M. (2004). Removing some 'A' from AI: Embodied cultured networks. In Iida, F., Pfeifer, R., Steels, L., & Kuniyoshi, Y. (Eds.), *Embodied Artificial Intelligence* (pp. 130–145). New York: Springer. doi:10.1007/978-3-540-27833-7_10.

Banker, G. A., & Cowan, W. M. (1977). Rat hippocampal neurons in dispersed cell culture. *Brain Research*, *126*, 397–342. doi:10.1016/0006-8993(77)90594-7 PMID:861729.

Brooks, R. (1986). A robust layered control system for a mobile robot. *IEEE Journal on Robotics and Automation, 2*, 14–23. doi:10.1109/JRA.1986.1087032.

Chao, Z. C., Bakkum, D. J., & Potter, S. M. (2007). Region-specific network plasticity in simulated and living cortical networks: Comparison of the center of activity trajectory (CAT) with other statistics. *Journal of Neural Engineering, 4*, 1–15. doi:10.1088/1741-2560/4/3/015 PMID:17409475.

Chao, Z. C., Bakkum, D. J., & Potter, S. M. (2008). Shaping embodied neural networks for adaptive goal-directed behavior. *PLoS Computational Biology, 4*(3). doi:10.1371/journal.pcbi.1000042 PMID:18369432.

Cozzi, L., D'Angelo, P., Chiappalone, M., Ide, A. N., Novellino, A., Martinoia, S., & Sanguineti, V. (2005). Coding and decoding of information in a bi-directional neural interface. *Neurocomputing, 65*, 783–792. doi:10.1016/j.neucom.2004.10.075.

DeMarse, T. B., Wagenaar, D. A., Blau, W. P. A., & Potter, S. M. (2001). The neurally controlled animat: Biological brains acting with simulated bodies. *Autonomous Robots, 11*, 305–310. doi:10.1023/A:1012407611130 PMID:18584059.

Gross, G. W., Rieske, E., Kreutzberg, G. W., & Meyer, A. (1977). A new fixed-array multi-microelectrode system designed for long term monitoring of extracellular single unit neuronal activity in vitro. *Neuroscience Letters, 6*, 101–105. doi:10.1016/0304-3940(77)90003-9 PMID:19605037.

Gross, G. W., Williams, A. N., & Lucas, J. H. (1982). Recording of spontaneous activity with photo etched microelectrode surfaces from mouse spinal neurons in culture. *Journal of Neuroscience Methods, 5*(1–2), 13–22. doi:10.1016/0165-0270(82)90046-2 PMID:7057675.

Hebb, D. O. (1949). *The organization of behaviour: A neuropsychological theory*. New York: John Wiley & Sons.

Jimbo, Y., Robinson, H. P., & Kawana, A. (1993). Simultaneous measurement of intracellular calcium and electrical activity from patterned neural networks in culture. *IEEE Transactions on Bio-Medical Engineering, 40*, 804–810. doi:10.1109/10.238465 PMID:8258447.

Jimbo, Y., Tateno, T., & Robinson, H. P. C. (1999). Simultaneous induction of pathway-specific potentiation and depression in networks of cortical neurons. *Biophysical Journal, 76*, 670–678. doi:10.1016/S0006-3495(99)77234-6 PMID:9929472.

Kawato, M. (2008). Brain controlled robots. *HFSP Journal, 2*(3), 136–142. doi:10.2976/1.2931144 PMID:19404467.

Kiyohara, A., Taguchi, T., & Kudoh, S. N. (2011). Effects of electrical stimulation on autonomous electrical activity in a cultured rat hippocampal neuronal network. *IEEJ Transactions on Electrical and Electronic Engineering, 6*(2), 163–167. doi:10.1002/tee.20639.

Kohonen, T. (2001). *Self-organizing maps* (3rd ed.). New York: Springer. doi:10.1007/978-3-642-56927-2.

Kudoh, S. N., Hosokawa, C., Kiyohara, A., Taguchi, T., & Hayashi, I. (2007). Biomodeling system—Interaction between living neuronal networks and the outer world. *Journal of Robotics and Mechatronics, 19*(5), 592–600.

Kudoh, S. N., Nagai, R., Kiyosue, K., & Taguchi, T. (2001). PKC and CaMKII dependent synaptic potentiation in cultured cerebral neurons. *Brain Research, 915*(1), 79–87. doi:10.1016/S0006-8993(01)02835-9 PMID:11578622.

Kudoh, S. N., & Taguchi, T. (2003). Operation of spatiotemporal patterns in living neuronal networks cultured on a microelectrode array. *Journal of Advanced Computational Intelligence and Intelligent Informatics, 8,* 100–107.

Kudoh, S. N., & Taguchi, T. (2003). Operation of spatiotemporal patterns in living neuronal networks cultured on a microelectrode array. *Journal of Advanced Computational Intelligence and Intelligent Informatics, 8,* 100–107.

Kudoh, S. N., Tokuda, M., Kiyohara, A., Hosokawa, C., Taguchi, T., & Hayashi, I. (2011). Vitroid—The robot system with an interface between a living neuronal network and outer world. *International Journal of Mechatronics and Manufacturing Systems, 4*(2), 135–149. doi:10.1504/IJMMS.2011.039264.

Mamdani, E. H. (1974). Application of fuzzy algorithms for the control of a simple dynamic plant. *Proceedings of the IEEE, 121*(12), 1585–1588.

Murata, M., Ito, H., Taenaka, T., & Kudoh, S. N. (2011). Modification of activity pattern induced by synaptic enhancements in a semi-artificial network of living neurons. In Proceedings of the International Symposium on Micro-NanoMechatronics and Human Science (MHS), (pp. 250–254). MHS.

Novellino, A., D'Angelo, P., Cozzi, L., Chiappalone, M., Sanguineti, V., & Martinoia, S. (2007). Connecting neurons to a mobile robot: An in vitro bidirectional neural interface. *Journal of Computational Intelligence and Neuroscience.* doi: 10.1155/2007/12725

Oka, H., Shimono, K., Ogawa, R., Sugihara, H., & Taketani, M. (1999). A new planar multielectrode array for extracellular recording: Application to hippocampal acute slice. *Journal of Neuroscience Methods, 93,* 61–67. doi:10.1016/S0165-0270(99)00113-2 PMID:10598865.

Pfeifer, R., & Scheier, C. (1999). *Understanding intelligence.* Cambridge, MA: The MIT Press.

Pine, J. (1980). Recording action potentials from cultured neurons with extracellular microcircuit electrodes. *Journal of Neuroscience Methods, 2,* 19–31. doi:10.1016/0165-0270(80)90042-4 PMID:7329089.

Potter, S. M. (2008). How should we think about bursts? In *Proceedings of the 6th International Meeting on Substrate-Integrated Microelectrodes.* ISBN 3-938345-05-5

Robinson, H. P., Kawahara, M., Jimbo, Y., Torimitsu, K., Kuroda, Y., & Kawana, A. (1993). Periodic synchronized bursting and intracellular calcium transients elicited by low magnesium in cultured cortical neurons. *Journal of Neurophysiology, 70*(4), 1606–1616. PMID:8283217.

Rolston, J. D., Wagenaar, D. A., & Potter, S. M. (2007). Precisely-timed spatiotemporal patterns of neural activity in dissociated cortical cultures. *Neuroscience, 148,* 294–303. doi:10.1016/j.neuroscience.2007.05.025 PMID:17614210.

Van Pelt, J., Vajda, I., Wolters, P. S., Corner, M. A., & Ramakers, G. J. (2005). Dynamics and plasticity in developing neuronal networks in vitro. *Progress in Brain Research, 147,* 173–188. doi:10.1016/S0079-6123(04)47013-7 PMID:15581705.

Wagenaar, D. A., Pine, J., & Potter, S. M. (2006). An extremely rich repertoire of bursting patterns during the development of cortical cultures. *BMC Neuroscience, 7*(11). doi: doi:10.1186/1471-2202-7-11 PMID:16464257.

Wolters, P. S., Rutten, W. L., Ramakers, G. J., Van Pelt, J., & Corner, M. A. (2004). Long term stability and developmental changes in spontaneous network burst firing patterns in dissociated rat cerebral cortex cell cultures on multielectrode arrays. *Neuroscience Letters, 361,* 86–89. doi:10.1016/j.neulet.2003.12.062 PMID:15135900.

Chapter 4
Novel Swimming Mechanism for a Robotic Fish

Sayyed Farideddin Masoomi
University of Canterbury, New Zealand

Stefanie Gutschmidt
University of Canterbury, New Zealand

XiaoQi Chen
University of Canterbury, New Zealand

Mathieu Sellier
University of Canterbury, New Zealand

ABSTRACT

Efficient cruising, maneuverability, and noiseless performance are the key factors that differentiate fish robots from other types of underwater robots. Accordingly, various types of fish-like robots have been developed such as RoboTuna and Boxybot. However, the existing fish robots are only capable of a specific swimming mode like cruising inspired by tuna or maneuvering inspired by labriforms. However, for accomplishing marine tasks, an underwater robot needs to be able to have different swimming modes. To address this problem, the Mechatronics Group at University of Canterbury is developing a fish robot with novel mechanical design. The novelty of the robot roots in its actuation system, which causes its efficient cruising and its high capabilities for unsteady motion like fast start and fast turning. In this chapter, the existing fish robots are introduced with respect to their mechanical design. Then the proposed design of the fish robot at University of Canterbury is described and compared with the existing fish robots.

1. INTRODUCTION

Recent advances in robotics have enabled underwater robots to replace humans in oceanic supervision, aquatic life-form observation, pollution search, undersea operation, military detection and so on (Junzhi, Min, Shuo, & Erkui, 2004). Accordingly, a number of underwater vehicles and robots such as Remotely Operated Vehicles (ROVs) and Autonomous Underwater Vehicles (AUVs) have been developed so far (Griffiths & Edwards, 2003; Wernli, 2000; Williams, 2004).

Among underwater vehicles, biomimetic swimming robots inspired by various types of fishes have shown superior performance over other types of underwater robots (Paulson, 2004). These fish-mimetic robots are highly efficient, manoeuvrable and noiseless in marine environment

DOI: 10.4018/978-1-4666-4225-6.ch004

(Hu, Liu, Dukes, & Francis, 2006). For instance, the propulsion system for some fishes is up to 90 percent efficient, while a conventional screw propeller has an efficiency of 40 to 50 percent (Yu & Wang, 2005).

Fish robots could be defined as fish-like aquatic vehicles whose motion is generated through undulatory and/or oscillatory motion of either body or fins (Hu et al., 2006).

The first fish robot, RoboTuna, was built at MIT in 1994 (Triantafyllou & Triantafyllou, 1995). As its name indicates, RoboTuna was inspired by a bluefin tuna. Developing this robot was a successful project of mimicking a fish robot; although, it has several deficiencies like being carriage mounted and non-autonomous. Three years later, at Charles Stark Draper Laboratory, the Vorticity Control Unmanned Undersea Vehicle (VCUUV) was developed based on the RoboTuna with some improvement and more capabilities such as being autonomous, capable of avoiding obstacles and having up-down motion (Anderson & Chhabra, 2002; Liu & Hu, 2004). However, VCUUV is more appropriate for a specific mode of swimming mainly cruising.

Accordingly, a number of robotic fishes for different mode of swimming were developed like lamprey robots capable of backward swimming (Ayers, Wilbur, & Olcott, 2000), MARCO inspired by boxfish suitable for maneuvering (Kodati, Hinkle, & Deng, 2007) and so on. Yet the state of the art in robotic fish shows that the robots built so far cannot have excellent performance in several swimming modes. For instance, tuna-mimicking robots are good at cruising while boxfish-mimicking robots are more suitable for hovering and maneuvering.

To have a skilled robot for two modes of swimming, the Mechatronics Group at University of Canterbury is developing a fish robot with novel mechanical design and control system. The novelty of the robot roots in its actuation system and causes its efficient cruising and its high capabilities for unsteady motion like fast start and fast turning. In the following chapter, the proposed mechanical design of the robot is introduced and discussed with respect to state of the art in the robotic fish.

The structure of the chapter is as follows. Section 2 introduces two main fish swimming categories and their subcategories. In this section, the state of the art in robotic fish is presented with respect to their swimming mechanisms. Section 3 discusses the capabilities and deficiencies of previously developed fish robots with respect to three main aspect of their mechanical design. Section 4 introduces the robotic fish designed at University of Canterbury. Section 5 concludes the chapter by comparing the mechanical design introduced in section 4 and fish robots developed by other institutes and universities.

2. LITERATURE REVIEW

Fish robots generally do not mimic the same fish motion in nature and, hence, have different swimming mechanisms. The main element which distinguishes fish robots from other types of underwater vehicles is their propulsion system. Fishes propel through undulation or oscillation of different parts of their body or fins called propulsors. When a fish passes a propulsive wave by its body or its fins in the opposite direction of its movement at a faster speed than swimming speed, its swimming method is referred to as undulation. On the other hand, in oscillation mode, fish generates propulsive waves by oscillating a certain part of its body around its base (Sfakiotakis, Lane, & Davies, 1999). Figure 1 presents some basic terminologies used in this chapter.

Taking fish propulsors into account, fish swimming modes are categorized into two main groups. Some fishes swim using their Body and/or Caudal Fins (BCF), whereas others apply their Median and/or Paired Fin (MPF). Figure 2 demonstrates the aforementioned swimming mode (Sfakiotakis et al., 1999).

Figure 1. Fish terminologies used in this chapter (Sfakiotakis, et al., 1999)

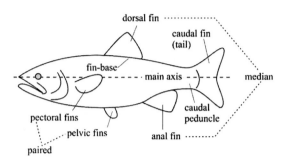

2.1. Body and/or Caudal Fin

As Figure 2(a) shows, BCF mode could be further distinguished by undulatory and oscillatory swimming to five subcategories: (1) anguilliform, (2) subcarangiform, (3) carangiform, (4) thunniform, and (5) osctraciiform.

2.1.1. Anguilliforms

The *anguilliforms* like eel and lamprey are the most undulatory fishes. These fishes need to be modeled by a long, slender, and flexible structure.

Anguilliforms are maneuverable enough to pass through narrow water area like coral reef. Besides, they are able to have backward swimming.

Anguilliform-like robot has also a big advantage in comparison with other BCF undulatory fish-like robot. Provided that the anguilliform structure consists of several equally distributed propulsors along the body, at least a complete wavelength is formed by propulsors. Accordingly, at least two lateral and thrust forces are created by the wave. The lateral forces have equal magnitude but counteract each other whereas the thrust forces act in the same direction. Then the fish robot does not oscillate around its center of mass or move laterally.

Considering the aforementioned advantages, anguilliforms are suitable robot for motion planning (McIsaac & Ostrowski, 2003). Despite the advantages, anguilliforms are not as fast and efficient swimmers as carangiforms or thunniforms. Some examples of the anguilliform robot are given in the following paragraphs.

First, Ayers et al. (Ayers et al., 2000) have developed a biomimetic lamprey robot shown in Figure 3(a). This robot includes a rigid head,

Figure 2. Fish swimming modes: (a) BCF mode and (b) MPF mode (Sfakiotakis, et al., 1999)

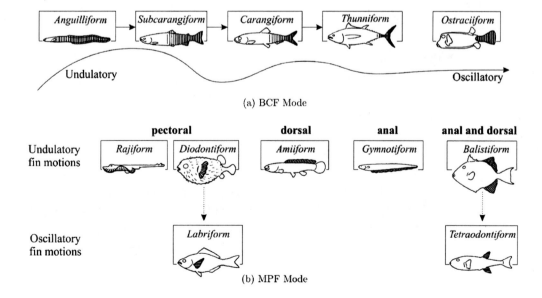

Figure 3. Two fish robot inspired by anguilliform and subcarangiforms: (a) lamprey robot and (b) trout robot (Salumäe, 2010)

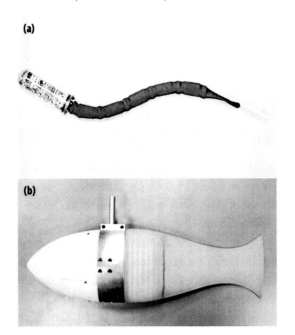

(a)

(b)

a flexible body and a passive tail. The flexible body is actuated by shape memory alloy artificial muscles. The artificial muscles are propagated on either side of the body. Actuation of these muscles in sequence provides the rhythmic lateral undulation of the robot that causes the swimming of the robot. The lamprey robot could also propel backward by reversing its rhythmic undulation.

Recently, Boyers et al. (Boyer, Chablat, Lemoine, & Wenger, 2009) have started to develop an eel-like swimming robot using parallel mechanism. This robot includes a head, 12 vertebrae with 3 Degrees Of Freedom (DOF) of rotation for each vertebra, and a tail. All vertebrae have identically parallel architectures but are mounted in series together. Each vertebra and the robot are actuated by electric motors. For the skin of the robot, rubber rings with intermediate rigid sections are used in order to provide easy distortion and also resist the pressure of the water around it.

2.1.2. Subcarangiforms

In comparison with anguilliforms, if the amplitude of the undulation wave decreases in the anterior part of the fish body and increases in the posterior part, the mode of swimming changes to *subcarangiform* mode.

The trout robot is an example of subcarangiform-mimetic robot (Salumäe, 2010) shown in Figure 3(b). The 0.5-meter artificial trout has three main parts: a nose cover, a middle flange and a silicone tail. All the electronics and actuation components are inside nose cover and the flange. The propulsion is made in the posterior part of the robot using a rotational actuator. In the other words, a DC motor is connected to a plate by two flexible steel cables while the plate is casted inside the tail. A sinusoidal rotation of the motor causes undulation of the tail to propel the robot.

As it has been described, Salmuae has not followed the traditional procedure to develop a swimming robot using multiple linkages. Instead, Salumae has implemented the work done by Alvardo (Alvarado, 2007) to develop a flexible composite model as the propulsor of the robot mimicking the morphology of a real fish, in particular, the geometry, stiffness and stiffness distribution of the body, and the caudal fin.

2.1.3. Carangiforms and Thunniforms

The next group of BCF mode is *carangiform*. In this swimming mode, the undulation is just by the last third of the body length in a way that thrust is generated more by the caudal fin. Carangiforms like mackerel usually have narrow peduncle and a tall forked caudal fin (Colgate & Lynch, 2004).

Usually the fastest and the most efficient carangiforms are categorized as *thunniforms*. Although, this is not a fully representative definition for thunniform since they apply different kind of hydrodynamic propulsion in comparison with carangiforms. The high speed and efficiency of thunniform in cruising is due to quite streamlined and optimized body of thunniforms, their crescent shape caudal fin and their lift-based method of propulsion. These fishes are up to 90% efficient (Sfakiotakis et al., 1999). Nevertheless, thunniforms are notorious for not being maneuverable. Because maneuverability needs quick changes of the acceleration while lift-based systems are not advantageous for those changes (Vogel, 1996).

Note that cetaceans have reasonably similar swimming mechanisms of thunniforms. The only difference between cetaceans and thunniforms is the shape of their tail. Cetaceans have horizontal fluke while thunniforms have vertical caudal fin. Due to the similarity in swimming mechanisms, dolphin-like robots are discussed in this section.

In addition, since the swimming mechanisms of carangiforms and thunniforms are moderately similar, a number of works done on thunniforms is called carangiforms in literature. In fact, in many cases the robots mimicking carangiforms and thunniforms are not distinguishable. Accordingly, in the following paragraphs the examples of both categories are discussed together.

The first fish robot, RoboTuna, built at MIT in 1994 is the prominent example of thunniform-mimetic robots (Triantafyllou & Triantafyllou, 1995). Six years later, RoboTuna was improved as RoboTuna II (Beal, 2003; Jakuba, 2000). Being carriage mounted and using external power support, these robots mimic the swimming mechanism of a bluefin tuna. RoboTuna has eight linkages while six links are independently actuated by six powerful servo motors. Figure 4 shows RoboTuna and RoboTuna II.

At MIT, a carangiform-like robot, RoboPike, was built as well (Triantafyllou, Kumph, et al., 2000). As its name indicates, this robot is inspired by a pike. On the contrary of previous robots, this robot is a free swimming robot consisted of three bending segments. The robot has three DOF, two of which are for producing undulation and one DOF is for changing the angle of caudal fin. RoboPike is actuated by DC servo motors. While

Figure 4. Tuna-like robots built at MIT: (a) RoboTuna I and (b) RoboTuna II

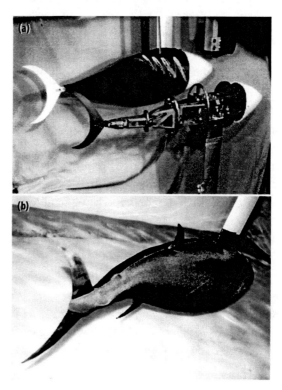

the undulation is to propel the robot, the pectoral fins are employed as rudders. The robot has a spring-wound fiberglass exoskeleton and a skin made of silicon rubber.

Inspired by RoboTuna, Vorticity Control Unmanned Undersea Vehicle (VCUUV) was developed in 1997. Similar to RoboPike, VCUUV is a free swimming robot that applies its tail for propulsion and pectoral fins for steering. Using hydraulic power unit, VCUUV has four active links to create undulation. This fish robot is able to go up-down underwater and avoid obstacles (Anderson & Chhabra, 2002; Kodati et al., 2007).

The mackerel-mimetic robot, BASEMACK1, developed by Lee et al. (Lee, Park, & Han, 2007) is another example of carangiform robots. BASEMACK1 has three links forming tail peduncle. The first link is attached to its front body (head) and the last one is attached to the caudal fin. The caudal fin which is shaped based on real mackerel tail is made from 1 mm thick flexible metal. All the links are actuated by DC servo-motors. BASEMACK1 has mimicked mackerel geometrically.

Regarding carangiforms, several fish robots have been developed at University of Essex (Liu, Dukes, & Hu, 2005; Liu, Dukes, Knight, & Hu, 2004). Although the robots built in this university are divided into two groups of G and MT series, they all produce undulatory motion using multi-linkages. For more information the reader is referred to (Jindong Liu, 2006). Figure 5 illustrates the fish robots model G9 and MT1.

At University of Washington, Morgenson et al. (Morgansen, 2003; Morgansen, Triplett, & Klein, 2007) have also developed a group of robotic fishes that have characteristic of a carangi-

form robot; however, they could use their pectoral fins for surfing and diving. As can be seen in Figure 6(a), this robot does not mimic the body shape of a real fish.

At Beihang University, several robotic fishes, SPC series, have been developed (Liang, Wang, & Wen, 2011; Wang, Liang, Shen, & Tan, 2005). These torpedo body shape robots swim using 2-link caudal fin. Beside caudal fin, the robot make benefit of two fixed dorsal and anal fins for stabilization. SPC series robots are built to study the performance of tail fin. Robofish and SPC-III developed at University of Washington and Beihang University are illustrated in Figure 6.

First dolphin-like robot is developed by Nakashima and Ono (2002). They built a three linkages robot. Using spring and damper at joints, they made the robot self-propelled. Soon

Figure 5. Two robots developed at the University of Essex: (a) model G9 and (b) model MT1 (Dag, 2009)

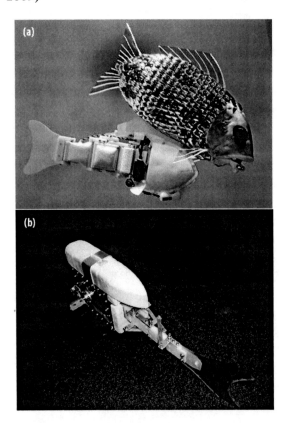

Figure 6. Fish robot built at (a) University of Washington, RoboFish (Morgansen, et al., 2007), and (b) Beihang University, SPC-III

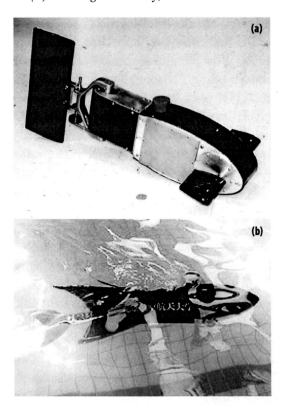

after, Nakashima et al. (Nakashima, Takahashi, Tsubaki, & Ono, 2004) built a new dolphin robot that use DC motor in the first joint while the next joint is passive. The new robot was able to have three-dimensional motion.

In 2005, Dogangil et al. built a pneumatically driven four-link dolphin robot (Dogangil, Ozcicek, & Kuzucu, 2005). Afterwards, Yu et al. built five-link dolphin robot actuated by five DC servo motors (Yu, Hu, Fan, Wang, & Huo, 2007). However, in order to increase the efficiency Yu et al. (Junzhi Yu, Hu, Huo, & Wang, 2007) designed a two-motor driven scotch mechanism for undulation of the tail. The length of crank in this mechanism is adjustable. Figure 7 shows uncoated dolphin robot built by Yu et al. (2007).

Carangiform and thunniform robots built thus far are not limited to what have been mentioned in the previous paragraphs. However, similar works are described in this section. For instance, the works done by Mason and Burdick (2000), Saimek and Li (2004), and Kim et al. (Kim, Lee, & Kim, 2007) are not described since similar fish robots whose undulation is caused by multi-linked actuation are introduced.

2.1.4. Ostraciiform

Ostraciiform, as the last propulsion mode of BCF, is related to fishes which only use their caudal fin to provide thrust by simply oscillating around a point at the end of their rigid body. This mode is the only one which generates thrust merely by oscillation of its fins. Note that sometimes ostraciiforms are categorized in MPF swimming mode since some ostraciiforms like boxfish also use their median or paired fin to swim. For instance, in (Colgate & Lynch, 2004) ostraciiforms are introduced as MPF swimmers.

To name an ostraciiform robot, Micro Autonomous Robotic Ostraciiform (MARCO) designed and fabricated by Kodati et al. (Kodati, Hinkle, Winn, & Deng, 2008) could be cited. MARCO is inspired by a boxfish. This fish robot shown in Figure 8 has a pair of 2-DOF pectoral fins and a single DOF caudal fin. The design of pectoral fins is according to actual boxfish shape while

Figure 8. MARCO inspired by a boxfish as an ostraciiform (Kodati, et al., 2008)

Figure 7. Uncoated dolphin robot built by Yu et al. (2007).

hydrodynamic experiments are considered for the design of tail shape. MARCO uses its pectoral fins for steering the motion while the caudal fin propels the robot. Noting that the robot mimics a real boxfish, MARCO has also a body shape quite similar to its corresponding actual fish in nature. The robot is highly manoeuvrable by making benefit of its pectoral fins. The pectoral fins also work as lifting surfaces for the robot which is specifically useful for up-down motion.

2.2. Median and/or Paired Fins

MPF mode, illustrated in Figure 2(b), is also categorized further based on the source of propulsion like pectoral, dorsal and anal fins.

2.2.1. Rajiforms

The first group is *rajiforms*. Rajiforms like cownose rays and stingrays use their wide pectoral fins for propulsion either in undulation mode or oscillation. In undulation mode, the amplitude of undulation is increasing from anterior to posterior to create wave. In oscillation mode, however, the fins behave like flapping wings of birds to create a wave with higher amplitude. Low and Willy (Low & Willy, 2005; Willy & Low, 2005) have developed a stingray-like robot which has undulation mode, while Gao et al. (Gao, Bi, Li, & Liu, 2009; Gao, Bi, Xu, & Liu, 2007) have developed BHRay-I and BHRay-II inspired by manta ray with oscillatory motion of fins. In the both aforementioned modes, wide pectoral fins help the fish robot to be maneuverable and noiseless. However, those wide fins do not allow the robot to pass through narrow water areas.

To create the undulation, Low and Willy (Low & Willy, 2005; Willy & Low, 2005) designed two flexible pectoral fins. The fins include fin rays where they are separately controlled. To actuate the fins, ten servo motors are used while a crank is attached to the end of each motor to play the

role of fin rays. All the rays are connected together using a flexible membrane made of thin acrylic sheet between each two rays. Figure 9 illustrates the swimming mechanism developed by Low and Willy.

In order to design BHRay-I, two 1-DOF fins actuated by a servo motor are employed. The pectoral fins are made of carbon fiber pipe, a silicone rubber board and reinforcing aluminum.

Figure 9. Model of fin mechanism mimicking rajiforms: (a) CAD model for two fin rays connected together and (b) rajiform robot made of nine lateral fins (Low & Willy, 2005)

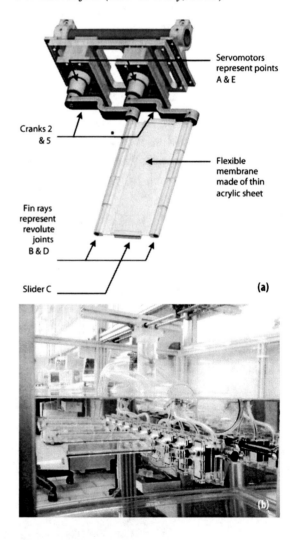

The carbon fibers placed at the leading edge of the fins are actuated by motors. Then the flexible silicone rubber passively generates phase difference, which is critical for an efficient thrust production of the fins. However, due to the non-adjustable flapping parameters, amplitude and frequency, Gao et al. (2009) enhanced their previous design using two servomotors working for the fins individually. They called the new version, BHRay-II. BHRay-I, and BHRay-II are shown in the Figure 10.

2.2.2. Labriforms

Similar to rajiforms, *labriforms* also use their pectoral fins for propulsion. As the most important MPF swimmer, labriforms like angelfish are highly maneuverable. This is due to their ability in controlling pectoral fins independently and producing backward thrust. Added to that, labriforms are capable of station keeping since they do not need caudal fin for propulsion. However, labriforms are low efficient swimmers in comparison with carangiforms and thunniforms. Several labriforms have been developed like Bass II (Kato, 2000), Bass III (Kato, Wicaksono, & Suzuki, 2000) which is enhanced version of Bass II, and BoxyBot (Lachat & Ijspeert, 2005).

BASS II mimics the swimming mode of a real black bass using two-motor driven mechanical pectoral fins (2MDMPF). These pectoral fins could have feathering motion and lead-lag motion. The combination of those two types of motions enables the robot to have forward and backward swimming, and also turning in horizontal plane. Substituting 2MDMPF with 3-motor driven mechanical pectoral fin (3MDMPF) in design of BASS III provides flapping motion for the pectoral fins. Flapping motion is to create the vertical swimming of the robot. BASS II and BASS III are shown in Figure 11. Regarding three aforementioned types of motion for pectoral fins, Low et al. (Low, Prabu, Yang, Zhang, & Zhang, 2007)

Figure 10. Fish robots inspired by manta ray: (a) BHRay-I (Gao, et al., 2007) and (b) BHRay-II (Gao, et al., 2009)

Figure 11. Fish robot mimicking black bass: (a) BASS II and (b) BASS III

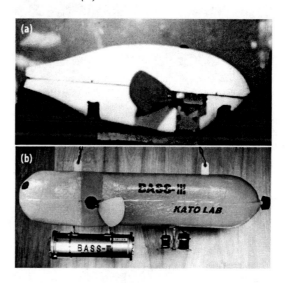

have designed a mechanism using a single motor and planetary gear assembly to provide all those motion for a labriform fish robot.

BoxyBot, showing in Figure 12, is another fish robot from labriform category. The fish robot has a pair of 1-DOF pectoral fins and one tail. The combination of motion of those fins and tail provide forward and backward swimming as well as turning motion of the robot. Although swimming mode of the robot is inspired by the boxfish; the robot does not mimic the shapes of boxfish fins and tail.

It should be noted that BoxyBot is capable of having different swimming modes. For instance, using the tail as a rudder and the fins as propeller, BoxyBot is employing labriform swimming mode. Conversely, using the fins as steering tools and the tail as propeller, BoxyBot has switched to ostraciiform swimming mode. Besides the capability of the robot in switching its swimming modes, the high maneuverability of the robot is considerable. Yet the robot is a planar robot and does not have the ability of diving.

2.2.3. Amiiforms and Gymnotiforms

Amiiforms and *gymnotiforms* are MPF swimmers which utilize a long-based undulatory fin, dorsal or anal, respectively. Several fish robots and mechanisms are invented to mimic swimming method of fishes from both aforementioned categories. For instance, Gymnarchus niloticus as an amiiform inspires the undulating fin, RoboG-nilos, designed and fabricated by Hu et al. (Hu, Shen, Lin, & Xu, 2009). The mechanism consists nine fin rays connected to an individual motor. The motors are independent to have adjustable amplitude, frequency and phase. All the rays are connected together by a membrane surface. Figure 13(a) shows RoboGnilos.

Regarding gymnotiforms, Epstein et al. (Epstein, Colgate, & MacIver, 2006) have developed a fish robot inspired by black ghost knifefish.

Figure 13. Two undulating fin mechanisms: (a) RoboGnilos (Hu, et al., 2009) and (b) undulating fin developed by Epstein et al. (2006)

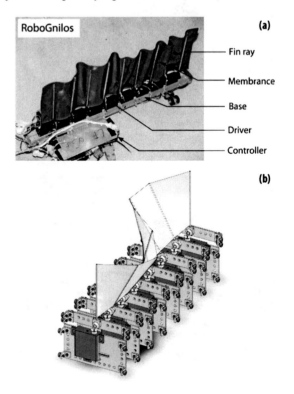

Figure 12. BoxyBot as a labriform-like and an ostraciiform-like robot (Lachat & Ijspeert, 2005)

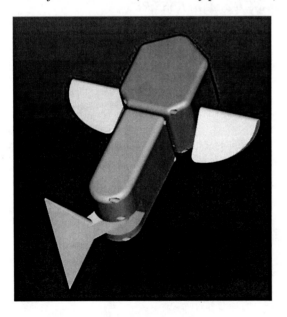

Their mechanism shown in Figure 13(b) has eight fin rays. Each ray is connected to a mitre gear actuated individually by a radio-controlled servo motor. The fin rays are connected together by a thin sheet of latex.

Low and Willy (2005) have also applied their undulating fin mechanism described in rajiforms section to mimic the swimming mode of a gymnotus carapo fish or a black ghost knifefish called later on Nanyang Knifefish (NKF-I) robot. On the contrary of the robot described in (Morgansen et al., 2007), the rotational axis in NKF-I is perpendicular to longitudinal wave direction since the fin rays are directly actuated by motors. Low and Yu improved their NKF-I presented in Figure 14(a) to a modular and reconfigurable robot called NKF-II. This version of NKF has three main parts: buoyancy tank, motor compartment and undulating fin module. For more information about the improvement the reader is referred to (Low, 2009; Low & Yu, 2007).

Siahmansouri et al. (Siahmansouri, Ghanbari, & Fakhrabadi, 2011) have also built a robot with six fin rays that has improved NKF robot series using two separate servomotors that could control the depth and the direction of the robot. Figure 14(b) illustrates Siahmansouri et al.'s work.

The robots inspired by amiiforms and gymnotiforms are manoeuvrable. Nevertheless, those robots are not fast enough as carangiforms and thunniforms. Their speed is even worse when diving and surfing.

Notice that some types of MPF swimming modes that do not have corresponding developed fish robot—to the best of the authors' knowledge—are not explained in the chapter.

3. DISCUSSION

3.1. Swimming Time-Based Features

Taking time into consideration, there are two groups of swimming motion including periodic and transient motion. Periodic motion or steady

Figure 14. (a) NKF-I (Low, 2009) and (b) its improved version by Siahmansouri et al. (2011)

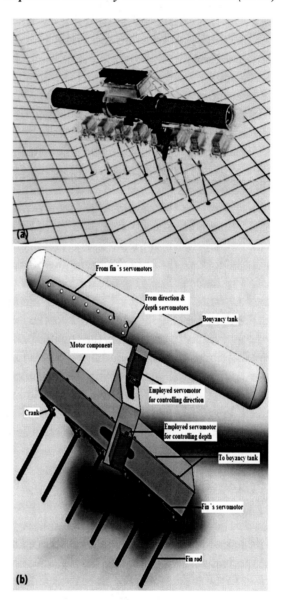

motion like cruising continues in a long period of time whereas transient or unsteady motion like fast start and sharp turn takes a short period of time (Sfakiotakis et al., 1999). Mechanically speaking, fishes could usually be optimal for one type of those motions, periodic or transient motion.

Fishes whose swimming motions are based on an oscillatory propulsor are not suitable for steady motion. This is due to the fact that their oscillatory propulsors need to have higher oscillatory speed in comparison with undulatory propulsors to cover the same distance. The oscillation-based rajifroms like manta ray are exceptions since their pectoral fins are wide and could provide sufficient thrust in each oscillation. Among the fish swimming modes, thunniforms like tuna are the most appropriate fish for periodic motion with efficiency up to 90 percent (Yu & Wang, 2005). This is essentially due to the lift-based undulatory motion as well as the hydrodynamically efficient shape of thunniforms.

On the contrary, undulation-based swimmers are not appropriate for transient movement as they have a long body shape while their main body is inflexible, shown in Figure 2, and undulation produces a small thrust in the beginning of motion. The latter is due to gradual production of traveling wave in undulatory swimming mode. Since the whole body of anguilliform is flexible and contributing to undulation, the anguilliforms are exceptional undulation-based swimmers that have a significant maneuvering performance. Considering the body shape and all swimming modes, labriforms and ostraciiforms like boxfish are the most proper swimming modes for transient locomotion.

3.2. Actuation System

The next key aspect of the robot mechanical design is its actuation system. The actuation source and the mechanism engaged to transfer the actuation power to the propulsors need to be detected.

Among fish robots described in the previous section, four types of actuation sources could be found including hydraulic power system (Anderson & Chhabra, 2002), Pneumatic power system (Dogangil et al., 2005), and electric motors, for instance (Hu et al., 2009; Kodati et al., 2008; Liang et al., 2011; Liu et al., 2005; Low & Willy, 2005; Morgansen et al., 2007; Triantafyllou et al., 2000; Triantafyllou & Triantafyllou, 1995), as conventional actuation systems, as well as Shape Memory Alloy (SMA) (Ayers et al., 2000).

Hydraulic and pneumatic systems could execute a high rate of energy with respect to their weight; however, the main concern about using these systems for fish robots are their large size and the lag in their control systems (Mavroidis, Pfeiffer, & Mosley, 2000).

In comparison with hydraulic and pneumatic actuators, electric motors produce smaller torque. Nevertheless, due to relatively small size, easier controllability, especially for maneuvering robots, and also the easy storage of their energy medium including recharging batteries, the electric motors are preferred to be used for fish robots (Mavroidis et al., 2000).

As an alternative to conventional actuators, SMA has also been employed in fish robots since they are small, light and easy adaptable to any type of mechanisms. Yet the low efficiency of SMA does not allow them to replace electric motors in large robots (Mavroidis et al., 2000). Therefore, the absolute majority of fish robots have electric motors in their actuation system.

Another key aspect of swimming mechanism is the actuation mechanism. Depending on the fish swimming modes, fish robots have different actuation mechanisms. Oscillatory fish robots have simple mechanism as their propelling fins are directly actuated by the motor. On the contrary, undulatory fish robots could have various mechanisms like using artificial muscle or linkages.

Artificial muscles could be used only when SMA is applied (Ayers et al., 2000). But link-based mechanisms could be used with all conventional actuators. In BCF swimming mode, the undulation is produced by a number of links connected in series. Each link then could be actuated either directly or passively by a phase lag. The phase lag is to create traveling wave. In MPF mode, undulation is caused by several parallel links connected by a flexible membrane. In the other word, parallel links play the role of fin rays in an undulatory fin. In some of the developed fish robot, the flexibility is made by connection of each two rays by a flexible mechanism (Low & Willy, 2005) while others have all rays connected together by a flexible sheet, for instance (Epstein et al., 2006; Hu et al., 2009). In both BCF and MPF modes, the links are rigid except in (Triantafyllou et al., 2000) which three bendable links are employed.

Besides linkages and artificial muscles, the mechanism introduced in the previous section for trout robot has unique features. A motor is connected to a plate casted inside a composite tail. The sinusoidal rotation of the motor provides the undulation of the tail (Salumäe, 2010).

3.3. Body Shape

The purpose of producing robots inspired by fishes is to mimic their swimming mechanisms to have efficient underwater robots. Yet mimicking the body shape of aquatic animals has a crucial role in the enhancement of the efficiency and the performance of the robot. This is even more significant when the fish body is undulating.

Accordingly, many fish robots developed are mimicking the geometry of fish bodies such as (Boyer et al., 2009; Triantafyllou & Triantafyllou, 1995; Yu et al., 2007). In order to do that, the skeleton of fish robot is made by either a flexible spiral or some rigid rings around the undulating body such a way that there is a sufficient distance between each two rings. This distance causes the flexibility of the body whereas rigidity of rings increases the resistance of body against water pressure when the body is covered by a flexible material.

Beside geometry, in some cases like (Salumäe, 2010), the body stiffness of the real fish is also taken into account. In (Salumäe, 2010), the work done in (Alvarado, 2007) is applied to develop a flexible composite model as the propelling part of the robot. The composite model is capable of adjusting its size, geometry and stiffness based on real corresponding trout.

4. NOVEL FISH ROBOT

As fishes in nature have the potential for all type of swimming including cruising, turning, up-down motion and so forth, approximately all fish-mimicking robots have that potential. However, fishes have outstanding performance in one type of swimming depending on their habitats and natures. Accordingly, the robots inspired by one fish cannot be skilled for different swimming modes. For instance, VCUUV and BoxyBot are very well-known fish robots for their swimming performance. But the performance of VCUUV in steady motion is not comparable with BoxyBot while the latter is better than the former in unsteady motion.

This section aims at introducing a novel fish robot which is an efficient cruising swimmer, highly maneuverable and capable of 3D motion. In the other words, despite the fish robots developed so far, this robot is inspired by two different types of fishes and, accordingly, is highly capable of both steady and unsteady motion.

4.1. Swimming Mechanism

Steady and unsteady motions of fish robots with respect to fish swimming modes are previously described. Among all types of fishes, thunniforms and labriforms have shown superior performance in steady and unsteady motion, respectively. Despite, neither of thunniforms nor labriforms is capable of having both steady and unsteady motion together efficiently. Then to have a fish robot that is maneuverable and capable of fast cruising, one solution involves combination of the fish swimming mechanisms as long as they are not stifling each other's motion.

As explained previously, the propulsion in thunniforms are based on the undulation of caudal fin and tail peduncle. Theoretically, pectoral fins do not affect the swimming behavior of thunniforms considerably. On the other hand, labriforms propels by their pectoral fins whereas their caudal fin do not have any significant effect on their motion. Then a system which could make benefit of undulation of tail peduncle and caudal fin inspired by thunniforms, and oscillation of pectoral fins inspired by labriforms is proposed.

4.2. Mechanical Design

As shown in Figure 15, a robot mimicking the geometry of a tuna fish is designed. However, it makes the benefit of having two pectoral fins. The pectoral fins are designed such a way that they produce small drags during steady motion and could be actuated by only two DC motors. This robot has three main parts: main body including pectoral fins, tail peduncle and caudal fin.

The main body includes all the electronics such as microcontroller, batteries, sensors and buoyancy control tank system. This part of the body has also two rotating pectoral fins actuated independently by two servo motors. As this part is assumed to be inflexible, it is covered by a rigid material to save all sensitive components inside and make waterproofing easier.

The tail peduncle is also consisted of two parts or links. One of the links is activated by a powerful DC servo motor while the second part is actuated passively depending on the first link. On the contrary of the main body, tail peduncle is undulating and subsequently need to be flexible. Then several elliptical rings connected through a flexible sheet are covering tail peduncle. The flexible sheet is not shown in the Figure 15.

Figure 15. CAD model of proposed fish robot

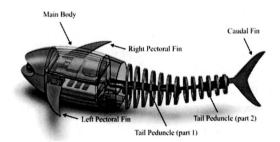

The last part of the fish robot is its caudal fin which is fixed to the second link of tail peduncle. Besides lateral motion, the caudal fin could change its direction vertically using a small DC servo motor located under the second link.

4.3. Swimming Strategy

Given that the fish robot is capable of both steady and unsteady motion, the robot has two corresponding swimming mechanisms. The first mechanism is during starting time and maneuvering and the second one is in cruising time. During these two mechanisms, the performance of caudal fin is similar whereas that of the pectoral fins is different.

In both mechanisms, the tail peduncle and caudal fin are undulating. If the robot is going to have forward motion, the caudal fin is aligned with tail peduncle. On the other hand, if the robot is going to have up or down motion, the caudal fin has downward or upward inclination, respectively. Besides, the buoyancy water tank system gets filled for diving and drained for surfing.

The pectoral fins, however, have different behavior in the two aforementioned mechanisms. For steady motion, the pectoral fins are like gliding wings. To turn right, the right pectoral fin rotates upward, and vice versa. To dive and surf, both pectoral fins rotate downward and upward, respectively. In unsteady motions, the direction of the fins are the same as steady motion; although, they are oscillating continuously.

Regarding up-down motion, it should be noted that the robot could have different mechanisms as explained by Hirata (2001). One of the mechanisms is through using pectoral fins for diving or surfing as used in (Liu et al., 2005; Morgansen, 2003). However, the lift forces produced by pectoral fins are not capable of neutralizing large buoyancy or gravity force of large fish robots. Buoyancy tank system could be also employed (Low, 2006). Using water tank to change the buoyancy and gravity actively is an adjustable procedure that could be used for any system. Nevertheless, this mechanism is slow and not suitable for fast up-down motion. Then to improve the mechanical design of the proposed fish robot both aforementioned methods are used. Moreover, the direction of caudal fin is adjustable to have even faster vertical motion.

5. CONCLUSION

With the intention of developing a fish robot capable of both steady and unsteady motion, a novel mechanical design and swimming mechanism is introduced in the chapter. The robot is preferred over the existing fish robot owing to three main reasons.

First, the robot is capable of having both steady and unsteady motion. In fact, the robot which is inspired by thunniforms and labriforms could behave like a fast cruising robot as well as a maneuverable one. When periodic motion is needed, the robot swims analogous with thunniforms. In contrast, when the robot needs to have a transient motion, the robot swims moderately similar to labriforms using oscillating pectoral fins.

In addition, the robot design is efficient from energy consumption point of view. In comparison with other types of actuation system, electric motors are the most useful actuators for medium-sized robots as previously explained. However, instead of using several motor for different linkages which reduces the battery life, the mechanism allows usage of one DC motor in the main body for undulation.

The last but not the least, the proposed mechanical design provides the robot with agile motion even during up-down motion. This agility roots in using pectoral fins, caudal fin inclination and buoyancy water tank simultaneously. These three elements help the robot to dive and surf during both steady and unsteady motion.

REFERENCES

Alvarado, V. Y. (2007). *Design of biomimetic compliant devices for locomotion in liquid environments* (Vol. 68).

Anderson, J. M., & Chhabra, N. K. (2002). Maneuvering and stability performance of a robotic tuna. *Integrative and Comparative Biology, 42*(1), 118–126. doi:10.1093/icb/42.1.118 PMID:21708700.

Ayers, J., Wilbur, C., & Olcott, C. (2000). *Lamprey robots*.

Beal, D. N. (2003). *Propulsion through wake synchronization using a flapping foil.* Cambridge, MA: Massachusetts Institute of Technology.

Boyer, F., Chablat, D., Lemoine, P., & Wenger, P. (2009). The eel-like robot. *Arxiv preprint arXiv:0908.4464*.

Charlie I. (n.d.). Retrieved 2012 http://Web.mit.edu/towtank/www-new/Tuna/Tuna1/pictures.html

Colgate, J. E., & Lynch, K. M. (2004). Mechanics and control of swimming: A review. *IEEE Journal of Oceanic Engineering, 29*(3), 660–673. doi:10.1109/JOE.2004.833208.

Dag, G. (2009). *Fish robots search for pollution in the waters.* Retrieved from http://www.robaid.com/bionics/fish-robots-search-for-pollution-in-the-waters.htm

Dogangil, G., Ozcicek, E., & Kuzucu, A. (2005). *Design, construction, and control of a robotic dolphin.* Paper presented at the IEEE International Conference on Robotics and Biomimetics (ROBIO). New York, NY.

Epstein, M., Colgate, J. E., & MacIver, M. A. (2006). *Generating thrust with a biologically-inspired robotic ribbon fin*. Paper presented at the IEEE/RSJ International Conference on Intelligent Robots and Systems. New York, NY.

Gao, J., Bi, S., Li, J., & Liu, C. (2009). *Design and experiments of robot fish propelled by pectoral fins*. Paper presented at the IEEE International Conference on Robotics and Biomimetics (ROBIO). New York, NY.

Gao, J., Bi, S., Xu, Y., & Liu, C. (2007). *Development and design of a robotic manta ray featuring flexible pectoral fins*. Paper presented at the IEEE International Conference on Robotics and Biomimetics (ROBIO). New York, NY.

Griffiths, G., & Edwards, I. (2003). AUVs: Designing and operating next generation vehicles. *Elsevier Oceanography Series, 69*, 229–236. doi:10.1016/S0422-9894(03)80038-7.

Hirata, K. (2001). *Up-down motion for a fish robot*. Retrieved from http://www.nmri.go.jp/eng/khirata/fish/general/updown/index_e.html

Hu, H., Liu, J., Dukes, I., & Francis, G. (2006). *Design of 3D swim patterns for autonomous robotic fish*.

Hu, T., Shen, L., Lin, L., & Xu, H. (2009). Biological inspirations, kinematics modeling, mechanism design and experiments on an undulating robotic fin inspired by gymnarchus niloticus. *Mechanism and Machine Theory, 44*(3), 633–645. doi:10.1016/j.mechmachtheory.2008.08.013.

Jakuba, M. V. (2000). *Design and fabrication of a flexible hull for a bio-mimetic swimming apparatus*.

Junzhi, Y., Min, T., Shuo, W., & Erkui, C. (2004). Development of a biomimetic robotic fish and its control algorithm. *IEEE Transactions on Systems, Man, and Cybernetics. Part B, Cybernetics, 34*(4), 1798–1810. doi:10.1109/TSMCB.2004.831151.

Kato, N. (2000). Control performance in the horizontal plane of a fish robot with mechanical pectoral fins. *IEEE Journal of Oceanic Engineering, 25*(1), 121–129. doi:10.1109/48.820744.

Kato, N. (n.d.). *Fish fin motion*. Retrieved from http://www.naoe.eng.osaka-u.ac.jp/~kato/fin9.html

Kato, N., Wicaksono, B. W., & Suzuki, Y. (2000). *Development of biology-inspired autonomous underwater vehicle BASS III with high maneuverability*. Paper presented at the International Symposium on Underwater Technology. New York, NY.

Kim, H., Lee, B., & Kim, R. (2007). *A study on the motion mechanism of articulated fish robot*. Paper presented at the International Conference on Mechatronics and Automation (ICMA). New York, NY.

Kodati, P., Hinkle, J., & Deng, X. (2007). *Micro autonomous robotic ostraciiform (marco): Design and fabrication*.

Kodati, P., Hinkle, J., Winn, A., & Deng, X. (2008). Microautonomous robotic ostraciiform (MARCO), hydrodynamics, design, and fabrication. *IEEE Transactions on Robotics, 24*(1), 105–117. doi:10.1109/TRO.2008.915446.

Lachat, D., & Ijspeert, A. J. (2005). *BoxyBot, the fish robot: Project report*. Lusanne, France: Biologially Inspired Robotics Group, School of Computer and Communication Sciences at Ecole Polytechnique Federale de Lausanne.

Lee, S., Park, J., & Han, C. (2007). Optimal control of a mackerel-mimicking robot for energy efficient trajectory tracking. *Journal of Bionics Engineering, 4*(4), 209–215. doi:10.1016/S1672-6529(07)60034-1.

Liang, J., Wang, T., & Wen, L. (2011). Development of a two-joint robotic fish for real-world exploration. *Journal of Field Robotics, 28*(1), 70–79. doi:10.1002/rob.20363.

Liu, J. (2006). *Welcome! Essex robotic fish*. Retrieved from http://dces.essex.ac.uk/staff/hhu/jliua/index.htm#Profile

Liu, J., Dukes, I., & Hu, H. (2005). *Novel mechatronics design for a robotic fish*. Paper presented at the IEEE/RSJ International Conference on Intelligent Robots and Systems (IROS). New York, NY.

Liu, J., Dukes, I., Knight, R., & Hu, H. (2004). Development of fish-like swimming behaviours for an autonomous robotic fish. *Proceedings of the Control, 4*.

Liu, J., & Hu, H. (2004). A 3D simulator for autonomous robotic fish. *International Journal of Automation and Computing, 1*(1), 42–50. doi:10.1007/s11633-004-0042-5.

Low, K. H. (2006). *Maneuvering and buoyancy control of robotic fish integrating with modular undulating fins*. Paper presented at the IEEE International Conference on Robotics and Biomimetics, ROBIO '06. New York, NY.

Low, K. H. (2009). Modelling and parametric study of modular undulating fin rays for fish robots. *Mechanism and Machine Theory, 44*(3), 615–632. doi:10.1016/j.mechmachtheory.2008.11.009.

Low, K. H., Prabu, S., Yang, J., Zhang, S., & Zhang, Y. (2007). *Design and initial testing of a single-motor-driven spatial pectoral fin mechanism*. Paper presented at the International Conference on Mechatronics and Automation (ICMA). New York, NY.

Low, K. H., & Willy, A. (2005). *Development and initial investigation of NTU robotic fish with modular flexible fins*. Paper presented at the IEEE International Conference on Mechatronics and Automation. New York, NY.

Low, K. H., & Yu, J. (2007). *Development of modular and reconfigurable biomimetic robotic fish with undulating fin*. Paper presented at the IEEE International Conference on Robotics and Biomimetics (ROBIO). New York, NY.

MT1 Gallery. (n.d.). Retrieved from http://dces.essex.ac.uk/staff/hhu/jliua/images/gallery/MT1/P1010049.JPG

Mason, R., & Burdick, J. (2000). Construction and modelling of a carangiform robotic fish. *Experimental Robotics, 6*, 235–242.

Mavroidis, C., Pfeiffer, C., & Mosley, M. (2000). 5.1 conventional actuators, shape memory alloys, and electrorheological fluids. *Automation, Miniature Robotics, and Sensors for Nondestructive Testing and Evaluation, 4*, 189.

McIsaac, K. A., & Ostrowski, J. P. (2003). Motion planning for anguilliform locomotion. *IEEE Transactions on Robotics and Automation, 19*(4), 637–652. doi:10.1109/TRA.2003.814495.

Morgansen, K. A. (2003). Geometric methods for modeling and control of a free-swimming carangiform fish robot. In *Proceedings of the 13th Unmanned Untethered Submersible Technology*. UUST.

Morgansen, K. A., Triplett, B. I., & Klein, D. J. (2007). Geometric methods for modeling and control of free-swimming fin-actuated underwater vehicles. *IEEE Transactions on Robotics, 23*(6), 1184–1199. doi:10.1109/LED.2007.911625.

Nakashima, M., & Ono, K. (2002). Development of a two-joint dolphin robot. *Neurotechnology for Biomimetic Robots*, 309.

Nakashima, M., Takahashi, Y., Tsubaki, T., & Ono, K. (2004). Three-dimensional maneuverability of the dolphin robot (roll control and loop-the-loop motion). *Bio-Mechanisms of Swimming and Flying*, 79-92.

Northeastern Marine Science Center - Biomimetic Robots. (n.d.). Retrieved from http://www.expo21xx.com/automation21xx/17600_st2_university/default.htm

Paulson, L. D. (2004). Biomimetic robots. *Computer, 37*(9), 48–53. doi:10.1109/MC.2004.121.

Robotic Fish SPC-03, BUA - CASIA, China. (n.d.). Retrieved from http://www.robotic-fish.net/index.php?lang=en&id=robots#top

RoboTuna II. (n.d.). Retrieved from http://Web.mit.edu/towtank/www-new/Tuna/Tuna2/tuna2.html

Saimek, S., & Li, P. Y. (2004). Motion planning and control of a swimming machine. *The International Journal of Robotics Research, 23*(1), 27–53. doi:10.1177/0278364904038366.

Salumäe, T. (2010). *Design of a compliant underwater propulsion mechanism by investigating and mimicking the body a rainbow trout (oncorhynchus mykiss)*. Tallinn University of Technology.

Sfakiotakis, M., Lane, D. M., & Davies, J. B. C. (1999). Review of fish swimming modes for aquatic locomotion. *IEEE Journal of Oceanic Engineering, 24*(2), 237–252. doi:10.1109/48.757275.

Siahmansouri, M., Ghanbari, A., & Fakhrabadi, M. M. S. (2011). Design, implementation and control of a fish robot with undulating fins. *International Journal of Advanced Robotic Systems, 8*(5), 61–69.

Triantafyllou, M., & Kumph, J. M. et al. (2000). *Maneuvering of a robotic pike*. Cambridge, MA: Massachusetts Institute of Technology.

Triantafyllou, M. S., & Triantafyllou, G. S. (1995). An efficient swimming machine. *Scientific American, 272*(3), 64–71. doi:10.1038/scientificamerican0395-64.

Vogel, S. (1996). The thrust of flying and swimming. In Life in moving fluids: The physical biology of flow.

Wang, T., Liang, J., Shen, G., & Tan, G. (2005). *Stabilization based design and experimental research of a fish robot*. Paper presented at the IEEE/RSJ International Conference on Intelligent Robots and Systems (IROS). New York, NY.

Wernli, R. L. (2000). *Low cost UUV's for military applications: Is the technology ready?* DTIC Document.

Williams, C. (2004). AUV systems research at the NRC-IOT. *An update*.

Willy, A., & Low, K. H. (2005). *Development and initial experiment of modular undulating fin for untethered biorobotic AUVs*. Paper presented at the IEEE International Conference on Robotics and Biomimetics (ROBIO). New York, NY.

Yu, J., Hu, Y., Fan, R., Wang, L., & Huo, J. (2007). Mechanical design and motion control of a biomimetic robotic dolphin. *Advanced Robotics, 21*(3-4), 499–513. doi:10.1163/156855307780131974.

Yu, J., Hu, Y., Huo, J., & Wang, L. (2007). *An adjustable scotch yoke mechanism for robotic dolphin*. Paper presented at the IEEE International Conference on Robotics and Biomimetics (ROBIO). New York, NY.

Yu, J., & Wang, L. (2005). *Parameter optimization of simplified propulsive model for biomimetic robot fish*.

Chapter 5
Efficient Evolution of Modular Robot Control via Genetic Programming

Tüze Kuyucu
Doshisha University, Japan

Ivan Tanev
Doshisha University, Japan

Katsunori Shimohara
Doshisha University, Japan

ABSTRACT

In Genetic Programming (GP), most often the search space grows in a greater than linear fashion as the number of tasks required to be accomplished increases. This is a cause for one of the greatest problems in Evolutionary Computation (EC): scalability. The aim of the work presented here is to facilitate the evolution of control systems for complex robotic systems. The authors use a combination of mechanisms specifically designed to facilitate the fast evolution of systems with multiple objectives. These mechanisms are: a genetic transposition inspired seeding, a strongly-typed crossover, and a multiobjective optimization. The authors demonstrate that, when used together, these mechanisms not only improve the performance of GP but also the reliability of the final designs. They investigate the effect of the aforementioned mechanisms on the efficiency of GP employed for the coevolution of locomotion gaits and sensing of a simulated snake-like robot (Snakebot). Experimental results show that the mechanisms set forth contribute to significant increase in the efficiency of the evolution of fast moving and sensing Snakebots as well as the robustness of the final designs.

DOI: 10.4018/978-1-4666-4225-6.ch005

1. INTRODUCTION

Genetic Programming (GP) is a methodology specifically developed for the evolution of computer programs that perform predefined tasks (Koza, 1994). It is a powerful representation for automatically creating computational systems via simulated evolution, and it has produced many human-competitive results (Koza, 2003). The use of GP for the automated design of artificial systems is a popular and promising method; it provides a valuable approach to achieving designs that are too complex or alien for human designers (such as the automated control of a racing car (Schichel & Sipper, 2011)).

Several examples of real-life designs obtained via GP can be found in literature, however, as we try to tackle harder problems with GP the inherent weaknesses in GP design become more apparent. The exponential growth of the search space with increasing complexity of the target design, and poor locality are only a couple of these weaknesses that present a challenge to GP design of real-life systems. Hence the design of systems with the use of GP is generally a time consuming process. The expensive evaluation times of real-life problems, such as physics, 3D graphics, and electronic circuit simulations, when multiplied by the large population sizes and the number of evolutionary generations required to obtain a result makes GP design a slow process. Furthermore, the designs obtained via GP can often end up overly fit, low quality solutions that fail to generalize. If used in real-world applications, a design must be reliable in terms of accurate functionality, and ensuring this can be a costly process in GP; multiple GP runs may be required until designs that are reliable can emerge. Therefore, the convergence of GP runs to a desired solution within a given time is important. Situations where the aim is to come up with designs with multiple features can especially present a challenge; the GP may fail to identify and preserve sub-solutions, and successfully build upon these partial solutions in order to provide the optimal design.

Poor scalability is a common problem in the simultaneous evolution of multiple features of a target system, this is because of the growing search space of evolution which increases faster than linearly with the increase of the number of simultaneously evolved features. In order to tackle the scalability problem, several approaches to the evolutionary design of systems with multiple features or objectives have been devised. However, achieving good scalability in the evolution of complex systems still remains to be one of the great challenges (McConaghy, Vladislavleva, & Riolo, 2010).

On the other hand, ensuring the evolution of reliable results in a timely manner is another persisting challenge present in GP design. Due to the opportunistic nature of evolution, the resulting designs achieved via GP can often be something other than what is desired. The design approach used by evolutionary designs exploit all the available properties of an evolutionary platform and environment. Due to this, the results can sometimes perform unexpectedly (good or more often bad (Bird & Layzell, 2002; Kuyucu, Trefzer, Greensted, Miller, & Tyrrell, 2008)).

In this paper we introduce mechanisms that will help to tackle these two great challenges. We show the effects of these new mechanisms by evolving designs for a challenging application, which will be described in Section 2. The computational effort of GP could be improved in several ways, such as, incorporating a domain-specific knowledge into the key attributes of GP (e.g., genetic representation, genetic operations, etc.), imposing problem-specific syntax constrains (i.e., grammar) on the evolved genetic programs, employing probability-distribution models, etc. These approaches are usually intended to steer the simulated evolution towards the most promising areas in the explored (presumably rugged) fitness landscapes. Another approach of improving computational effort of GP stems from the assumption that, the main genetic operations (crossover and mutation), due to their randomness, can be damaging the already existing building blocks of the solution. Thus, the destructive effects of these operations could be limited if they are occasionally allowed to operate on the genetic code that is irrelevant to the quality of the corresponding genetic program. Moreover, such an irrelevant, or neutral code, might offer the simulated evolution with a "playground" where it can experiment with developing either novel-, or better-than-existing genotypic traits without the risk of damaging the already evolved ones.

A biologically inspired mechanism based on Genetic Transposition (GT) in order to facilitate incremental evolution in achieving well scaling GP solutions by introducing a more efficient use of existing sub-solutions will be described in Section 3. A specifically constrained crossover mechanism for GP using the GT-inspired mechanism that is aimed to facilitate the quick evolution of more reliable designs will be described in Section 4. Section 6 will provide the details on the evolutionary framework, the experimental setup and GP parameters used for the experimental data presented in the following section; Section 7 will present experiments that demonstrate the effects of mechanisms described in Sections 3 and 4. The effects of using a MultiObjective (MO) optimization approach, namely Nondominated Sorting Genetic Algorithm 2 (NSGA-II) (Deb, Pratap, Agarwal, & Meyarivan, 2002), in solving a control problem with multiple features via GP will also be briefly explored via the experiments provided. Section 8 will finalize the paper with a brief summary and conclusions.

The objective of our work is to investigate the effects of the outlined mechanisms on the efficiency of GP employed for simulated incremental evolution of locomotion of sensing snake-like robot (Snakebot) in a challenging environment with obstacles. The successful bots should feature the evolved (emergent) know-how about how to clear a narrow corridor by (1) moving fast, (2) following the walls of the corridor, (3) overcoming a number of randomly scattered small boxes, and (4) circumnavigating large obstacles.

2. SIDEWINDING AND SENSING SNAKE-LIKE MODULAR ROBOT

Snake-like robots present potential robustness characteristics beyond the capabilities of most wheeled and legged vehicles, such as: the ability to traverse challenging terrain and insignificant performance degradation when partial damage is inflicted. Some useful features of snake-like robots include smaller size of the cross-sectional areas, stability, ability to operate in difficult terrain, good traction, and complete sealing of the internal mechanisms (Dowling, 1999; Hirose, 1993). Moreover, due to the modularity and homogeneity of their design, the snake-like robots have high redundancy, which in turn provides it with inherent fault tolerance and adaptability properties (Tanev, Ray, & Buller, 2005). Robots with such properties can be valuable for applications that involve exploration, reconnaissance, medicine and inspection.

Designing a controller that can achieve the optimal locomotion of a modular Snakebot is a challenging task due to the large number of degrees of freedom in the movement of segments. The locomotion gait of such bots is often seen as an emergent property; observed at a higher level of consideration of complex, non-linear, hierarchically organized systems, comprising many relatively simply-defined entities (morphological segments). In such complex systems the higher-level properties of the system and the lower-level properties of comprising entities cannot be directly induced from each other (Morowitz, 2002). Therefore even if an effective incorporation of sensing information in fast and robust locomotion gaits might emerge from intuitively defined sensing morphology and simple motion patterns of morphological segments, neither the degree of optimality of the developed code nor the way of how to incrementally improve this code is evident to the human designer (Koza, Keane, Yu, Bennett, & Mydlowec, 2000).

The earlier research demonstrates that the control for a fast-moving modular robotic organism could be automatically developed through various nature-inspired paradigms, based on models of learning and evolution. The work, presented in (Tanev, Ray, & Buller, 2005) demonstrates the use of GP (Koza, 1994) for evolution of sensorless sidewinding Snakebots in various environmental conditions. Furthermore, the coevolution of active sensing and the control of the locomotion gaits was demonstrated to be achievable (Tanev & Shimohara, 2008). The morphology of the sensors, attached to the segments of the bot, coevolve with the way to incorporate the sensory readings into the control of locomotion of the bot. The genetically optimized morphological traits of the bot include the initial orientation, the timing of switching on, and the range of the simulated Laser Range Finders (LRF)–an LRF is a sensor used to detect the objects nearby and their distance from the sensor, in the Snakebot we use this information to evolve wall avoidance behaviour–attached to each of the segments of the bot. The emergent features of the evolved gaits include both the contact and contactless wall-following navigation accomplished via adaptive, sensory-controlled differential steering of the fast moving sidewinding bot. Despite the evidence on the feasibility of coevolution of active sensing and the locomotion, the resulting wall-following behaviour is achieved in an environment that is too simple, and therefore too distant from the real-world applications: a simple curved corridor with an obstacle-free, smooth surface.

In this paper we investigate the coevolution of the active sensing and locomotion control of sidewinding Snakebot in a more complex environment that, in addition to a narrow corridor, includes several large obstacles and many randomly placed small obstacles constituting a rugged terrain within this challenging environment. The sensors on the Snakebot used in this paper follow the same model as proposed in (Tanev & Shimohara, 2008): each segment of the Snakebot is provided with a fixed, immobile LRF with

evolvable initial orientation, range and timing of firing. The most efficient locomotion gaits of Snakebot are not necessarily associated with the forward, rectilinear motions (and sidewinding might emerge as a fast and robust locomotion). Therefore, the eventual fusion of the readings of many sensors mounted in all the segments would provide Snakebot with the capability to perceive the features of surrounding environment along its whole body. In addition to the widening of the area of the perceived surroundings, multiple sensors offer the potential advantages of robustness to damage of some of them, dependability of the sensory information, and an ability to perceive the spatial features of the surrounding environment due to the motion parallax.

Scalable approaches that can handle multiple tasks are important in the evolution of Snakebot, as the complexity of evolving a controller for the described set up can become an important issue. The size of the evolutionary search space is a multiplication of the sizes of the search spaces of the following interdependent evolutionary subtasks:

- **Evolution of the Control of Locomotion:** The time patterns of turning angles of actuators that result in a fast locomotion of the bot.
- **Evolution of the Morphology of the Active Sensing:** Initial orientation of the sensors, their range, and timing of their activation.
- **Evolution of the Incorporation of the Sensor Signals:** Into the control of locomotion of the bot.

The evolution of both the morphology and the incorporation of the signals from many sensors face the challenge of dealing with the uncertain sensor readings as they move synchronously with the coupled segments. Figure 1 illustrates how the initial orientation of the axes of the internal coordination systems of the segments of a bot dramatically differs from a sample instant orientation of

Figure 1. Orientation of the axes of the internal coordination systems of the central segment at two different snakebot positions

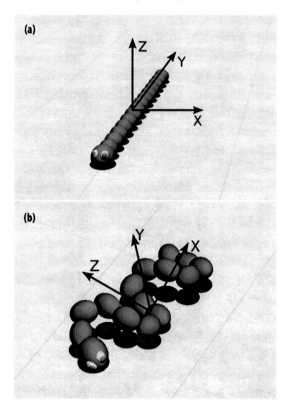

these axes in a moving bot. A sensor fixed to the segment of a moving Snakebot would constantly change its spatial orientation, and consequently it might alternatively perceive no signal, a signal from the ground surface or from another segment of the snake (in both cases the sensory reading should be ignored), or eventually from an obstacle. Moreover, in the targeted environment the obstacle could be either a wall (to be followed), a large box (to be circumnavigated), or a small box (to be overcome).

The large search space of the evolution of the considered Snakebot results in an intractable computational effort, and as it will be demonstrated in Section 7, canonical GP with Automatically Defined Functions (ADF) is unable to effectively explore the emerging search space in a reasonable time frame: ADF, a method of gene reuse during the evolution of genetic programs, has been shown to speed up evolution and improve the achievable complexity via GP (Koza, 1994). In the following two sections, we will outline two simple mechanisms that can transform a complex problem such as the aforementioned Snakebot behaviour modelling, into an amenable task for GP.

3. GENETIC TRANSPOSITION IN GP

Discovered by Barbara McClintock in maize (Zea mays), the transposons (jumping genes) are sequences of DNA that can move around to different positions within the genome of a single cell, in a mechanism called transposition (McClintock, 1950). In the process, they can cause mutations and change the amount of DNA in the genome. It is recognized that the transposons, facilitate the evolution of increasingly complex forms of life by providing the creative playground for fast mutations where the latter could experiment with developing novel genetic structures without the risk of damaging the already existing, well-functioning genome (Nowacki, Higgins, Maquilan, Swart, Doak, & LandWeber, 2009; Strand & McDonald, 1985). Hence, the transposons enhance neutrality in the genome, which is known to facilitate the evolution of successful traits (Huynen, Stadler, & Fontana, 1996; Wilke, Wang, Ofria, Lenski, & Adami, 2001).

The transposition-inspired research in EC was initially started by the work of Simoes et al. (Simoes, Costa, Sim, & Costa, 1999; Sim & Costa, 2000) on the favorable effect of transposition on the performance of Genetic Algorithms (GA). The first of their methods is intended to enhance the crossover operation in GA by exchanging only the genetic material that is specifically marked as a transposon (Sim & Costa, 2000). Their second approach, (termed "asexual transposition") models the mutation of GA as a "cut and paste" operation observed in biological genetic transposition (Sim

& Costa, 2000). Chan et al demonstrate a successful implementation of a GT inspired mechanism in MO optimization, which is shown to have superior performance in achieving Pareto-optimal solutions in comparison to MO optimization without the GT mechanism (Chan, Man, Tang, & Kwong, 2008). Liu et al employ a similar GT inspired mechanism in a clonal selection algorithm, which is shown to provide improved performance in automatic clustering problems (Liu, Sheng, & Jiao, 2009). In a related research, McGregor and Harvey use a mechanism similar to transposition which they termed as "plagiarism" (McGregor & Harvey, 2005). The "plagiarism" operation copies one part of a genotype into another: replacing the latter completely. The authors demonstrate that the proposed mechanism improves the performance of the evolution of solutions to the Boolean logic problems (such as the design of digital logic circuits and counting-ones problem) (McGregor & Harvey, 2005). Spirov et al. also develop an original implementation of artificial transposition, used as a form of mutation operator for the simulated evolution of evolving a finite state machine as a solver for the artificial ant problem (Spirov, Kazansky, Zamdborg, Merelo, & Levchenko, 2009).

In our model, we implement an operation similar to the "copy and paste" action found in transposons. We use previously evolved GP trees as transposons that create copies of themselves to expand their own genotype. Although the source of inspiration is the same, the implementation of the proposed model here differs significantly with the existing (GT inspired) mechanisms in EC, i.e. not a genetic operation such as crossover and mutation. Our model of the GT inspired mechanism is an approach to introducing neutrality and compartmentalization in incremental evolution via seeding. Inspired by genetic transposition, we use a seed to create only part of a new individual. Using the GT inspired mechanism, we introduce new genetic code in addition to the seeding genome to make up the whole genome of an individual Snakebot. We believe that, similar to the nature, the latter

would offer the opportunity to preserve the genetic make-up of the seeding individual. In our case this constitutes of preserving the generic locomotion features intact, while incrementally "upgrading" it with the coevolution of the sensing abilities of the bot. Therefore in contrast to aforementioned works, as well as biology, where genetic transposition can occur frequently during the evolutionary cycle (just like other common genetic operations, such as crossover), the proposed method in this paper executes genetic transposition only once for each "seeding phase" (which is one in the experiments presented), and not for any other time during an evolutionary run. For the rest of the paper, we will refer to the GT inspired mechanism introduced here as Genetic Transposition (GT) for simplicity and succinctness. The rationale for proposing such an approach is based on the observations documented by Tanev et al (Tanev & Shimohara, 2008), who suggest that the evolved fast moving Snakebots with sensory abilities exhibit some of the locomotion traits that are pertinent to the generic, sensorless sidewinding locomotion. We speculate that it might be more efficient first to evolve these generic features in a sensorless bot moving in a smooth, plain terrain (corresponding to a narrow search space for the evolution), and then to incorporate the genotypes of these bots into the evolution of the morphologically more complex bots (with sensors) in a more challenging environment. The proposed mechanism of incorporation of these generic features of locomotion is based on seeding the initial population of GP (employed for the evolution of the bot with sensors) via the GT inspired mechanism.

In our work we demonstrate the effects of the proposed mechanisms via the coevolution of locomotion gaits and sensing of the simulated Snakebot. At the initial stage of the proposed GT inspired approach, we evolve a pool of generic fast-moving sidewinding bots in a flat, smooth terrain. Then, during the second stage, we use the genotypes of these Snakebots to seed the initial population of the bots that are further subjected to coevolution of their locomotion control and sensing in a more challenging environment. During the seeding process the generic, fast moving, sensorless bots are subjected to genetic retro-transposition (i.e., duplicated within the same genome). The resulting transposon (connected with the seeding genome via a randomly initialized control gene) is subjected to 100% random mutation (i.e. it is a randomly created genome that is same size as the seed genome) in order to allow

for the incorporation of the sensing information into the locomotion control of the bot. The schematic parse tree of the genotype of an example Snakebot created during the initialization of GP via GT is illustrated in Figure 2.

Seeding of the initial population via existing solutions (full or partial) has been shown to be an effective way of improving the efficiency of simulated evolution. For example, Nolfi et al. (Nolfi, Floreano, Miglino, & Mondada, 1994)

Figure 2. The mechanism of the proposed GT inspired seeding in GP (Stage 2b) and the typical seeding process (Stage 2a). Both of these cases need to make use of a preliminary seed, and in the proposed approach this seed comes from a previously evolved sensorless Snakebot (Stage 1) that achieves fast locomotion on a smooth open terrain (shown in Figure 6c). In either of the Stages 2a and 2b, the resulting genome from Stage 1 is used as a seeding individual and further evolved, with additional sensory abilities (illustrated by the terminal symbol LRF) in a more challenging terrain (shown in Figures 6a and 6b). For Stage 2a the seed from Stage 1 makes up the whole genome of the initial Snakebot. For Stage 2b, the seed from Stage 1 is only a part (Transposon A) of the initial genome of the Snakebot. The rest of it contains a clone of the seed that has gone 100% mutation (Transposon B), and a randomly initialized group of control gene (Transposon C) which connects Transposons A and B.

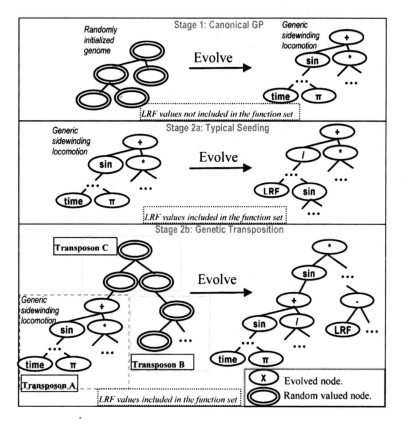

evolve the controller of simulated robot and then re-evolve (or adapt) the obtained results on real robots to accelerate the evolutionary process. Other examples of successful seeding include the work of Vassilev et al (Vassilev, Job, & Miller, 2000) on the optimization of the existing digital circuit design; Thomsen et al (Thomsen, Fogel, & Krink, 2002) on the use of solution obtained from a domain-neutral algorithm as a seed to evolve an even better performing solution; Langdon et al (Langdon & Nordin, 2000) on seeding the evolutionary population with hand-coded solutions that allow a better generality of the evolved results. In addition, by utilizing the previously evolved solutions, seeding has also been applied to improve the performance of evolution of solutions from scratch. This technique, termed by Perry as "population enrichment" (Perry, 1994), has been demonstrated to be more efficient in discovering solutions in GP. "Population enrichment" is a form of seeding that is closest to the GT technique described in this paper. The main difference in these methods is the form of initialization, where in the "population enrichment" the seed is used to create a complete individual (See Stage 2a in Figure 2), while in GT the seeded genotype only forms a part of the genetic make-up of the newly created individual in the initial evolutionary population (See Stage 2b in Figure 2). This way we hope to provide a neutral playground for genetic operations, and encourage the preserving of the functional part of the genome. Neutrality has been shown to be beneficial in complex, rugged landscapes with multiple optima (Beaudoin, Verel, Collard, & Escazu, 2006; Doerr, Gnewuch, Hebbinghaus, & Neumann, 2007; Lobo, Miller, & Fontana, 2004). Hence, we hope that the increased neutrality in the genetic programs via the use of GT will enhance the evolutionary performance in designing sensing and fast moving Snakebots. In the rest of this document we refer to "population enrichment" as typical seeding or seeding.

4. STRONGLY-TYPED CROSSOVER

Strongly Type Genetic Programming (STGP) is a version of GP that enforces data type constraints, by which the evolutionary performance can be improved due to the decrease in search time and the increase in the quality of the solutions found (Montana, 1995). The crossover operations in STGP are processed to ensure that always valid branches of GP trees are created as a result of recombination.

The term strongly-typed crossover used here presents a different approach to the one introduced in (Montana, 1995). Our method, specifically designed to facilitate the evolution of reliable systems when GT is used, applies constraints to the crossover operation to prevent the emergence of not only illegal tree branches, but also the merging of separate strains of genome. The aim is to preserve an already functional genetic code while evolving a new feature in another part of the genome. When conventional crossover is used, the crossover point is randomly selected within the genotype–this means that in the case of the GT mechanism described in Section 3, the genetic code of the seed will be mixed with the newly created genes for some of the offspring created; i.e. the genes from different "transposons" (as shown in Figure 2, Stage 2b) can be crossed-over. The strongly-typed crossover mechanism aims to protect the already well-functioning parts of the genome by avoiding a crossover between different transposons. The purpose of this mechanism is to promote the evolution of objectives as distinct blocks of genes, and protect the already functional genetic blocks: therefore improving the locality of crossover operations during the evolution phase of GP. In the strongly-typed crossover approach, the genetic code from the initial seed (Figure 2, Stage 2b, Transposon A) is not allowed to recombine with the genetic code from the newly created transposons (Figure 2, Stage 2b, transposons B and C). Hence the crossover operation is only carried out within the individual transposons dedicated for the same objective (i.e. Transposon A with Transposon A, B with B and C with C only). This way compartmentalization, the division of tasks into separate strands of genome, is encouraged.

5. MULTIOBJECTIVE FITNESS FUNCTION

Since the mechanisms introduced in this paper are aimed at tackling issues raised by the evolution of a design with multiple features (that are also partly conflicting), we also study the effects of using MO optimization. MO algorithms optimize towards several objectives by searching Pareto-optimal solutions, and their use in GP has been shown to enhance performance in cases such as reducing bloat (uncontrolled growth of genotype in GP) (Bleuler, Brack, Thiele, & Zitzler, 2001).

The MO method chosen was based on the popular NSGA-II algorithm (Deb, Pratap, Agarwal, & Meyarivan, 2002). NSGA-II uses a non-dominated sorting algorithm with a crowding distance operator, ranking the individuals in the population into a number of Pareto-optimal fronts based on the number of objectives, and allows an unbiased trade-off between the objectives. Therefore, the individuals which do well on some objectives while performing worse on other objectives are preserved in the evolutionary process; allowing a wide spread of results in the phenotypic landscape. An example of Pareto-optimal fronts for a two objective problem is shown in Figure 3.

In our case, there are 2 objectives that the Snakebots need to be optimized for; forward movement, and collision avoidance with the large obstacles and walls. When NSGA-II is utilized, the population will converge to a Pareto-optimal front known as the Pareto-optimal. The results produced by NSGA-II along the Pareto-optimal front provides solutions with phenotypic as well as genotypic diversity, which should be good for evolving well performing Snakebots. One of the

Figure 3. Illustration of the formation and presence of different pareto-optimal fronts for an optimization problem with two objectives. The aim of an MO optimization algorithm is to encourage diversity of solutions while looking for convergence to the optimal solutions. The candidate solutions are categorized into different fronts, and the ones in the upper front are considered to have higher ranks (i.e. fitness value).

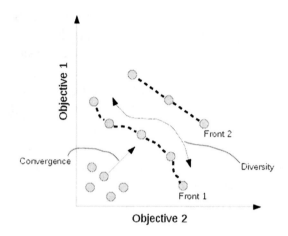

key factors in the decision to introduce MO optimization was the observations of the deteriorating performance of single objective GP runs due to stagnating runs. One of the most common reasons for unsuccessful GP runs to evolve Snakebots with sensing capabilities were observed by the authors to be the stagnation of the runs, where the optimization algorithm would cease to improve the fitness of the best individuals over a large number of generations (i.e. getting stuck in local optima).

6. EVOLUTIONARY FRAMEWORK AND THE SIMULATION ENVIRONMENT

For the experiments presented in this work we employ the Document Object Model (DOM) and the eXtensible Markup Language (XML) for the implementation of the genetic programs

(Tanev, 2004): we represent evolved genotypes of simulated Snakebot as DOM-parse trees featuring equivalent flat XML-text. Both (1) the calculation of the desired turning angles during fitness evaluation and (2) the genetic operations are performed on DOM-parse trees using API of off-the-shelf, platform-, and language neutral DOM-parsers. The corresponding XML-text representation (rather than S-expression) is used as a flat file format, feasible for migration of genetic programs among the computational nodes in an eventual distributed implementation of the GP. This approach means a significant reduction in time in the usually slow software engineering of GP, and offers a generic way to facilitate the reduction of computational effort by limiting the search space of GP by handling only semantically correct genetic programs. The concept is accomplished through strongly typed GP, in which the use of W3C-recommended standard XML schema is proposed as a generic way to represent and impose the grammar rules in strongly typed GP. Furthermore, the evolved solutions are stored in a human-friendly, text-based representation.

Although using a DOM/XML-based representation has the disadvantage of slower traversal of the GP programs than a traditional approach. The performance degradation caused by the traversal of the DOM/XML-based representation of genetic programs during fitness evaluation is negligible for the overall performance of GP. The performance profiling results indicate that fitness evaluation routine consumes more than 99% of GP runtime, however, even for relatively complex genetic programs featuring a few hundred tree nodes, most of the fitness evaluation runtime at each time step is associated with the relatively enormous computational cost of the physics simulation of the simulated Snakebot.

In Section 6.2 we discuss the details of the evolutionary framework. However, first, Section 6.1 will introduce the representation of the evolved Snakebot.

6.1. Representation of the Snakebot

We employ Open Dynamics Engine (ODE) as a simulation platform for the Snakebot. ODE is a free, industrial quality software library for simulating articulated rigid body dynamics (29). It is fast, flexible and robust, and it has built-in collision detection. Therefore, ODE is suitable for a realistic simulation of the physics of an entire Snakebot when applying actuating forces to its segments. The main ODE related parameters of the simulated Snakebot are summarized in Table 1.

Snakebot is simulated in ODE as a set of 15 identical spherical morphological segments ("vertebrae"), linked together via universal (Cardan) joints; i.e. the joints have 2 degrees of freedom (Figure 4). All joints feature identical angle limits and each joint has two attached actuators ("muscles"). A single LRF sensor, with a limited range is rigidly attached to each of the segments.

Table 1. ODE-related parameters of simulated snakebot

Parameter	Value
Number of modules in Snake	15
Module shape	Sphere
Radius of a module	3cm
Overlap between modules	25%
Snakebot length	66cm
Weight of a module	100g
Joint type between modules	Universal
Actuators per joint	2 (Horizontal and Vertical)
Actuator operation mode	dAMotorEuler
Actuator max torque	12000gcm
Max angular velocity of actuators	100degrees/s
Actuator stops (angular limits)	±50
Coefficient of Friction (μ)	0.5
Friction model	Pyramid approx. of Coloumb friction model
Simulation sampling frequency	20Hz

Figure 4. Morphological segments of snakebot are linked via universal joint. Horizontal and vertical actuators attached to the joint perform rotation of the segment #i-1 in vertical and horizontal planes respectively. A single LRF is attached to each of the segments in the plane of the axes of the universal joint.

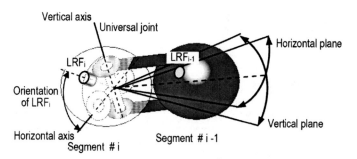

6.2. Algorithmic Paradigm

The functionality of the LRF can be defined by the values of the following set of parameters: (1) orientation, measured as an angle between the longitudinal axis of the sensor and the horizontal axis of the joint, (2) range of the sensor (in cm), and (3) the timing of activation, expressed as a threshold value of the turning angle of the horizontal actuator. The reading of LRF is a scalar value which corresponds inversely to the distance between the sensor and an object (if any within the sensor's range), measured along the longitudinal axis of the LRF. In the initial standstill position of Snakebot (as depicted in Figure 1a) the rotation axes of the actuators are oriented vertically (vertical actuator) and horizontally (horizontal actuator) and perform rotation of the joint in the horizontal and vertical planes respectively.

Considering the representation of Snakebot, the task of designing the fastest locomotion can be rephrased as developing temporal patterns of desired turning angles of horizontal and vertical actuators of each segment that result in fastest overall locomotion of Snakebot. The proposed representation of Snakebot as a homogeneous system comprising identical morphological segments is intended to significantly reduce the size of the search space of the GP.

For the evolution of the Snakebot, the genotype is represented as a triple consisting of a linear chromosome containing the encoded values of the three relevant parameters of LRF, and two parse trees corresponding to the algebraic expressions of the temporal patterns of the desired turning angles of both the horizontal and vertical actuators, respectively (Figure 5).

The Snakebot is genotypically homogeneous so the same triple is applied for the setup of the LRF and for the control of actuators of all morphological segments. The encoding of the parameters of LRF is as described in Figure 5. The same figure also illustrates the function set and the terminal set of the GP, employed to evolve the control sequences of both actuators. Since the locomotion gaits by definition are periodical, the periodic functions sine and cosine are included in the function set of GP in addition to the basic algebraic functions. Terminal symbols include the variables time, segment_ID, an ADF, the reading of the sensor (LRF), and two constants: pi, and a random constant within the range (0, 2). The incorporation of the terminal symbol segment_ID (a unique index of morphological segments of Snakebot) provides GP with an effective way to specialize (by phase, amplitude, frequency, etc.) the genetically identical motion patterns of actuators of each of the morphological segments of the Snakebot.

Figure 5. Genotype of the snakebot; represented as a triple containing the values of the parameters of LRF and two algebraic expressions of the temporal patterns of the desired turning angles of horizontal and vertical actuators, respectively. The genotype of snakebot is homogeneous: therefore, all segments feature the same triple.

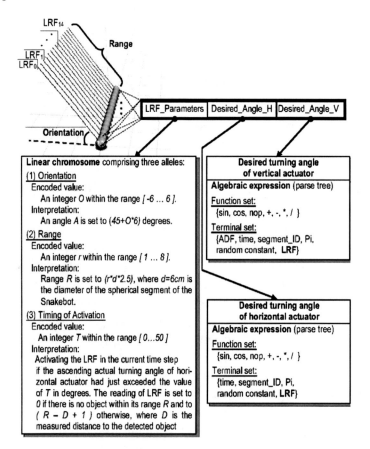

The rationale of employing ADFs is based on the observation that the evolvability of straightforward, independent encoding of desired turning angles of both horizontal and vertical actuators is rather poor. Even without ADFs, GP is able to adequately explore the potentially large search space and ultimately discover the areas that correspond to fast locomotion gaits in the solution space. However, it was observed in the previous work of Tanev et al (Tanev, Ray, & Buller, 2005) that not only the motion patterns of adjacent segments are correlated, but the motion patterns of horizontal and vertical actuators of each segment in fast locomotion gaits are highly correlated too. Moreover, discovering and preserving such

correlation by GP is associated with enormous computational effort. ADFs, which provide a way of introducing modularity and reuse of code in GP (Koza, 1994), are employed in our approach to allow GP to explicitly evolve the correlation between motion patterns of horizontal and vertical actuators as shared fragments in algebraic expressions of desired turning angles of both actuators. Furthermore, we observed that the best results are obtained by: (1) allowing the use of ADF as a terminal symbol in algebraic expression of desired turning angle of vertical actuator only, and (2) evaluating the value of ADF by equalizing it to the value of currently evaluated algebraic expression of desired turning angle of horizontal actuator.

6.2.1. Genetic Operations

The main GP (hence the GA) parameters are summarized in Table 2. The genotype is formed of three parts: two GP trees to control the Snakebot movement, and a set of parameters to determine the sensor settings. The population size is set to 200, and an elitist GA with 4 elite individuals is utilized: 4 best individuals are passed down to the next generation without any alterations. To create the rest 196 individuals of a new generation, we employ a binary tournament selection: two individuals are picked at random, and 90% of the time a new individual is created via single point crossover (reproduction) and 10% of the time the fittest of the two is chosen to be passed on to the next generation. The crossover point is randomly selected between the three components of the genotype (as shown in Figure 5), unless stated otherwise. The mutation randomly alters 1% of the newly created individuals (all except the four elite): either a value of an allele in the linear chromosome representing the parameters of LRF, or a sub-tree in one of the two parse tress that correspond to the temporal patterns of the control sequences of actuators. We have chosen to set the aforementioned parameters as described relying on

Table 2. Main parameters of GP

Category	Value
3*Genotype	LRF parameters (linear chromosome)
	Horizontal actuator control (parse tree)
	Vertical actuator control(parse tree)
Population Size	200 individuals
2*Selection	Binary Selection ratio: 0.1
	Reproduction ratio: 0.9
Elitism	4 individuals
Mutation Rate	0.01
Trial Interval	16s (400 time steps of 40ms per step)
Termination Criteria	(Fitness=120) or (Tot. No. of generations=80)

our experience with previous experimental runs of evolving GP for the control of a fast moving and sensing Snakebot (Tanev, Ray, & Buller, 2005; Tanev & Shimohara, 2008). Furthermore, we also kept the parameters same as the previous studies for consistency and a fair comparison between the different versions.

6.2.2. Fitness Evaluation

The fitness function is based on the average velocity of Snakebot, which is estimated from the distance travelled during the trial. The real values of the raw fitness, which are usually within the range (0, 2) are multiplied by a normalizing coefficient in order to deal with integer fitness values within the range (0, 200). A normalized fitness of 100 is equivalent to a velocity that displace the Snakebot a distance equal to twice its length. The fitness value is penalized by $\frac{1}{10}$ times the size of the genotype in order to control bloat. The fitness value is also penalized if the evolved Snakebot is not using sensor information by reducing the achieved fitness score by 98%. As we shall elaborate later in Section 6.3, the confined environment that the Snakebot need to clear during the trial is simulated by a narrow corridor covered with obstacles of various sizes (Figure 6). The velocity of locomotion needed to reach the end of the corridor for the given time of the trial (16s) corresponds to the fitness value of 100. The evolution is terminated if the bot reaches a fitness of 120 (or higher) or if the maximum number of generations is reached.

6.2.3. Experimental Cases

We present six experimental cases that will be studied in Section 7. These cases are aimed to demonstrate the effects of the proposed mechanisms on the performance of GP. The environment is described later in Section 6.3.

Figure 6. The experimental scenes used. The main environment for evolving snakebots with sensors (a and b), and the obstacle-free environment for evolving Snakebots with no sensors (c).

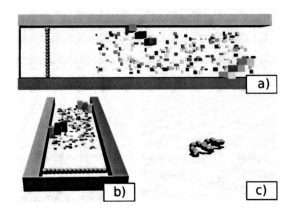

1. **Canonical GP (Single Stage Approach):** In this case the evolution of the Snakebot controller is done from scratch; i.e. evolution starts with a population of randomly created individuals and optimizes these individuals to satisfy the target fitness. The limit of the evolutionary generations of GP is set to 80. This experimental case aims to demonstrate the performance of canonical GP with ADF in solving the described design problem.

2. **Seeding (Two-Staged Approach):** The genotypes of six best sensorless Snakebots that have already been evolved to achieve fast sidewinding locomotion in a plain, smooth terrain (Figure 2, Stage 1), are used to create the initial population. These evolved genotypes are used as elite individuals to seed the initial population, where the exact copies of these six sensorless bots are used to form a small part of the initial evolutionary population. The remaining part of the population (194 bots) is randomly generated. This seeded population is then evolved to fully satisfy the target fitness. The limit of the generations of both stages of evolution is set to 40 (totaling 80).

3. **GT (Two-Staged Approach) with Conventional Crossover:** The first stage of the GT approach is identical to that of the seeding method. Similarly, the six best sensorless genotypes are used as elite individuals in the initial population of the second evolutionary stage. To create the remaining 194 bots of the initial population, however, we use GT on a randomly chosen seeding individual (out of the 6) as described in Section 3. The created population is evolved to fully satisfy the target fitness. Similar to the seeding, the limit of the generations of both stages of evolution is set to 40.

4. **GT with Strongly-Typed Crossover:** GT is applied as explained in the previous case, however in this case the crossover operation differs in the constraints applied to the available genetic pool for a crossover operation. Crossover operation is only carried out between the two transposons that are dedicated for the same task. In the implementation of this crossover technique, the genes belonging to a certain transposon group are given the same identification numbers at the start of the evolutionary phase in order to allow convenient classification of transposons. For the experiments presented in this paper, there are three transposons: formed by the initial seed, the newly created transposon, and the control gene that connects the two previously mentioned transposons (See Figure 2, Stage 2b).

5. **MO:** In this experimental case, the evolution of the Snakebot is carried out in a similar way as the first case (Canonical GP), but instead of utilizing the fitness function described in Section 6.2.2, the distance travelled (to be maximized) and the number of collisions (to be minimized) are optimized as distinct objectives via the techniques outlined by NSGA-II. Therefore, the main difference that arises in this case is the fitness evaluation

which is altered to treat the two target features of the Snakebot as distinct objectives: (1) moving forward as fast as possible, and (2) not hitting any of the large obstacles. We use the same selection mechanism described earlier, but the ranking of individuals differs due to the change in fitness evaluation as well as the ranking used in NSGA-II (Deb, Pratap, Agarwal, & Meyarivan, 2002). The maximum number of evolutionary generations is limited to 80.

6. **MO with GT and Strongly-Typed Crossover:** This has the same implementation as the previous case, but also uses GT and strongly-typed crossover as in case 4. Therefore, MO in this case is only used for stage 2; stage 1 is same as other experimental cases utilizing a seed. The limit of evolutionary generations is set to 40 for each stage, i.e. 80 generations in total.

6.3. Experimental Setup

The experimental environment (Figure 6) is formed of a straight narrow corridor (the width is the same as the length of the Snakebot) that has two groups of tall boxes which protrude to about 40% of the width of the corridor. In addition, part of the corridor is covered by many, randomly located and sized, small boxes that are designed to create a rough terrain and noisy environment for the sensors. The length of the corridor is set to seven times the length of the bot. Starting from the end of the corridor, the aim of the bot is to reach the other end within the given time-span. We designed this environment with the intention to encourage the evolving bot to develop the following abilities: (1) fast locomotion (long enough corridor), that is (2) not hindered by rugged terrain (small boxes), (3) following of obstacles that cannot be overcome (walls), and (4) circumnavigating obstacles that cannot be overcome (tall boxes).

The Snakebot is initialized with 15 modules, on full stretch at the dead end of the corridor with its longitudinal axis perpendicular to the intended direction of movement. Initially, the rough terrain is not present to facilitate the evolution of basic locomotion on smooth terrain. After a fitness value of 60 is reached (i.e. the first set of obstacles cleared), a large portion of the corridor is filled with randomly initialized boxes (random size and location). The initial orientation of the Snakebot and the corridor is influenced by the previous work suggesting that sidewinding is the fastest and most robust locomotion gait for a Snakebot. Therefore, the Snakebot would be expected to enter a corridor featuring a similar orientation.

7. EXPERIMENTAL RESULTS

The evolution of a sensing and fast moving Snakebot is tested for the experimental cases and conditions described in Section 6. For each approach we executed 38 independent runs. The average fitness convergence characteristics of these runs are shown in Figure 8, and the results are summarized in Table 3.

As Figure 8 depicts, the canonical GP features average fitness (over all independent runs) of about 40, which roughly corresponds to the 40% of the length of the corridor, which is the starting position of the first set of tall obstacles. The pace of the improvement of the fitness for canonical GP runs is rather slow, with average value of about 30 at generation 40.

When six of the previously evolved sidewinder Snakebots (Figure 7) are incorporated via *seeding* into the initial population as a second stage of evolution (Figure 8), and allowed to evolve for 40 more generations, the average fitness value is 1.6 times higher than the result obtained by canonical GP (Table 3). The seeded runs reach the fitness value of 40 after 3 generations of the second stage of evolution, instead of a total of 80 generations

Table 3. Statistics of the experimental results. The fitness-based values of each experimental case are obtained from 38 runs. Out of the 38 runs, the snakebots that achieved a fitness value of 100 or higher are considered "successful," and the snakebots that could obtain a fitness value of 100 or more in an experimental environment with different configuration of small obstacles (rough surface) than the one used during evolution are considered "robust."

	Average Fitness	Median Fitness	Std Dev. of Avg. Fit.	Successful Runs	Robust Runs
Canonical GP	39.5	32	23	1	0
Seeding	69.3	67	27.2	3	1
GT	91.2	91	19.4	8	1
GT with Strongly-typed Crossover	86.9	88.5	21.7	10	3
MO	33	29	17.1	1	0
MO with GT and Str-typed Xover	92.6	94	17	14	6

Figure 7. Evolution of the seeds. The average fitness convergence with error bars displaying the standard deviation of the runs from the "first stage" of the experiments using a seed (i.e. a GP with ADFs is used for these runs)—the data used for this plot are also from the runs in Tanev, Ray, and Buller (2005). The target of these runs were to evolve a snakebot that can achieve fast sidewinding locomotion in an environment with no obstacles or walls, the experimental environment is shown in Figure 6c. The fitness evaluation here is as described in Section 6.2.2, with a fitness value of 100 as the goal.

Figure 8. The average fitness convergence plots of each experimental case are displayed in the plot. The results of the runs utilizing a 2 stage approach (i.e., a seed) are plotted from generation 40 onwards; the first 40 generations of these runs take place in a different environment with different conditions. Therefore they can not be part of this plot, instead they are shown in Figure 7. Small degradations in fitness can be observed in some places, when the fitness is above 60, this is due to the randomization of the small obstacles during the map (as explained in Section 6.3).

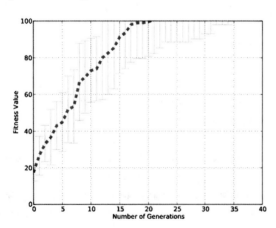

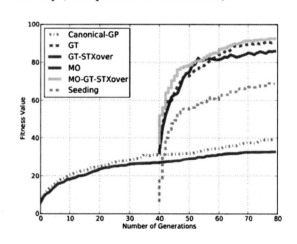

required by the canonical GP runs; demonstrating a speed up of almost 2 fold (43 generations vs 80 generations) in achieving this fitness. The performance of evolution is even better when *GT* is used–the average final fitness is higher than 90, with 8 successful runs, and a smaller deviation in the fitness values achieved (Figure 8 and Table 3). When GT is used, the average fitness value achieved via seeding runs (69.5) after 40 generations of the second stage run, is reached in approximately 5 generations.

From the results obtained, a significant change in the GP performance was not observed with different crossover mechanisms (GT vs. GT with strongly-typed crossover): in both cases, Snakebots that achieve similar fitness scores are found, and the number of successful runs increase by 2 (from 8 to 10 out of 38 runs) when strongly-typed crossover is used.

Two more experiments were undertaken using MO optimization; one with canonical GP, and another utilizing GT with strongly-typed crossover. The fitness progression of the MO runs are shown side by side with the other runs for comparison. Only for this purpose, the individuals from MO runs are assigned fitness values using the fitness function described in Section 6.2.2 for the plots in Figure 8 and Table 3. In both cases of MO runs, the use of MO optimization provided a smaller standard deviation in comparison to all the other experimental cases. However, the use of a MO approach did not affect the overall performance of canonical GP in evolving Snakebots positively, in fact a lower average fitness was obtained in this case. On the other hand, the use of MO optimization in conjunction with GT and strongly-typed crossover provided a significant boost in the number of successful runs achieved, while maintaining a lower standard deviation and higher average fitness. The overall best performance was achieved via MO with GT and strongly typed crossover: the highest average final fitness, lowest deviation, and the highest success rate (with 14 successful, 6 of them robust) runs are achieved.

Furthermore, the experiments using MO with GT and strongly typed crossover achieved the average final value achieved by typical seeding runs of 69.3 in 2 generations instead of 40.

It is important to note that the fitness based comparison made between the MO runs and the other experimental cases are not completely fair since the MO runs had different objectives to optimize (as described in Section 5) than the rest of the cases. However, the results from MO runs are re-assigned a fitness using the same fitness function as the other cases, and plotted side-by-side for a simple and convenient comparison in order to achieve a basic idea on how well the MO optimization performs. The more reliable measure in deciding whether a MO optimization provides genotypes that achieve significantly better Snakebot control is by comparing the number of successful and robust robots shown in Table 3. In fact, since the main objective is to design a robust controller for modular Snakebots, the quality of an algorithm is best quantified in terms of the number of robust Snakebot controllers achieved. We consider an algorithm that can achieve the design of a single robust Snakebot controller as successful, since we only need a single good controller to be implemented on the actual Snakebot modules. Thus, a good way to compare the performance of two algorithms is to quantify how successful these algorithms are in the given task.

The fitness values achieved for each case are also shown as box and whisker plots in Figure 9. There are two mechanisms that can be observed to cause significant improvements (i.e. the notches in the interquartiles do not overlap) over canonical GP with ADFs: seeding and GT, where seeding runs show a significantly better median fitness than canonical GP runs and all the GT runs have significantly better median fitness than the non-GT runs. As mentioned earlier, the fitness based plots of MO runs can not reflect the true performance of these runs, but they are included in this plot for completeness.

Figure 9. Box and whisker plots of the fitness values achieved by each experimental case over 38 runs. The three best cases utilize GT, and they are all significantly better than the other cases including classical seeding.

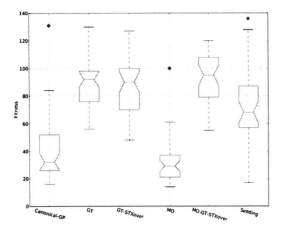

Fitness convergence plots displaying all 38 runs for each experiment can be found at the end of the paper as a source of more detailed information, Figures 11 and 12.

7.1. Analysis and Discussion of Experimental Results

We can observe from the results that GP performs best when a combination of MO with GT and strongly-typed crossover mechanisms are used, and performs worst when no incremental evolutionary approach is utilized. The sole use of canonical GP with ADFs is unable to discover bots that even demonstrate generic locomotion gaits that can result in the movement of the Snakebot. The large search space of the evolution, caused by the need to integrate the sensing information to locomotion and the method of incorporating the sensing signals into the locomotion control are some of the reasons for poor efficiency. Another reason is the challenging environment imposed by the walls and various obstacles, which means that the fitness landscape of evolution features fewer (compared to the simpler environment in

Figure 6c, and the previously tested cases (Tanev, Ray, & Buller, 2005; Tanev & Shimohara, 2008)) and narrower optimal areas. Indeed, even if fast locomotion emerges during the initial stages of evolution, its survival value could be easily underestimated by evolution because the Snakebots get stuck at the first obstacle.

Furthermore, the results of the first stage of both the seeding and GT (Figure 7) indicate that the evolution of the locomotion of a Snakebot is more efficient when relieved from the burden of dealing with the sensors and the sophisticated environment, and utilization of an MO optimization alone is not sufficient to deal with the complexities of the more sophisticated environment. In the obstacle-free environment, a velocity (equivalent to a fitness value of 100) sufficient to clear the large obstacles (shown in Figures 6 a and b) is easily achievable within 10 to 36 generations. In the final results we see that the use of incremental GP yields significantly better Snakebots than GP without seeding. However, there is still a significant gap in the evolutionary performance when seeding is compared to the GT-inspired seeding mechanism. The proposed seeding mechanism, GT, allows the evolution to experiment with the way of processing the sensory signals without the risk of damaging the already evolved, fast locomotion control. Therefore, GT is able to facilitate the protection of the already evolved beneficial building blocks from the destructive effects of genetic operations. Conversely, since the locomotion control comprises 100% of the genotype of the bots created via seeding, any incorporation of the sensing information as a result of a genetic operation would most likely result in a damage of this control.

The use of strongly-typed crossover with GT, however, does not seem to significantly influence the performance of GP; in comparison to GT with regular crossover, a moderate increase in the number of successfully evolved Snakebots can be observed. On the other hand, when the successfully evolved Snakebots are more closely studied, it is observed that the solutions achieved via the runs using strongly-typed crossover are more robust in comparison (Table 3). The runs marked by evolution as "successful" are not always robust, and the number of Snakebots that are fully functional, general solutions (i.e. robust) are significantly different from the number of solutions marked as "successful" for each case. The strongly-typed crossover provides focus in a fruitful area of the optimization landscape drawing the GA away from areas of the optimization space with low quality results. Due to this constrained approach, the number of solutions that GA can come up with did not diminish (on the contrary improved), but more importantly the quality of the found solutions increased significantly. These results indicate that it is beneficial to prevent the inherited genotypes from mixing with newly created genes in order to preserve their functionality.

As mentioned earlier, the use of a MO optimization method to evolving Snakebot controllers appeared to boost the number of successful runs when utilized with GT. Even though there were only two objectives, this increase in evolutionary performance suggests that an equal evolutionary pressure to "avoiding collision" with "reaching the end of the corridor" was beneficial in finding a Pareto-optimal solution. The use of MO did not have any significant effect on the evolutionary performance when GT is not used, because of the difficulties presented by the challenging environment to a completely randomized Snakebot. Although GT is aimed at improving evolutionary performance in solving problems with multiple objectives, a similar task with MO, the effects and the main contribution to the evolutionary performance of GT is different from MO, and the best GP performance in the experiments presented is achieved by the use of these two techniques together. Moreover, the use of MO with GT and strongly typed crossover significantly improved the number of robust runs as well, Table 3.

All the runs marked as "successful" from every experimental case are re-run on an identical map to the one used during evolution, except this time without a rough terrain. Even though all the runs marked "successful" can clear the corridor when the rough terrain used during evolution is again utilized, some of these runs fail to clear the corridor without a rough terrain, see Table 3. This is due to the exploitation of the rugged features of the terrain by the evolved snakes as a form of guide through the map (i.e. over-fitting the solution to the environmental landscape). From the successful runs in each experiment, only 1 run from the experiment with seeding, and GT with conventional crossover are able to clear the map with smooth terrain, whereas 3 runs from the experiments using GT with strongly-typed crossover are able to clear the map: demonstrating a higher robustness in the results obtained via the use of strongly-typed crossover.

We further extend our investigation to robust Snakebots by testing the successful runs for a longer run time (of 800 time steps instead of 400) in a larger corridor with more of the large obstacles (illustrated in Figure 10). The best Snakebots from each experimental case are shown in Figure 10. There are no Snakebots from canonical GP experiments (with or without MO) displayed due to the best of the Snakebots from these runs not even being able to clear the original smaller map when there is no rough terrain. For the experimental cases that make use of strongly-typed crossover, the Snakebots can clear the whole of the extended map with the Snakebot using MO travelling the furthest. Snakebot from the GT with conventional crossover experiments is able to clear all the obstacles but not the whole corridor, and the Snakebot from seeding experiments can only clear the second set of obstacles in the given time frame. Once more, the performance of the evolved Snakebots observed in the latter experiments show that the Snakebots evolved are more

Figure 10. The moving trajectory of the central segment and the center of gravity (COG) of the sample best-of run snakebots on an extended map. Best snakebots of runs with seeding (a), GT-inspired seeding (b), GT with strongly-typed crossover (c), and GT with strongly-typed crossover using MO optimization (d). Runs from canonical GP and MO are not shown due to the best snakebots from these runs not being able to clear even the original map with no obstacles.

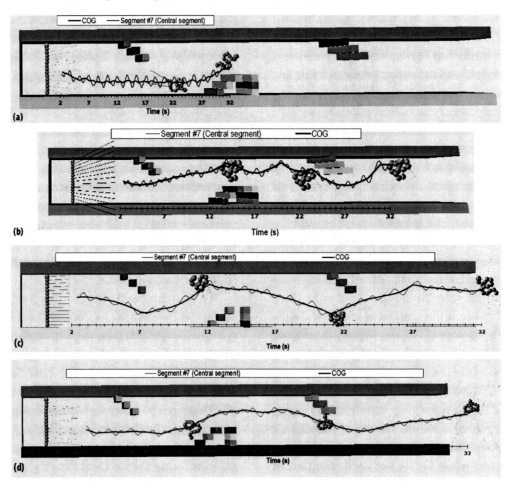

robust when strongly-typed crossover is used with GT. Also, the use of MO optimization techniques is shown to improve the quality of the resulting Snakebot as well as the number of successfully evolved Snakebots.

We also list the genotypes of each of the fittest runs from every experimental case in Table 4. These genotypes are the evolved controls for the sensing and sidewinding Snakebot. In addition the genotype of a simple sidewinding Snakebot is also shown (marked as "Sidewinder" in Table

4). The genotypes of the most robust Snakebots, which are all from experiments utilizing GT, had recognizable parts that correspond to sidewinding locomotion $\left(sin(id + t + C)\right)$, however this was not the case with the rest.

Since the locomotion control comprises 100% of the genotype of the bots created via seeding, and for runs without any seeding the locomotion control has to be coevolved at the same time with sensing abilities, parts of their genotype that resemble the generic form found by the sidewinding

Table 4. Example resulting genotypes from the experiments. The values LRF, t and ID are, respectively, the input values from the sensor mounted on the module, the time from the internal clock

	Genotype
Canonical GP	$(\frac{\pi}{2} + 4t) \times cos(\frac{\pi}{2} ID + 7t - \frac{cos(t)cos(cos(LRF))}{cos(cos(LRF \times t))})$
Seeding	$$\frac{\frac{\pi}{2} cos(\frac{463 + 154ID - t}{\frac{2}{\pi} t - 9.8ID})}{1043 + (3 + cos((16.4 + \frac{\pi}{2}(3 + LRF)^2) \times \sqrt{\frac{13 - \sqrt{LRF} - \frac{LRF + 1^4}{3}}{cos(sin(t \times LRF))}}))^2}$$
Genetic Transposition	$(\pi + \sqrt{sin(LRF)}) \times \frac{1}{8} sin(\sin(415t - 103ID))$
GT with Strongly-typed Crossover	$-1.01 \times (2sin(4 + LRF) + 7 + LRF) \times \sin(399t - 099ID)$
MO GP	$tcos(LRF + ID - 8t + 03)$
MO with GT and Strongly-typed Xover	$(\frac{7}{3} + 115 \times 10^{-6} \times LRF) \times (084 - \cos(097t + 039ID))$
Sidewinder	$cos(4t - \frac{ID}{13} + \pi)$

locomotion experiments do not exist. Instead large, complicated equations that are hard to comprehend are the results. As a matter of fact these genotypes are not able to fulfill the task of controlling a sidewinding Snakebot that can avoid walls using sensors; as demonstrated in Figure 10. Therefore, it can be said that using GT-inspired seeding mechanism not only improves the success rates of GP but also the emergence of compact genotypes, as well as reliable evolution of complex computational systems. The latter are true more so in the case of strongly-typed crossover. A plot of the average genotype sizes in order to show the progression of bloat in GP trees for each experimental case is shown in Figure 13 for the interested reader (attached to the end of this document).

We believe the improvement in the efficiency of GP due to the use of GT is for the following three major reasons:

1. A wider spread of the initial seed into the population (than the typical seeding) of genotype that features generic ability to move.
2. A better value of the initial fitness of the bots as they already feature the generic ability to move in their genotypes.
3. A separation of the sensing and locomotion parts of the genome, which may create a more efficient control mechanism for the bot.

Figure 11. Fitness convergence characteristics of 38 independent evolutionary runs are displayed for non-seeded experimental cases. The dashed horizontal line (fitness = 100) corresponds to the fitness required by the Snakebot to reach the end of the corridor for the allocated time of the trial (16s). The thick line of fitness plot over evolutionary time line is the average fitness achieved by all evolutionary runs.

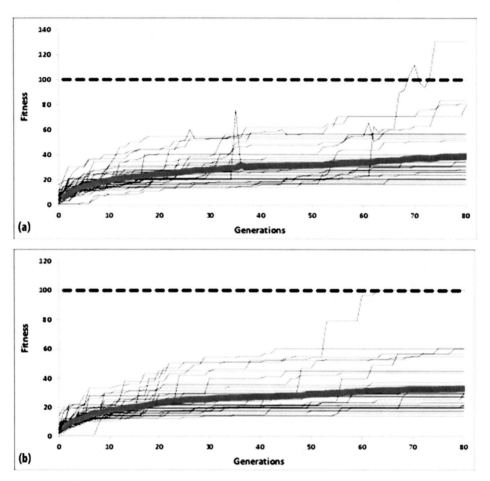

We would like to point out that latter of the before mentioned arguments might provide a further insight into the design of robotic control systems and their sensorimotor control. The locomotion property of the Snakebot can be viewed as a continuous process that needs to be applied regularly under normal conditions, and the sensing property of the Snakebot can be viewed as a reflex that only needs to affect the actions of the bot when an event occurs. Such a concept might be seen as analogous to the reactive behaviour related to the reflexes observed in biological organisms. For example, the collision-free flight of locusts in a crowded swarm is recognized to be

achieved by direct, real-time input of the sensory signals into the wings muscles. The latter serve as a mediator for both the (1) default oscillating signals (generated by a Central Pattern Generator (CPG)) and (2) the visual sensors (Uvarov, 1977).

From another viewpoint, our results can be seen as an evidence of the computational benefits of mimicking the neurobiological concept of achieving complex navigation behaviors of species in nature through sensory-controlled modulation of CPG. The moving trajectory of the sample best of run bots (Figure 10) illustrate the emergence of the following abilities of the Snakebots: (1) fast locomotion (clearing the corridor), that is

Figure 12. Fitness convergence characteristics of 38 independent evolutionary runs are displayed for seeded experimental cases

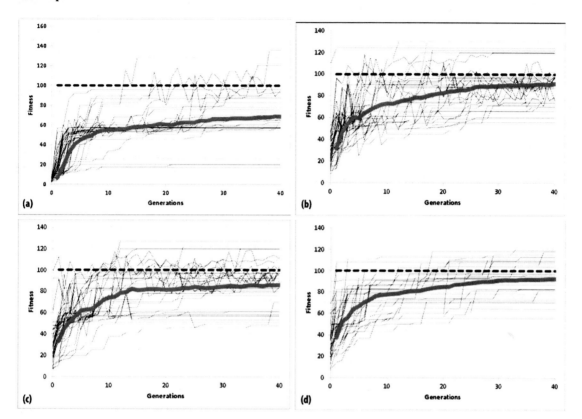

(2) not hindered by rugged terrain (overcoming small boxes), (3) following obstacles that cannot be overcome (walls), and (4) circumnavigating obstacles that cannot be overcome (two groups of tall boxes).

The quality of the obtained solutions from an evolutionary run determines the amount of post-processing required by the experimenter in filtering out the overly-fit solutions that are not able to satisfy the expected general behaviour. Therefore, the significant increase in the quality of the produced results (which corresponds to a reduction in time and effort required in achieving perfect solutions) due to the use of the combination of all the mechanisms presented (GT, strongly-typed crossover, MO) signifies an important step in creating higher reliability in the genetic programming of complex computational systems.

8. CONCLUSION

The results obtained here once more highlight the importance of achieving reliable results when an optimization algorithm is used to aid the design of a computational system. A weakly defined fitness function, complex evolutionary environment, and evolution with limited or no constraints will always have holes that a Genetic Algorithm (GA) can exploit in the design of a computational system, and when this happens the solutions presented via the algorithm are most often not the solution the experimenter desires.

In this paper, we tackled the design of a complex application that deals with interactions within an intricate environment. Although a simpler version of this application (where only a single feature locomotion is controlled) is shown to be perfectly

Figure 13. The average genotype lengths (tree sizes) over 38 runs for the final 40 generations of each experimental case are plotted. Each experimental case used the same parsimony pressure of a fitness penalty for the $\frac{1}{10}$ times the genotypes size. Although there was no other objective encouraging smaller genotypes, the average tree size of the solutions achieved are smallest for runs utilizing MO optimization. Seeding and Canonical-GP runs achieved solutions with similar average tree sizes, whereas a 25% increase in the average tree sizes can be seen when GT is used (in comparison to canonical GP).

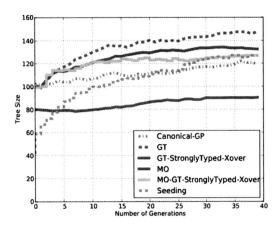

within the design capabilities of Genetic Programming (GP) with Automatically Defined Functions (ADF), the addition of only one more feature and an increase in the complexity of the environment render canonical GP with ADFs ineffective. We solve this challenge with the use of incremental evolution; however see that a typical seeding mechanism still performs poorly.

A Genetic Transposition (GT) inspired seeding mechanism is introduced that provides a safer approach to genetic alteration of existing seeds in incremental evolution by explicitly introducing neutral spaces in the genome. The GT-inspired mechanism aims to provide a safe playground for the genetic operators; sharing a similar purpose with the biological GT, albeit in a different form of implementation. Furthermore, the GT inspired seeding mechanism introduces modularity via partitioning of the genome (GP tree) to the number of dedicated objectives. The success of GT is demonstrated to significantly surpass the performance shown by typical seeding (also known as population enrichment (Perry, 1994)), and by canonical GP.

A further improvement in the execution of genetic crossover is demonstrated, in the hope of further improving the evolvability of a seeded system by protecting the already well-functioning areas of the genome. This mechanism, named strongly-typed crossover, is demonstrated to further improve the reliability of the results achieved, which is very important in real-life applications. The results obtained are analysed in detail via re-runs and the study of the resulting genotypes. The studies show that the best results are achieved via the protection of the seeding genotypes in the final solution, which is the aim of the strongly typed crossover in GT.

Finally, the popular MultiObjective (MO) optimization (Nondominated Sorting Genetic Algorithm 2) is used to accompany the canonical GP and GT with strongly-typed crossover in the evolution of genetic programs for the target Snakebots. When the NSGA-II algorithm is used for the evolution of Snakebots with sensors with canonical GP and ADFs, no significant change in the performance can be seen. Thus, the use of an MO does not provide any benefits in this case. However, when used in conjunction with GT, it is shown that MO optimization significantly improves the evolutionary performance in terms of average fitness achieved, number of high quality Snakebots achieved, as well as the number of robust Snakebots. It is shown that the use of MO optimization effectively boosts the evolutionary performance when used together with GT, but is unable to solve the given task without the help of an incremental evolution method.

The work presented demonstrates the benefits of utilizing a biologically inspired mechanism that introduces compartmentalization and the neutrality in the genotype of an incremental GP used for automatically designing computational systems. Future directions include further investigations into the effects of neutrality on the effects of GP performance. The positive effects of neutrality in the evolutionary performance is documented in evolutionary algorithms, and there are some solid studies that contribute to the understanding of neutrality. Neutrality has been shown to be beneficial in complex, rugged landscapes with multiple optima (Beaudoin, Verel, Collard, & Escazu, 2006; Doerr, Gnewuch, Hebbinghaus, & Neumann, 2007; Lobo, Miller, & Fontana, 2004). However, the role of neutrality in GP is not as widely accepted and more controversial, with conflicting studies (Galv, Poli, Kattan, O'Neill, & Brabazon, 2011). We hope to conduct studies that can help understand the role of neutrality within GP in certain practical situations.

REFERENCES

Beaudoin, V. Collard, & Escazu. (2006). Deceptiveness and neutrality: The ND family of fitness landscapes. In *Proceedings of the 2006 Conference on Genetic and Evolutionary Computation*. GECCO.

Bird & Layzell. (2002). The evolved radio and its implications for modelling the evolution of novel sensors. In *Proceedings of the Evolutionary Computation on 2002. CEC '02*, (pp. 1836–1841). Washington, DC: IEEE Computer Society.

Bleuler, B. Thiele & Zitzler. (2001). Multiobjective genetic programming: Reducing bloat using spea2. In *Proceedings of the 2001 Congress on Evolutionary Computation*, (vol. 1, pp. 536–543). IEEE.

Chan, Man, Tang, & Kwong. (2008). A jumping gene paradigm for evolutionary multiobjective optimization. *IEEE Transactions on Evolutionary Computation, 12*, 143–159.

Deb, Pratap, Agarwal, & Meyarivan. (2002). A fast and elitist multiobjective genetic algorithm: Nsga-ii. *IEEE Transactions on Evolutionary Computation, 6*(2), 182–197. doi:10.1109/4235.996017.

Doerr, G. Hebbinghaus, & Neumann. (2007). A rigorous view on neutrality. In Proceedings of Evolutionary Computation, (pp. 2591–2597). IEEE.

Dowling. (1999). Limbless locomotion: Learning to crawl. In *Proceedings of Robotics and Automation*, (vol. 4, pp. 3001–3006). IEEE.

Galv, Poli, Kattan, O'Neill, & Brabazon. (2011). Neutrality in evolutionary algorithms: What do we know? *Evolving Systems, 2*, 145–163. doi:10.1007/s12530-011-9030-5.

Hirose. (1993). *Biologically inspired robots: Snake-like locomotors and manipulators*. Oxford, UK: Oxford University Press.

Huynen, Stadler, & Fontana. (1996). Smoothness within ruggedness: The role of neutrality in adaptation. *Proceedings of the National Academy of Sciences of the United States of America, 93*, 397–401. doi:10.1073/pnas.93.1.397 PMID:8552647.

Koza, Keane, & Yu, Bennett, & Mydlowec. (2000). Automatic creation of human-competitive programs and controllers by means of genetic programming. *Genetic Programming and Evolvable Machines, 1*, 121–164. doi:10.1023/A:1010076532029.

Koza. (1994). *Genetic programming II: Automatic discovery of reusable programs*. Cambridge, MA: MIT Press.

Koza. (2003). *Genetic programming IV: Routine human-competitive machine intelligence*. Boston: Kluver Academic Publishers.

Kuyucu, Trefzer, Greensted, Miller, & Tyrrell. (2008). Fitness functions for the unconstrained evolution of digital circuits. In *Proceedings of the 9th IEEE Congress on Evolutionary Computation (CEC08)*, (pp. 2589–2596). Hong Kong: IEEE.

Langdon & Nordin. (2000). Seeding genetic programming populations. In *Proceedings of the European Conference on Genetic Programming*, (pp. 304–315). London, UK: Springer-Verlag.

Liu, Sheng, & Jiao. (2009). Gene transposon based clonal selection algorithm for clustering. In *Proceedings of the 11th Annual Conference on Genetic and Evolutionary Computation*, (pp. 1251–1258). IEEE.

Lobo, Miller, & Fontana. (2004). Neutrality in technological landscapes. *Santa Fe Working Paper*.

McClintock. (1950). The origin and behaviour of mutable loci in maize. *Proceedings of the National Academy of Sciences of the United States of America, 36*, 344–355. doi:10.1073/pnas.36.6.344 PMID:15430309.

McConaghy, Vladislavleva, & Riolo. (2010). Genetic programming theory and practice. In *Genetic programming theory and practice 2010: An introduction*, (pp. vii–xviii). New York: Springer.

McGregor & Harvey. (2005). Embracing plagiarism: Theoretical, biological and empirical justification for copy operators in genetic optimisation. *Genetic Programming and Evolvable Machines, 6*, 407–420. doi:10.1007/s10710-005-4804-9.

Montana. (1995). Strongly typed genetic programming. *Evolutionary Computation, 3*(2), 199–230.

Morowitz. (2002). *The emergence of everything: How the world became complex*. Oxford, UK: Oxford University Press.

Nolfi, F. Miglino, & Mondada. (1994). How to evolve autonomous robots: Different approaches in evolutionary robotics. In *Proceedings of the 4th International Workshop on Artificial Life*. Boston: MIT Press.

Nowacki, Higgins, & Maquilan, Swart, Doak, & LandWeber. (2009). A functional role for transposases in a large eukaryotic genome. *Science, 324*(5929), 935–938. doi:10.1126/science.1170023 PMID:19372392.

Perry. (1994). The effect of population enrichment in genetic programming. In *Proceedings of Evolutionary Computation*, (pp. 456–461). IEEE.

Shichel & Sipper. (2011). Gp-rars: Evolving controllers for the robot auto racing simulator. *Memetic Computing, 3*, 89–99. doi:10.1007/s12293-011-0056-9.

Sim & Costa. (2000). Using genetic algorithms with asexual transposition. In *Proceedings of the Genetic and Evolutionary Computation Conference* (pp. 323–330). San Francisco, CA: Morgan Kaufmann.

Simoes, C. Sim, & Costa. (1999). Transposition: A biologically inspired mechanism to use with genetic algorithms. In *Proceedings of the Fourth International Conference on Neural Networks and Genetic Algorithms (ICANNGA'99)*, (pp. 612–619). Berlin: Springer-Verlag.

Smith. (2004). *Open dynamics engine*.

Spirov, Kazansky, Zamdborg, Merelo, & Levchenko. (2009). *Forced evolution in silico by artificial transposons and their genetic operators: The John Muir ant problem*. Technical Report arXiv:0910.5542, Oct 2009. Comments: 33 pages.

Strand & McDonald. (1985). Copia is transcriptionally responsive to environmental stress. *Nucleic Acids Research, 13*(12), 4401–4410. doi:10.1093/nar/13.12.4401 PMID:2409535.

Tanev, Ray, & Buller. (2005). Automated evolutionary design, robustness and adaptation of sidewinding locomotion of simulated snake-like robot. *IEEE Transactions on Robotics, 21*, 632–645. doi:10.1109/TRO.2005.851028.

Tanev & Shimohara. (2008). Co-evolution of active sensing and locomotion gaits of simulated snake-like robot. In *Proceedings of the 10th Annual Conference on Genetic and Evolutionary Computation*, (pp. 257–264). New York, NY: ACM.

Tanev. (2004). Dom/xml-based portable genetic representation of the morphology, behavior and communication abilities of evolvable agents. *Artificial Life and Robotics, 8*, 52–56.

Thomsen, Fogel, & Krink. (2002). A clustal alignment improver using evolutionary algorithms. In *Proceedings of Evolutionary Computation, 2002,* (vol. 1, pp. 121–126). IEEE..

Uvarov. (1977). *Grasshoppers and locusts.*

Vassilev, Job, & Miller. (2000). Towards the automatic design of more efficient digital circuits. In *Proceedings of the 2nd NASA/DoD workshop on Evolvable Hardware*. Washington, DC: IEEE Computer Society.

Wilke, Wang, & Ofria, Lenski, & Adami. (2001). Evolution of digital organisms at high mutation rates leads to survival of the flattest. *Nature, 412*, 331–333. doi:10.1038/35085569 PMID:11460163.

Chapter 6

Awareness–Based Recommendation:
Toward the Human Adaptive and Friendly Interactive Learning System

Tomohiro Yamaguchi
Nara National College of Technology, Japan

Takuma Nishimura
Nara National College of Technology, Japan & NTT WEST, Japan

Keiki Takadama
The University of Electro-Communications, Japan

ABSTRACT

This chapter describes the interactive learning system to assist positive change in the preference of a human toward the true preference. First, an introduction to interactive reinforcement learning with human in robot learning is given; then, the need to estimate the human's preference and to consider its changes by interactive learning system is described. Second, requirements for interactive system as being human adaptive and friendly are discussed. Then, the passive interaction design of the system to assist the awareness for a human is proposed. The system behaves passively to reflect the human intelligence by visualizing the traces of his/her behaviors. Experimental results show that subjects are divided into two groups, heavy users and light users, and that there are different effects between them under the same visualizing condition. They also show that the system improves the efficiency for deciding the most preferred plan for both heavy users and light users.

DOI: 10.4018/978-1-4666-4225-6.ch006

INTRODUCTION

Interactive Reinforcement Learning with Human

In field of robot learning (Kaplan 2002), interactive reinforcement learning method, reward function denoting goal, is given interactively and has worked to establish the communication between a human and the pet robot AIBO. The main feature of this method is the interactive reward function setup, which was a fixed and built-in function in the main feature of previous reinforcement learning methods. So the user can sophisticate reinforcement learner's behavior sequences incrementally.

Shaping (Konidaris 2006; Ng 1999) is the theoretical framework of such interactive reinforcement learning methods. Shaping is to accelerate the learning of complex behavior sequences. It guides learning to the main goal by adding shaping reward functions as subgoals. Previous shaping methods (Marthi 2007; Ng 1999) have three assumptions on reward functions as following:

- Main goal is given or known for the designer.
- Subgoals are assumed as shaping rewards those are generated by potential function to the main goal (Marthi 2007).
- Shaping rewards are policy invariant, it means not affecting the optimal policy of the main goal (Ng 1999).

However, these assumptions will not be true on interactive reinforcement learning with an end-user. Main reason is that it is not easy to keep these assumptions while the end-user gives rewards for the reinforcement learning agent. It is that the reward function may not be fixed for the learner if an end-user changes his/her mind or his/her preference. However, most of previous reinforcement learning methods assumes that the reward function is fixed and the optimal solution is unique, so they will be useless in interactive reinforcement learning with an end-user.

Table 1 shows the characteristics on interactive reinforcement learning. In reinforcement learning, an optimal solution is decided by the reward function and the optimality criteria. In standard reinforcement learning, an optimal solution is fixed since both the reward function and the optimality criteria are fixed. On the other hand, in interactive reinforcement learning, an optimal solution may change according to the interactive reward function. Furthermore, in interactive reinforcement learning with human, various optimal solutions will occur since the optimality criteria depend on human's preference.

Then the objective of this research is to recommend preferable solutions of each user. The main problem is "how to guide to estimate the user's preference?". Our solution consists of two ideas. One is to prepare various solutions by *every-visit-optimality* (Satoh 2006), another is the *coarse to fine recommendation* strategy (Yamaguchi 2008).

Requirements for Interactive System as Being Human Adaptive and Friendly

Recent years there are many researches on recommender systems to present information items that are likely of interest to the user. A recommender system is one of the major intelligent and interactive systems with a human user. However, it sometimes seems to be officious for the user. Main problem is that it is active but less intelligent (Yamaguchi 2012).

Table 1. Characteristics on interactive reinforcement learning

Type of reinforcement learning	an optimal solution	reward function	optimality criteria
standard	fixed	fixed	fixed
interactive	may change	interactive	fixed
interactive with human	various optimal	may change	human's preference

For examples, Office Assistant (Office97-2003) and the recommender system by Amazon.com are typical intelligent & interactive systems. Office Assistant is to assist users by offering advice based on Bayesian algorithms. However, the feature evoked a strong negative response from many users. On the other hand, the recommender system by Amazon.com is based on collaborative filtering (Schafer 2001). After a user views some items in the amazon's Web site, the recommender system shows information to the user such as "What Do Customers Ultimately Buy After Viewing This Item?" or "Customers Who Bought This Item Also Bought ...." However, it sometimes seems to be officious. Why? The key point is that these intelligent & interactive systems are active but less intelligent.

Useful suggestion is from social psychology in 1970s. In the *human performance theory* (Anderson 1974), a human's performance depends on both his ability and his motivation. Figure 1 shows the human performance theory applied to an agent. Figure 1(a) shows a human's performance relation. The better cases are when the ability is high. It is beneficial if high motivation, or it is potentially hopeful if low motivation. On the other hand, if the ability is low, the bad case is not low motivation (harmless) but high motivation (harmful). Therefore, the worst case is the combination of low ability and high motivation.

Then we define that an agent's performance depends on both its intelligence and its activity. Figure 1(b) shows an agent's performance relation.

Note that the agent's intelligence axis is the ability axis in Figure 1(a), and that the agent's activity axis is the motivation axis in Figure 1(a). Such as the human's performance relation in Figure 1(a), the worst case in Figure 1(b) is the combination of low intelligence and high activity. This is the reason why active but less intelligent agents seem to be officious.

To solve this problem, less active but more intelligent agent is desirable. Therefore, objective of the research is to realize harmless or potentially desirable interactive recommendation for a human user by our less active recommendation agent. There are several merits. First, it allows a user's preference shift since it performs less active recommendation. Second, it can assist the user for its convergence of his preference. Therefore, we aim to design to assist the awareness for a user.

How to Keep the Current Preference of a User?

Another problem is to keep the current preference of a user. Typically, a recommender system compares the user's profile to some reference characteristics, and seeks to predict the "rating" that a user would give to an item they had not yet considered. These characteristics may be from the information item (the content-based approach) or the user's social environment (the collaborative filtering approach) (Schafer 2001; Riecken 2000; Balabanovic 1997).

Figure 1. The human performance theory applied to an agent: (a) human's performance relation; (b) agent's performance relation

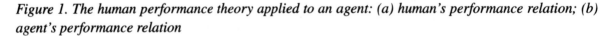
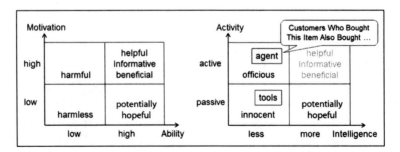

One of the main research issues is building the user's profile. There are two ways whether explicit data collection or implicit data collection. The explicit data collection is asking a user that he/she likes, or to choose the best one. The implicit data collection is observing behaviors of a user, for example the record of items that the user views or chooses. However, there is a problem that the collected user's profile is not same as the user's current preference, we called it the preference change problem (Yamaguchi 2009). In other words, these systems often recommend irerelevant items for the user since most previous researches do not directly specify the user's current preference. Consequently, it is necessary to treat the ambiguity, shift, or changes of a user's preference.

Passive Interaction Design to Assist the Awareness for a Human

To solve these problems, first we define the model of a user's *preference shift* for the recommendation space by two axes, preference *reduction* and preference *extension*. Then the system prepares various recommendation plans according to the goals that the user selected by our extended reinforcement learning method (Konda 2002a, 2002b; Satoh 2006). Second, to assist the user's awareness for his/her *preference shift*, we propose the user-centered recommendation by visualizing both the recommendation space with prepared recommendation plans and the user's preference trace as the history of the recommendation in it. Note that the recommendation space visualizes the possible *preference shift* of the user. We consider the interaction between a user and a recommender system to be the cooperative learning in human agent interaction. We assume that the process to make clear the user's preference is to *be aware the true preference* of himself. To realize this process, we try to support the user's awareness (Yamaguchi 2009; Takadama 2012). Note that

supporting the awareness means not to prevent it but to assist it. The experimental results show that our system improves the efficiency for deciding the most preferred plan.

BACKGROUND

This section describes an overview of our plan recommendation system. First, we introduce the plan recommendation task and the model of a user's *preference shift*. Then our plan recommendation system based on our extended reinforcement learning called LC-learning (Satoh 2006) is described. As for details, please refer (Yamaguchi 2011).

The Round-Trip Plan Recommendation Task

The task for a user is to decide the most preferred round-trip plan after selecting four cities to visit among eighteen cities. The task for the system is to estimate the preferable round-trip plans to each user and to recommend them sequentially. The way of generating these plans by LC-learning will be described in subsection "Overview of the interactive plan recommendation system."

The plan recommendation procedure is as follows:

Step 1: A user selects four cities to visit. These are called goals.
Step 2: Various round-trip plans including at least one goal are recommended.
Step 3: The user decides the most preferred round-trip plan among them.

The Model of a User's Preference Shift

Figure 2 shows the model of a user's preference shift. It is defined by two-axes space. Figure 2(a) illustrates the model. Comparing the previous

Figure 2. The model of a user's preference shift: (a) the model; (b) an illustrated example of the model

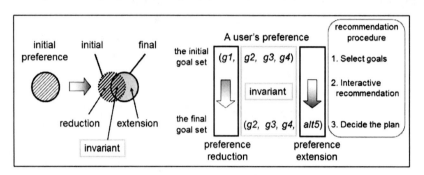

preference set and the current preference set, the common set is the invariant preference, the reduction set is called *preference reduction*, and the addition set is called *preference extension*.

Next, we explain the significance of our model on user profiling. There are two ways whether the data collection is explicit or implicit. From the viewpoint of the explicit data collection, our model aims for realizing not similar but various recommendations with a set of goal cities by a user. From the viewpoint of the implicit data collection, our model makes the user aware of his preference shift by visualizing the user's preference trace.

Now we describes an illustrated example of the model of a user's preference shift as shown in Figure 2(a) according to the plan recommendation procedure in previous subsection. Figure 2(b) shows the illustrated example of the model of a user's preference shift. To begin with, a user selects four goals (g1, g2, g3, g4) as the initial goal set in the first step as shown in previous subsection. After the second step of interactive recommendation, the user decides the most preferred plan including the final goal set as (g2, g3, g4, alt5) in the third step. By comparing the final goal set with the initial goal set, the invariant preference is (g2, g3, g4), the preference reduction is (g1), and the preference extension is (alt5).

Overview of the Interactive Plan Recommendation System

This subsection describes an overview of the interactive plan recommendation system. When a user input several goals to visit constantly as his/her preference goals, they are converted to the set of rewards in the plan recommendation block for the input of interactive LC-learning (Satoh 2006) block. Note that a goal is a reward to be acquired, and a plan means a cycle that acquires at least one reward in a policy.

Interactive LC-Learning block is based on extended model-based Average Reward reinforcement learning (Mahadevan 1996). In that, the simple MDP model (Puterman 1994) is used. Our learning agent consists of three blocks those are model identification block, optimality criterion block and policy search block. The novelty of our method lies in optimality criterion as every-visit-optimality and the method of policy search collecting various policies (Yamaguchi 2011).

After various policies are prepared, each policy is output as a round plan for recommendation to the user. The user comes into focus on his/her preference criteria through the interactive recommendation process. The interactive recommendation will finish after the user decides the most preferred plan.

Overview of the Passive Interactive Recommendation

This subsection describes the interactive recommendation space and the passive recommendation strategy by our system. In the recommendation space, the user can view and select various plans actively. The recommendation space consists of two dimensions, the preference reduction axis and the preference extension axis, in that, various plans are arranged in a plane. As the passive recommendation strategy, we introduce coarse to fine recommendation strategy (Yamaguchi 2008) that consists of two steps, coarse recommendation step along the preference reduction axis and fine recommendation step along the preference extension axis. Figure 3 shows coarse to fine recommendation strategy in the recommendation space.

Grouping Various Plans by the Visited Goals

After preparing various round plans, they are merged into group at the round-trip plan recommendation block. Figure 3(a) shows grouping various plans by the number of included goals. When three goals g1, g2, and g3 are input by a user, then, Group1 in Figure 3(a) holds various plans including all goals, g1, g2, and g3. Group2 holds various plans including all but one among g1, g2, or g3, and Group3 holds various plans including all but two among g1, g2, or g3.

Coarse Recommendation Step

For the user, the aim of this step is to select a preferable group according to the preference reduction axis. To support the user's decision, the system recommends a representative plan in each selected group to the user. Figure 3(a) shows a coarse recommendation sequence when a user changes his preferable group as Group1, Group2, and Group3 sequentially. When the user selects a group, the system presents a representative plan in the group as recommended plan.

Fine Recommendation Step

For the user, the aim of this step is to decide the most preferable plan in the selected group in previous step according to the preference extension axis. To support the user's decision, the system recommends plans among his/her selected group to the user. Figure 3(b) shows a fine recommendation sequence after the user selects his/her preferable group as Group2. In each group, plans are ordered according to the length of a plan.

AWARENESS-BASED RECOMMENDATION

To assist the user's awareness for his/her preference shift, we propose the user-centered recommendation by visualizing both the recommendation

Figure 3. Coarse to fine recommendation strategy in the recommendation space: (a) coarse recommendation step; (b) fine recommendation step

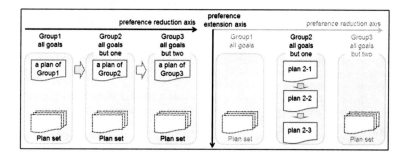

space with prepared recommendation plans and the user's preference trace as the history of the recommendation in it. The recommendation space visualizes the possible preference shift of the user.

Overview of the GUI

Figure 4 shows an overview of the GUI of our interactive recommendation system. A current recommendation plan in the Hokkaido map is displayed in the left side. In the right mid area, there is the control panel in which a current recommendation plan can move up, down, right or left in the recommendation space. The recommendation space is displayed in the rightmost. Each plan icon is arranged in a plane in the recommendation space. The boxed plan icon in it indicates the current recommendation plan.

Visualizing the Recommendation Space

This subsection describes an overview of the visualization of the *recommendation space*. The objective of this visualization is to inform a user of two kinds of information. First is that the recommendation space consists of two-axes. Second is that in each axis, groups or plans are ordered according to the recommendation order.

Figure 4. Overview of the GUI of our interactive recommendation system

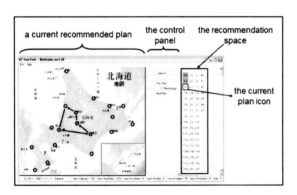

Figure 5 shows a flow of the preference trace in the recommendation space for four goal cities. Figure 5(a) shows an example of the beginning of the recommendation space of a user. Each plan icon in it has the plan-id x-y consists of group-id x and plan-number y. In Figure 5(a), after a user inputs four cities as goals, these are highlighted by red color. The plan 1-1 is origin of the recommendation space that is the minimum length plan visiting all four goals. The recommendation space is defined by two-axes as follows:

- Coarse Axis (Preference Reduction)
- Fine Axis (Preference Extension)

The horizontal axis is coarse axis that means the group-id for plans defined by equation (1). group-id is the horizontal distance between plan 1-1 at the origin and the group of the current plan. Note that N_0 is the number of initial goals (default is four), and N_i is the number of included goals in each recommended plan. A larger group-id is less familiar but fresh to the user since the plans in it include alternative cities without initial goal set of the user.

$$\text{group-id} = N_0 - N_i + 1 \qquad (1)$$

The vertical axis is fine axis that means the plan-number of each group. The order is according to the length of the plans from shorter to longer. Each plan-number y is the vertical distance between the current plan x-y of the group x and uppermost plan x-1, it is about the number of alternative cities in the current plan.

Visualizing the User's Preference Trace

This subsection illustrates the visualizing process of the user's *preference trace*. The objective of this visualization is to show the distribution and the degree of the user's preference to him/herself.

Figure 5. A flow of the preference trace in the recommendation space for four goal cities: (a) the beginning , (b) the middle, (c) the end, (d) visualizing rule

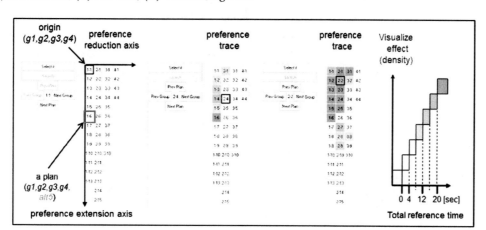

Figure 5(b) shows the middle of the preference trace of the user. The preference trace is visualized as the set of density of the background color of each plan icon. It indicates the total reference time of each plan by the user. In Figure 5(b), the preference trace is mainly distributed in a vertical direction, so that the user referred six plans in the Group 1 those are plans with four goals. Note that the current plan 2-4 is one of the plans with three goals within four goals. Within plan 1-1 to 1-6, plan 1-6 shows the highest density of the color so that the user watched this plan with the longest time.

Figure 5(c) shows the end of the preference trace of the user. The preference trace shows that the user has referred all recommended plans from plan 1-1 to 4-4. Then the distribution of the density of background color indicates the bias of the user's preference trace. By these visualizations, the user can notice the distribution or the bias of the reference time of each plan in the recommendation space.

Effect of Visualizing the Degree of Preference

Next we analyze the meaning of visualizing the reference time of each plan. If we assume that the reference time corresponds to the degree of preference of the user, the user can notice the distribution or the bias of the preference of each plan in the recommendation space as follows:

- The plan with highest density color suggests the (currently) most preferred plan as shown in Figure 5(b).
- Plans with same high-density color suggests that the user wavered in his/her preference as shown in Figure 5(c).
- The distribution of the plans with high-density color suggests the extent of the concern of the user.

Experiment

This subsection describes the experimental setup. To examine the effects our interactive recommendation system, we perform the comparative experiments. There are two objectives. First one is "dose the preference of a user change during recommendation process?" Second question is "what kind of effect occurs by our awareness based recommendation?"

A total of sixteen subjects are divided into two groups for comparative experiment. Figure 6 shows the experimental condition whether recommendation space is visualized or not during the experiment. Both no visualizing group and

Figure 6. The experimental condition on the recommendation space: (a) no visualizing condition, (b) visualizing condition

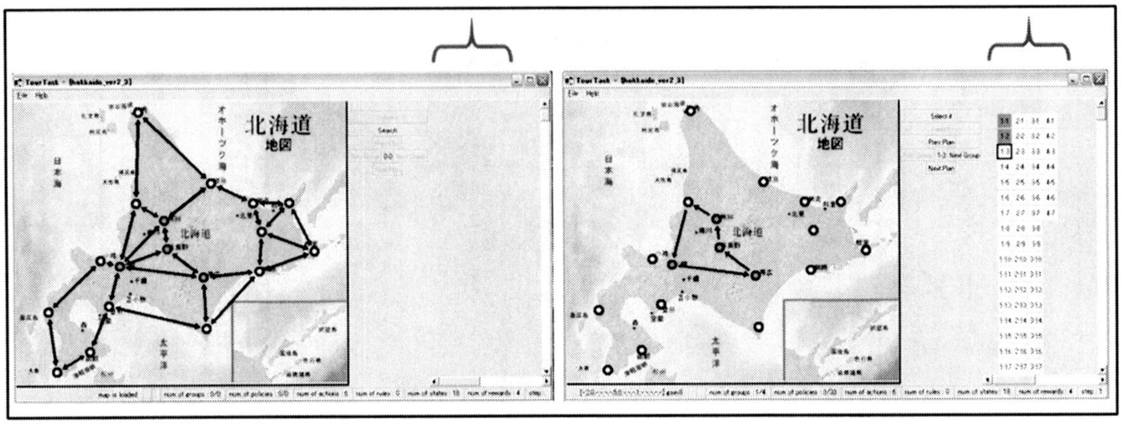

visualizing group are eight subjects each. During the interactive recommendation as described previously, a subject can select either coarse recommendation step or fine recommendation step until the subject decide the most preferred plan.

Solutions and Recommendations: Experimental Results

Does the Preference of a User Change during Recommendation Process?

Table 2 shows the number of subjects whether the preference has been changed or not. As the result, about 70% of subjects (11 subjects among 16 subjects) have changed their preferences. There is no significant difference whether visualizing

or not. This result suggests that our interactive recommendation method brings out the user's preference shift.

What Kind of Effect Occurs with Our Awareness-Based Recommendation?

To look into the effects of our recommendation, we analyze the subjects' behavior. Figure 7 shows the rate of referred plans during the interactive recommendation. Horizontal axis is ID of sixteen subjects. What is interesting is that 16 subjects are divided into two groups whether almost all plans have been referred or not. One group of 8 subjects is called *heavy users* who decide the

Table 2. The number of subjects with preference shift

Condition	Preference shift	No preference shift
No visualizing	6	2
Visualizing	5	3
Total	11	5

Figure 7. The rate of referred plans during the interactive recommendation

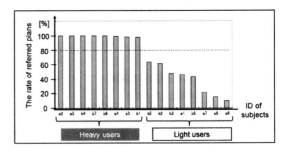

most preferred plan after watching almost all plans. Another group of 8 subjects is called *light users* who do not watch all plans since they stop watching when a preferred plan is found.

After this, Figure 8 shows the comparative analysis in each heavy users and light users. Figure 8(a) shows the total number of referred plans between heavy users and light users. Vertical axis is the total number of references, that is the number of watched plans. Horizontal line is classified result (heavy users or light users) and experimental conditions whether visualizing or not in each group. What is interesting is that by visualizing, heavy users reduced the number of reference. It suggests that heavy users improve the efficiency by visualizing support. Besides, the number of reference between heavy users and light users is homogenized by visualizing condition since the light users increase the number of reference. There is significant difference (5%) among light users.

Figure 8(b) shows the average reference time per plan between heavy users and light users. Horizontal axis is same as Figure 8(a). Vertical axis is the average reference time per plan. What's interesting is that by visualizing, light users reduced the average reference time. It suggests that light users improve the efficiency by visualizing support. On the other hand, there is no difference in heavy users in this metric. Besides, the average reference time between heavy users and light users is slightly homogenized by visualizing condition.

Summary of Visualizing Effects

Now we summarize the visualizing effects on our experiments. First, subjects are classified into two groups, heavy users and light users by their behaviors. Second, different effects occurred by visualizing condition. Heavy users reduced the number of reference and the total reference time. On the other hand, light users reduced the average reference time, and increased the number of reference. Third, both group improved the efficiency for deciding the most preferred plan.

FUTURE RESEARCH DIRECTIONS

Discussions

Significance of the User-Centered Recommendation

We discuss the major significance of the user-centered recommendation. First issue is "why an intelligent agent seems to be officious?" An agent is active and less intelligent than a human. The aspect of less intelligence can cause trouble to catch the human needs. Then the human might feel officious since the agent is active but needless. Second issue is "why passive recommendation is better?" The aspect of passive interaction results in ability for reflecting human intelligence. It covers less intelligence of the agent. Besides, it also brings out human awareness.

Figure 8. The comparative analysis in each heavy users and light users: (a) the total of referred plans; (b) the average reference time

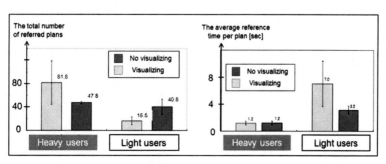

Awareness for the Preference Distribution

Now we discuss the effect of awareness by visualizing the distribution of the user's preference trace in the recommendation space. We present two kinds of awareness by coarse to fine recommendation as follows:

- Aware of the preference reduction by coarse axis: The distribution toward coarse axis of the user's preference trace suggests the shift of the preference. Since the leftmost Group1 is every-visit plans including all original goals, when the most preferred plan is in the Group1, the preference of the user dose not shifted by the recommendation. On the other hand, if the most preferred plan is far from the Group1 as the right side to coarse axis, the preference of user has shifted during the recommendation process.
- Aware of the preference extension by fine axis: The distribution toward fine axis of the user's preference trace suggests the extension of the preference. Since fine axis vertically means the degree of plan length, the upper side plans are short, in other words, the number of cities without goals is small in each group. Contrast to it, the lower side of plans is longer with many cities to be visit. When the most preferred plan is in the upper side, the preference of the user dose not been extended by the recommendation. On the other hand, if the most preferred plan is lower side toward fine coarse axis, the preference of user has been extended during the recommendation process since many alternative cities except goal cities are included.

Related Works on Recommender Systems

This subsection describes additional readings on the relations between our proposed solutions and current research issues on recommendation systems. The main feature of our recommendation system is interactive and adaptable recommendation for human users by interactive reinforcement learning. First, we describe two major problems on traditional recommenders. Second, interactive recommendation system called Conversational Recommender is summarized. At last, adaptive recommenders with learning ability are described.

Major Problems on Traditional Recommenders

Main objective of recommender systems is to provide people with recommendations of items, they will appreciate based on their past preferences. One of the major approaches (Adomavicius 2005) is collaborative filtering (Resnick 1994), whether user-based or item-based (Sarwar 2001) such as by Amazon.com (Linden 2003). The common feature is that similarity is computed for users or items, based on their past preferences.

However, there are two major issues. First issue is the similar recommendations problem (Ziegler 2005) in that many recommendations seem to be "similar" with respect to content. It is because of lack of novelty, serendipity (Murakami 2007) and diversity of recommendations. Second issue is the preference change problem (Yamaguchi 2009) that is inability to capture the user's preference shift during the recommendation. It often occurs when the user is a beginner or a light user. For the first issue, there are two kinds of previous solutions. One is topic diversification (Ziegler 2005) that is designed to balance and diversify personalized recommendation lists for user's full range of interests in specific topics. Another is visualizing the feature space (Hijikata 2006) for editing a user's profile to search the different items on it by the user. However, these solutions do not directly considering a user's preference shift. To solve this, this paper assumes a user's preference shift as two-axes space, coarse and fine axes.

Interactive Recommendation Systems

Traditional recommenders are simple and non-interactive since they only decide which product to recommend to the user. So it is hard to support for recommending more complex products such as travel products (Mahmood 2009b). Therefore, conversational recommender systems (Bridge 2005) have been proposed to support more natural and interactive processes. Typical interactive recommendation is the following two strategies (Mahmood 2008a):

Step 1: Ask the user in detail about her preferences.
Step 2: Propose a set of products to the user and exploit the user feedback to refine future recommendations.

A major limitation of this approach is that there could be a large number of conversational but rigid strategies for a given recommendation task (Mahmood 2008b).

Adaptive Recommenders with Learning Ability

There are several adaptive recommenders using reinforcement learning. Most of them observe a user's behavior such as products the user viewed or selected, then learn the user's decision processes or preferences (Preda 2009). To improve the rigid strategies for conversational recommenders, learning personalized interaction strategies for conversational recommender systems has been proposed (Mahmood 2008a, 2009a, 2009b).

Major difference from them, the feature of our approach is adaptable recommendation for human users by passive recommendation strategy called coarse to fine recommendation. Adaptable recommendation means that during our recommendation, a user can select these two steps (coarse step or fine step) as his/her likes before deciding the most preferable plan.

CONCLUSION

We described awareness-based recommendation toward the human adaptive and friendly interactive recommendation system. For considering preference shift of a user, we proposed user-centered interactive recommendation by visualizing both the recommendation space with prepared recommendation plans and the user's preference trace as the history of the recommendation in it. Experimental results showed that subjects are divided into two groups, heavy users and light users, and that there are different effects between heavy users and light users under the same visualizing condition. They also showed that our system improves the efficiency for deciding the most preferred plan for both heavy users and light users. Future work is ways to help to widen light users' perspectives through being exposed to different values.

ACKNOWLEDGMENT

The authors would like to thank Prof. Shimohara and Prof. Habib for offering a good opportunity to present this research. The authors would also like to thank Maiko Adachi, Kouki Takemori, and Tatsuya Muraoka for helping prepare the chapter. This work was supported by JSPS KAKENHI (Grant-in-Aid for Scientific Research (C)) Grant Number 23500197.

REFERENCES

Anderson, N. H., & Butzin, C. A. (1974). Performance = motivation x ability: An integration-theoretical analysis. *Journal of Personality and Social Psychology, 30*(5), 598–604. doi:10.1037/h0037447.

Balabanovic, M., & Shoham, Y. (1997). Fab: Content-based, collaborative recommendation. *Communications of the ACM, 40*(3), 66–72. doi:10.1145/245108.245124.

Kaplan, F., Oudeyer, P.-Y., Kubinyi, E., & Miklosi, A. (2002). Robotic clicker training. *Robotics and Autonomous Systems, 38*(3-4), 197–206. doi:10.1016/S0921-8890(02)00168-9.

Konda, T., Tensyo, S., & Yamaguchi, T. (2002). LC-learning: Phased method for average reward reinforcement learning - Analysis of optimal criteria. In Ishizuka & Sattar (Eds.), PRICAI2002: Trends in Artificial Intelligence (LNCS), (vol. 2417, pp. 198-207). Berlin: Springer.

Konda, T., Tensyo, S., & Yamaguchi, T. (2002). Learning: Phased method for average reward reinforcement learning - Preliminary results. In Ishizuka & Sattar (Eds.), PRICAI2002: Trends in Artificial Intelligence (LNAI), (vol. 2417, pp. 208-217). Berlin: Springer.

Konidaris, G., & Barto, A. (2006). Automonous shaping: Knowledge transfer in reinforcement learning. In *Proceedings of the 23rd International Conference on Machine Learning* (pp. 489-496). New York: ACM.

Mahadevan, S. (1996). Average reward reinforcement learning: Foundations, algorithms, and empirical results. *Machine Learning, 22*(1-3), 159–195. doi:10.1007/BF00114727.

Marthi, B. (2007). Automatic shaping and decomposition of reward functions. In *Proceedings of the 24th International Conference on Machine Learning*. ACM.

Ng, A. Y., Harada, D., & Russell, S. J. (1999). Policy invariance under reward transformations: Theory and application to reward shaping. In *Proceedings of the 16th International Conference on Machine Learning* (pp. 278-287). New York, NY: ACM.

Puterman, M. L. (1994). *Markov decision processes: Discrete stochastic dynamic programming*. New York, NY: John Wiley & Sons, Inc. doi:10.1002/9780470316887.

Riecken, D. (2000). Introduction: Personalized views of personalization. *Communications of the ACM, 43*(8), 26–28. doi:10.1145/345124.345133.

Satoh, K., & Yamaguchi, T. (2006). Preparing various policies for interactive reinforcement learning. In *Proceedings of the SICE-ICASE International Joint Conference 2006 (SICE-ICASE 2006)* (pp. 2440-2444). New York: Institute of Electrical and Electronics Engineers (IEEE).

Schafer, J. B., Konstan, J. A., & Riedl, J. (2001). E-commerce recommendation applications. *Journal of Data Mining and Knowledge Discovery, 5*, 115–153. doi:10.1023/A:1009804230409.

Takadama, K., Sato, F., Otani, M., Hattori, K., Sato, H., & Yamaguchi, T. (2012). Preference clarification recommender system by searching items beyond category. In *Proceedings of IADIS (International Association for Development of the Information Society) International Conference Interfaces and Human Computer Interaction 2012* (pp. 3-10). Lisbon, Portugal: IADIS Press.

Yamaguchi, T., & Nishimura, T. (2008). How to recommend preferable solutions of a user in interactive reinforcement learning? In *Proceedings of the International Conference on Instrumentation, Control and Information Technology (SICE2008)*, (pp. 2050-2055). Tokyo, Japan: The Society of Instrument and Control Engineers.

Yamaguchi, T., Nishimura, T., & Sato, K. (2011). How to recommend preferable solutions of a user in interactive reinforcement learning? In Mellouk, A. (Ed.), *Advances in Reinforcement Learning* (pp. 137–156). Rijeka, Croatia: InTech Open Access Publisher. doi:10.5772/13757.

Yamaguchi, T., Nishimura, T., & Takadama, K. (2009). Awareness based filtering - Toward the cooperative learning in human agent interaction. In *Proceedings of the ICROS-SICE International Joint Conference (ICCAS-SICE 2009)*, (pp. 1164-1167). Tokyo, Japan: The Society of Instrument and Control Engineers.

Yamaguchi, T., Nishimura, T., & Takadama, K. (2012). *Awareness based recommendation - Toward the cooperative learning in human agent interaction*. Paper presented at the International Conference on Humanized Systems 2012 (OS02_1003). Daejeon, Korea.

ADDITIONAL READING

Adomavicius, G., & Tuzhilin, A. (2005). Toward the next generation of recommender systems: A survey of the stateof-the-art and possible extensions. *IEEE Transactions on Knowledge and Data Engineering*, *17*(6), 734–749. doi:10.1109/TKDE.2005.99.

Bridge, D., Goker, M. H., McGinty, L., & Smyth, B. (2005). Case-based recommender systems. *The Knowledge Engineering Review*, *20*(3), 315–320. doi:10.1017/S0269888906000567.

Hijikata, Y., Iwahama, K., Takegawa, K., & Nishida, S. (2006). Content-based music filtering system with editable user profile. In *Proceedings of the 21st Annual ACM Symposium on Applied Computing (ACM SAC 2006)* (pp. 1050-1057). New York: ACM.

Linden, G., Smith, B., & York, J. (2003). Amazon. com recommendations: Item-to-item collaborative filtering. *IEEE Internet Computing*, *7*(1), 76–80. doi:10.1109/MIC.2003.1167344.

Mahmood, T., & Ricci, F. (2008a). Adapting the interaction state model in conversational recommender systems. In *Proceedings of the 10th International Conference on Electronic Commerce* (pp. 1-10). New York: ACM.

Mahmood, T., & Ricci, F. (2009a). Improving recommender systems with adaptive conversational strategies. In *Proceedings of the 20th ACM Conference on Hypertext and Hypermedia* (pp. 73-82). New York: ACM.

Mahmood, T., Ricci, F., & Venturini, A. (2009b). Improving recommendation effectiveness by adapting the dialogue strategy in online travel planning. *International Journal of Information Technology and Tourism*, *11*(4), 285–302. doi:10.3727/109830510X12670455864203.

Mahmood, T., Ricci, F., Venturini, A., & Hopken, W. (2008b). Adaptive recommender systems for travel planning. In O'Connor, Hopken, & Gretzel (Eds.), *Information and Communication Technologies in Tourism 2008: Proceedings of ENTER 2008 International Conference in Innsbruck* (pp. 1-11). New York: Springer.

Murakami, T., Mori, K., & Orihara, R. (2007). Metrics for evaluating the serendipity of recommendation lists. In Satoh, et al. (Eds.), *New Frontiers in Artifical Intelligence (LNAI)* (Vol. 4914, pp. 40–46). Berlin: Springer. doi:10.1007/978-3-540-78197-4_5.

Preda, M., Mirea, A. M., Teodorescu-Mihai, C., & Preda, D. L. (2009). Adaptive web recommendation systems. *Annals of University of Craiova*, *36*(2), 25–34.

Resnick, P., Iacovou, N., Suchak, M., Bergstrom, P., & Riedl, J. (1994). GroupLens: An open architecture for collaborative filtering of netnews. In *Proceedings of the 1994 ACM Conference on Computer Supported Cooperative Work* (pp. 175–186). New York, NY: ACM.

Sarwar, B., Karypis, G., Konstan, J., & Reidl, J. (2001). Item-based collaborative filtering recommendation algorithms. In *Proceedings of the 10th International Conference on World Wide Web* (pp. 285-295). New York: ACM.

Yamaguchi, T., Nishimura, T., & Takadama, K. (2009). Awareness based filtering - Toward the cooperative learning in human agent interaction. In *Proceedings of the ICROS-SICE International Joint Conference (ICCAS-SICE 2009)* (pp. 1164-1167). Tokyo, Japan: The Society of Instrument and Control Engineers.

Ziegler, C. N., Mcnee, S. M., Konstan, J. A., & Lausen, G. (2005). Improving recommendation lists through topic diversification. In *Proceedings of the 14th International Conference on World Wide Web (WWW2005)* (pp. 22-32). New York: ACM.

KEY TERMS AND DEFINITIONS

Awareness-Based Recommendation: It is the user-centered recommendation by visualizing both the recommendation space with prepared recommendation plans and the user's preference trace as the history of the recommendation in it. The recommendation space visualizes the possible *preference shift* of the user.

Heavy Users: Users of the our interactive recommendation system who decide the most preferred plan after watching almost all plans.

Human Adaptive and Friendly: Less active but more intelligent agent is desirable since it does not seem to be officious for the human.

Interactive Recommendation Space: In the recommendation space, the user can view and select various plans actively. The recommendation space consists of two dimensions; the preference *reduction* axis and the preference extension axis, in that, various plans are arranged in a plane.

Interactive Reinforcement Learning with Human: Reinforcement learning method in that reward function denoting goal is given interactively by a human. It is not easy to keep reward function being fixed while the human gives rewards for the reinforcement learning agent. It is that the reward function may not be fixed for the learning algorithm if an end-user changes his/her mind or his/her preference.

Light Users: Users of the our interactive recommendation system who do not watch all plans since they stop watching when a preferred plan is found.

Model of a User's Preference Shift: It is defined by two axes, preference reduction and preference extension. Comparing the previous preference set and the current preference set, the common set is the *invariant* preference, the reduction set is called preference *reduction*, and the addition set is called preference *extension*.

Preference Change Problem: It is a problem that the collected user's profile is not same as the user's current preference.

Visualizing the Recommendation Space: The objective of this visualization is to inform a user of two kinds of information. First is that the recommendation space consists of two-axes. Second is that in each axis, groups or plans are ordered according to the recommendation order.

Visualizing the User's Preference Trace: The objective of this visualization is to show the distribution and the degree of the user's preference to him/herself.

Section 2

Chapter 7
Design and Implementation of a Step–Traversing Two–Wheeled Robot

Huei Ee Yap
Waseda University, Japan

Shuji Hashimoto
Waseda University, Japan

ABSTRACT

In this chapter, the authors present the design and implementation of a step-traversing two-wheeled robot. Their proposed approach aims to extend the traversable workspace of a conventional two-wheeled robot. The nature of the balance problem changes as the robot is in different phases of motion. Maintaining balance with a falling two-wheeled robot is a different problem than balancing on flat ground. Active control of the drive wheels during flight is used to alter the flight of the robot to ensure a safe landing. State dependent feedback controllers are used to control the dynamics of the robot on ground and in air. Relationships between forward velocity, height of step, and landing angle are investigated. A physical prototype has been constructed and used to verify the viability of the authors' control scheme. This chapter discusses the design attributes and hardware specifications of the developed prototype. The effectiveness of the proposed control scheme has been confirmed through experiments on single- and continuous-stepped terrains.

1. INTRODUCTION

Mobile intelligent robots are expected to play an increasing role in aiding humans in various tasks. Advancements in robotic technologies have enabled increasing number of robots to be deployed in the fields of exploration, surveillance, health care, and entertainment. Robots that need to operate in an uncontrolled human environment will have to be able to navigate stepped and uneven terrain. Properly implemented control methods will ensure the usability and safe deployment of these robots.

DOI: 10.4018/978-1-4666-4225-6.ch007

Many existing stair climber robots have used rocker-bogie mechanisms or crawler mechanisms (Volpe, Balaram, Ohm, & Ivlev, 1996; Hayati et al., 1997; Saranli, Buehler, & Koditschek, 2001; Lamon, Krebs, Lauria, Siegwart, & Shooter, 2004; Guarnieri, Takao, Fukushima, & Hirose, 2007). Robots using these mechanisms rely on the static stability of the platform to perform step traversal. Relying on static stability has drawbacks of low traversal speed and increased complexity in the structure design. Any external perturbation which forces these robots outside their basin of stability will lead to loss of control and inability to recover.

Dynamically stable robots offer better agility and are more robust to external disturbances (Lauwers, Kantor, & Hollis, 2006; Kumagai & Ochiai, 2008; Xu & Au, 2004; Brown & Xu, 1997). Such advantages can be used to achieve rapid, stable transition through stepped terrain (Kikuchi et al., 2008). In our research, we focus specifically on using a two-wheeled robot to travel continuously through stepped terrain without losing balance. The problem of maintaining balance with a falling two-wheeled robot is highly non-linear. The nature of the balance problem changes as the robot is in different phases of motion. A free falling robot is a different control problem than a robot climbing a step or traversing flat ground.

Two-wheeled robots have been a popular research platform due to their simple design yet complex dynamic behavior. Most of the research literature available focuses on continuous ground balance (Teeyapan, Wang, Kunz, & Stilman, 2010; Stilman, Olson, & Gloss, 2010; Grasser, D'Arrigo, Colombi, & Rufer, 2002). The problem of balancing a two-wheeled robot through discontinuous terrain has received relatively little attention. Related research includes a reconfigurable hopping rover as proposed in Schmidt-Wetekam et al. (2007) and Schmidt-Wetekam et al. (2011). The hopping rover resembles a 3 dimensional reaction wheel pendulum with a set of orthogonally arranged drive wheels. The drive wheels are used to provide torque for attitude correction to re-orient the vehicle during flight and ground balance. The hopping action is provided by an extendable leg. The hopping robot exhibited good dynamic stability on a flat surface, but performance on a stepped surface was not evaluated.

In this chapter, we introduce the development of a step traversing two-wheeled pendulum robot (Yap & Hashimoto, 2012). We first present the theoretical analysis of the problem and derive the equations of motion of the system. Two independent feedback controllers are then designed to control the system on the ground and in the air. We also present the detailed design of the robot in terms of hardware, electronics, as well as the choice of sensors used to achieve step traversing. Experiments of the robot traversing various stepped terrains are conducted and the results are presented and discussed.

2. DYNAMICS OF THE ROBOT

In this section, we derive the planar equations of motion of the system on ground and in airborne phases to understand the dynamic behavior of the two-wheeled pendulum. When traversing stepped terrain the robot goes through several different dynamic phases, from on the ground to airborne to on the ground again. The changes in the dynamic phases cause discontinuities in the governing equations of motion. To simplify the analysis we have assumed the equations for different phases of motion are separable and can be derived and analyzed independently.

The dynamics during free-fall are modeled using projectile motion with the robot treated as a rigid body. The initial conditions are derived from the state of the robot just before it began falling. Unmodeled dynamics of the system are treated as disturbance forces and be compensated by feedback controllers. The robot is assumed to have reflective symmetry along its vertical axes. This allows the dynamics of the robot to be analyzed in two dimensions only instead of three.

One limitation to this simplification is that the roll dynamics of the robot is not taken into consideration. During all analysis the no-slip condition is assumed to hold.

2.1. Ground Model

Figure 1a describes the 2D ground model of the robot. Equations of motion of the system are derived using Euler/Lagrange's method. The Lagrangian is defined as the difference between the kinetic energy and potential energy of the system, $L = T - U$. The Euler/Lagrange equations of motions for the robot are given by

$$\frac{d}{dt}\left(\frac{\partial L}{\partial \dot{q}_i}\right) - \frac{\partial L}{\partial q_i} = Q_i \qquad i = 1,2,\ldots,n$$

where $q = (\theta, \psi)$ are the generalized coordinates wheel angle and body pitch angle for the ground model. Q_i denotes the generalized force for each coordinate. The physical parameters of the robot used in our analysis are listed in Table 1.

The kinetic energy of the body, T_b, and the kinetic energy for the wheels, T_w, are expressed as

$$T_b = \frac{1}{2}m_b R^2 \dot{\theta}^2 + m_b lR\,\dot{\theta}\,\dot{\psi}\cos\psi$$
$$+\frac{1}{2}(I_b + m_b l^2)\dot{\psi}^2 \tag{1}$$

$$T_w = \frac{1}{2}m_w R^2 \dot{\theta}^2 + \frac{1}{2}I_w \dot{\theta}^2 \tag{2}$$

The potential energy for the robot is given by

$$U = m_b gl \cos\psi \tag{3}$$

Figure 1. 2D schematic of the two-wheeled robot on ground and in air

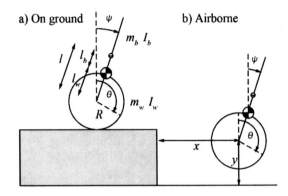

Table 1. List of physical parameters

Parameter	Unit	Description
g	[ms^{-2}]	Gravitational acceleration constant
m_b	[kg]	Body mass
m_w	[kg]	Wheel mass
R	[m]	Wheel radius
l	[m]	Length of wheel axis to body's center of mass
l_b	[m]	Length of body's centre of mass to robot's centre of mass
l_w	[m]	Length of wheel axis to robot's centre of mass
h	[m]	Height of robot's center of mass to ground
I_b	[kgm^2]	Body pitch inertia moment
I_w	[kgm^2]	Wheel inertia moment
I	[kgm^2]	Overall robot inertia moment
ψ	[rad]	Tilt angle of robot body
θ	[rad]	Rotational angle of wheels
τ	[Nm]	Motor torque

The Lagrangian of the system on ground can then be expressed as

$$L = \frac{1}{2}\left(m_b + m_w\right)R^2 \dot{\theta}^2 + m_b lR\,\dot{\theta}\,\dot{\psi}\cos\psi$$
$$+\frac{1}{2}\left(I_b + m_b l^2\right)\dot{\psi}^2 + \frac{1}{2}I_w \dot{\theta}^2 - m_b gl \cos\psi \tag{4}$$

Evaluating Euler/Lagrange Equation (1) for each of the coordinates θ and ψ gives the equations of motion of the system on ground

$$\left(I_w + \left(m_b + m_w\right)R^2\right)\ddot{\theta} + m_b lR \cos \psi \, \ddot{\psi}$$
$$-m_b lR \, \dot{\psi}^2 \sin \psi = \tau \tag{5}$$

$$\left(I_b + m_b l^2\right)\ddot{\psi} + m_b l \, R \cos \psi \, \ddot{\theta}$$
$$-m_b gl \sin \psi = -\tau \tag{6}$$

Equation (5) and (6) are the non-linear model of the robot on ground. The linear approximation of the robot system can be found by using the small angle approximation around the equilibrium point, $\psi = 0$. The linearised equations of motion can be rearranged into a state-space form as follows:

$$\dot{x}(t) = \mathbf{A}x(t) + \mathbf{B}u(t) \tag{7}$$

where

$$\mathbf{x} = (\theta, \psi, \dot{\theta}, \dot{\psi}), \; u(t) = \tau(t)$$

and

$$\mathbf{A} = \begin{bmatrix} 0 & 0 & 1 & 0 \\ 0 & 0 & 0 & 1 \\ 0 & a_{32} & 0 & 0 \\ 0 & a_{42} & 0 & 0 \end{bmatrix}, \quad \mathbf{B} = \begin{bmatrix} 0 \\ 0 \\ b_3 \\ b_4 \end{bmatrix}$$

$$den = I_w I_b + I_w m_b l^2 + I_b \left(m_b + m_w\right)R^2$$
$$+ m_b m_w R^2 l^2$$

$$a_{32} = -\frac{m_b^2 gRl^2}{den}$$

$$a_{42} = \frac{\left(I_w + \left(m_b + m_w\right)R^2\right)m_b gl}{den}$$

$$b_3 = -\frac{\left(I_b + m_b l^2\right) - m_b Rl}{den}$$

$$b_4 = \frac{\left(I_w + \left(m_b + m_w\right)R^2\right) - m_b Rl}{den}$$

2.2. Airborne Model

During airborne motion the robot can be modeled as a freefalling reaction wheel pendulum (Spong, Corke, & Lozano, 2001; Astrom, Block, & Spong, 2007) with a pivot at its center of mass (See Figure 1b). The dynamic equations of the robot in air and freefalling are separately derived. Motion of the robot in air derived here does not include the falling projectile motion and hence the kinetic energy and potential energy of the system are

$$T_b = \frac{1}{2}(I_b + m_b l^2)\dot{\psi}^2 \tag{8}$$

$$T_w = \frac{1}{2}I_w \dot{\theta}^2 + \frac{1}{2}m_w l_w^2 \dot{\psi}^2 \tag{9}$$

$$U = 0 \tag{10}$$

The Lagrangian of the system is then

$$L = \frac{1}{2}\left(I_b + m_b l_b^2 + m_w l_w^2\right)\dot{\psi}^2 + \frac{1}{2}I_w \dot{\theta}^2 \tag{11}$$

Using Equation (1) the equations of motion can be expressed as

$$I_w \ddot{\theta} = \tau \tag{12}$$

$$\left(I_b + m_b l_b^2 + m_w l_w^2\right)\ddot{\psi} = -\tau \qquad (13)$$

2.3. Projectile Motion

The trajectory of the robot falling from a step can be modeled as a simple projectile motion problem by treating the robot as a rigid body with initial horizontal velocity set to the velocity of the robot right before takeoff, and the initial vertical velocity set to zero. Once the robot loses contact with the ground it flies through the air affected only by gravity and hence horizontal acceleration is zero,

$$x(t) = \dot{x}_0 \ t \quad \dot{x}(t) = \dot{x}_0 = R\dot{\theta}_0 + l\cos\psi_0 \ \dot{\psi}_0 \qquad (14)$$
$$\ddot{x}(t) = 0$$

$$y(t) = h - \frac{1}{2}gt^2 \quad \dot{y}(t) = gt \quad \ddot{y}(t) = g \qquad (15)$$

where x and y are the horizontal and vertical displacements of the center of mass of the robot. Time t starts from the time the robot leaves the edge. , and are the wheel angular velocity, tilt angle and tilt angular velocity at the moment the robot transit into airborne phase. Equations (12) – (15) completely describe the motion of the robot while airborne.

2.4. Impact

At the moment of impact linear kinetic energy is converted into rotational kinetic energy. The rotational torque pushes the body of the robot forward upon landing and helps maintain the forward motion of the robot. To simplify the understanding of the dynamics of the robot we modeled the system as a falling rod (Cross, 2006). Assuming the non-slip condition, conservation of momentum and neglecting impulse reaction, the problem simplifies to

$$\left(m_b + m_w\right)(l_w + R)(\dot{x}\cos\psi + \dot{y}\sin\psi) = I\dot{\psi} \qquad (16)$$

$$\frac{1}{2}I\dot{\psi}^2 + (m_b + m_w)g(l_w + R)\cos\psi = constant \qquad (17)$$

3. DESIGN OF THE CONTROLLER

The motion of traversing through a step can be separated into on ground and airborne phases. During airborne motion the drive wheels of the robot are used as reaction wheels to create reaction torque to alter the tilt angle of the robot. The landing angle is chosen in relation to the pre-impact velocity to ensure successful landing. Two independent controllers are used to control the motion of the robot on ground and in air.

3.1. Ground Balance Control

Ground balance control is conducted using the state feedback control law $u_g = -\mathbf{K}_g \mathbf{x}$ which minimizes the quadratic cost function

$$J(u) = \int_a^b \mathbf{x}^T Q \mathbf{x} + u^T R u \, dx \qquad (18)$$

subject to the system defined in Equation (7). The values of weight matrix Q and R are chosen experimentally to yield a satisfactory feedback response. The final feedback control input u_g is given by

$$u_g = -\mathbf{K}_g \mathbf{x} + k_{g1}\theta_{ref} \qquad (19)$$

where $\mathbf{K}_g = [\ k_{g1}\ k_{g2}\ k_{g3}\ k_{g4}\ ]$ is the feedback gain matrix and θ_{ref} is the reference value for controlling the rotational angle of the wheel.

3.2. Airborne Attitude Control

From Equations (12) and (13), the torque generated by the rotation of the wheel will directly affect the tilt angle of the robot. Equating both equations yield the relationship between the wheel angular acceleration and the body angular acceleration as

$$\frac{\ddot{\psi}}{\ddot{\theta}} = -\frac{I_w}{\left(I_b + m_b l_b^2 + m_w l_w^2\right)} = constant \qquad (20)$$

The angular acceleration of the body is proportional to the angular acceleration of the wheel.

Control the tilt angle of the robot in mid-air is conducted using the state feedback control law

$$u_a = -\mathbf{K}_a \mathbf{x} + k_{a1}\psi_{ref} \qquad (21)$$

where $\mathbf{K}_a = [\, k_{a1}\, k_{a2}\, k_{a3}\, k_{a4}\,]$ and ψ_{ref} is the reference body tilt angle during airborne motion.

3.3. Impact Compensation and Landing Angle

If the impact force is sufficiently large, the robot will rebound on impact. However, during rebound the robot is effectively returned to airborne phase and hence the same airborne controller is used to control and anticipate the subsequent landing. Effectively, additional dynamics introduced from impact rebound will be mitigated by the airborne controller. Impact forces and rebound disturbances are therefore not taken into consideration in our analysis. Depending on the angle at which the robot hits the ground, angular momentum will cause the robot to fall forward or backward. By setting Equation (17) to $(m_b + m_w)g(l_w + R)$ we can calculate the landing angle $\psi_{landing}$ at which the angular momentum pushes the robot to upright neutral position. Setting the body tilt angle ψ_{ref} larger than $\psi_{landing}$ will ensure that the robot fall forward upon landing. The value of ψ_{ref} is determined experimentally.

3.4. Switching Control

A simple switching controller is used to select whether ground phase or airborne phase control is used,

$$u = \begin{cases} u_g, & |h| \le h_{threshold} \\ u_a, & |h| > h_{threshold} \end{cases} \qquad (22)$$

The value of $h_{threshold}$ is decided experimentally.

4. CONSTRUCTION OF THE ROBOT SYSTEM

A durable two-wheeled robotic platform was designed and constructed to explore the ability of a two-wheeled robot to maintain balance while traversing stepped terrain (Figure 2). The robot is equipped with customized electronics including a digital three-axis accelerometer with a minimum full-scale range of ± 3g and a digital three-axis gyro sensor with a full-scale range of $\pm 2000°$/sec. A Kalman filter is used to compensate the Gaussian noise from the accelerometer and gyro sensor drift and to improve accuracy of angle measurements.

The prototype two-wheeled robot is a self contained robotic system where electronics circuit, power supply, motor drivers and sensors are mounted onboard. The lack of wired connections is very important as it allows unconstrained motion of the robot. A 32-bit ARM microcontroller running at 96MHz is used to process the sensor

Figure 2. CAD drawing on the prototype two-wheeled robot and the image of the actual prototype

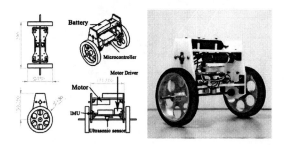

data and provide the outputs for real time control. The ARM microcontroller is useful in this application due to its low power consumption of 220mA at 3.3V.

The microcontroller communicates with the inertia measurement unit (accelerometer and gyro) through digital channel, I²C. Encoder interface is done through the embedded QEI module (Quadrature Encoder Interface) of the microcontroller. This eliminates the need to use any interrupt function of the microcontroller and hence reserve more computing power for more calculation intensive process such as Kalman filtering and balance control. An ultrasonic distance sensor is mounted on the bottom of the robot for height detection. The ultrasonic sensor interfaces with the microcontroller through a positive TTL pulse channel.

A serial Bluetooth communication channel is used to send raw data of the system to an external computer for data plotting and logging. The system is equipped with a USB Bluetooth interface for communication with a Wiimote controller for robot control. The microcontroller communicates with the Sabertooth motor driver through a digital Pulse Width Modulation (PWM) channel to control the drive wheels. A schematic of the structure of the on-board electronics components is shown in Figure 3. The attributes of the prototype system is summarized in Table 2.

Figure 2 shows the CAD design drawing and dimensions of the prototype robot. Each component and part of the robot is custom designed and manufactured. The robot needs to have a low mo-

Table 2. Prototype two-wheeled pendulum attributes

Parameter	Value
Overall size	(W)130x(L)230x(H)190mm
Overall Mass	1.53kg
Body COM Inertia	$4.873 \times 10^{-3} kgm^2$
Battery	11.1V 2200mAh Lipo
Microcontroller	96MHz ARM Cortex-M3
Sensors	Ultrasonic Distance Sensor
	ADXL345 3 axis accelerometer
	ITG3200 gyro sensor
	512 ppr Encoder
Motor driver	Sabertooth 2x5
Motor	Maxon 10W 12V
Gear Ratio	6.24:1
Max Torque	$92.88 \times 10^{-3} Nm$

ment of inertia on its pitch axis to easily rotate about its centre of mass easily while airborne. The wheels of the robot need sufficient moment of inertia to generate sufficient torque to change its attitude during flight. It would be desired to have the moment of the wheels as large as possible so that large torque can be generated. The moment of inertia is limited by the maximum torque output of the motor. Our final design incorporates the following design measurements:

- **Mass of the Wheel:** 0.535kg
- **Moment of Wheel:** $1.354 \times 10^{-3} kgm^2$
- **Mass of Body:** 0.995kg
- **Moment of Body:** $4.913 \times 10^{-3} kgm^2$

5. EXPERIMENTAL RESULTS

5.1. Single Step Terrain

In our first experiment the robot was commanded to fall from a 10cm step using conventional feedback controller. Figure 4a shows the corresponding snapshots of the motion. Conventional feedback

Figure 3. Schematic of electronics components

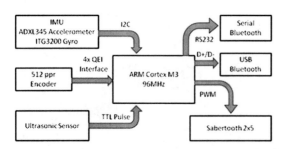

Figure 4. Snapshots of step traversing experiments: a) 10cm step traversing using conventional feedback controller; b) 10cm step traversing using proposed control scheme; c) 30cm step traversing using conventional feedback controller; d) 30cm step traversing using proposed control scheme

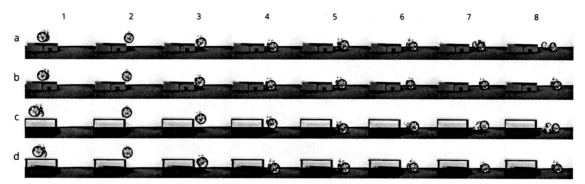

controller is designed under the assumption that the drive wheels of the robot are constantly in contact with the ground. During free fall the robot lost contact with the ground and was unable to generate balancing torque. This triggered the feedback controller to increase the feedback signal and spun the wheel faster. Due to low falling height, the robot landed with relatively small tilt angle and wheel velocity. Upon landing, non-zero initial conditions of the robot caused the controller

to apply large opposite torque in attempt to bring the robot back into balance. The robot fell forward with large overshoot. The motor was saturated and failed to generate sufficient torque to keep the robot in balance.

Next, the robot was commanded to fall from a 30cm step using conventional feedback controller. Figure 5a shows the experiment results of the free fall dynamics of the robot. The corresponding snapshots of the motion are shown in Figure 4c.

Figure 5. a) Experiment results of 30cm step traversing using conventional feedback control. Label 1~8 are points of time correspond to snapshot in Figure 4c. b) Experiment results of 30cm step traversing using proposed feedback control. Label 1~8 are points of time that correspond to snapshot in Figure 4d.

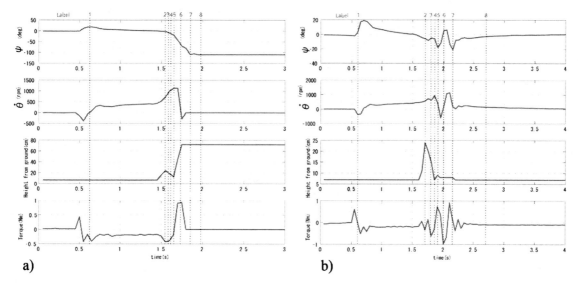

Height from ground against time plot in Figure 5a plots the relative height from ground measured using the ultrasonic sensor. It can be seen from Figure 5 that when the robot lost contact with the ground ($t \approx 1.55s$), wheel velocity $\dot{\theta}$ increased. This caused the robot to tilt further away from vertical position. On landing with fast spinning wheels and large tilt angle, the motor was saturated and unable to provide balance recovery torque. The robot lost balance and fell. At $t \approx 1.7s$, height from ground versus time plot increases dramatically. This is because when the robot fell, the sonar sensor did not have any reflecting plane to bounce back the sonar signal and hence gave large reading values. We have also experimented with turning off the motor during airborne motion. However, forward momentum pushed the robot forward, resulting in large tilt angle upon landing and the robot failed to recover.

Figure 3b shows the experiment results with our proposed control scheme activated. Corresponding snapshots of the motion are shown in Figure 2d. During the fall, the drive wheels were used as reaction wheels to generate reaction torque to correct its attitude. The reference body tilt angle used in the control system was pre-calculated before the experiment due to difficulties with accurately measuring linear velocity during the experiment. Also non-linearities such as friction forces, impact forces and air drag forces were not considered in our equations of motion, and hence the calculated reference body tilt angle and actual value used in experiments do not conformed exactly. The experimental results show that by setting the correct reference body tilt angle our approach enabled the robot to traverse stepped terrain while maintaining balance. It can be seen from Figure 3b that the robot started to fall at $t \approx 1.65s$. During falling, the wheels were actively controlled to keep the body tilt angle ψ small. Upon first landing at $t \approx 1.8s$, the robot experienced small amount of rebound. Airborne controller was activated to keep the tilt angle in control before

subsequent landing. Upon the second landing, the body tilt angle remained small and enabled the controller to keep the body upright without saturating the motors. The experiments were conducted multiple times and the robot had repeatedly demonstrated safe landing without falling with our proposed control scheme.

We have conducted our experiments with increased height up to 70cm. However it was observed that for higher step, rebound problem became obvious and caused imbalance in roll direction. The roll disturbance is caused by unequal weight distribution between left and right sides of the robot. This resulted in uneven landing of the wheels and caused the robot to fall. The maximum traversable step height is 50cm. The experiments were conducted in flat steps. In the case of sloped step, the reference tilt angle in air can be adjusted to compensate the rotational torque caused by the slope during impact and ensure a successful landing.

5.2. Continuous Step/Stair Terrain

The next experiment was to evaluate the effectiveness of the proposed control scheme by testing the behavior of the robot on a flight of stairs. A continuous snap shot of the motion is shown in Figure 6. From the figure we can see that the robot was able to traverse through a flight of stairs using the proposed control scheme without falling. The robot exhibited uncontrolled roll dynamics during falling due to unequal weight distribution between left and right sides. This can be seen from $t = 1197ms$ in Figure 6 where the robot landed on its left wheel before right. This caused the robot to misalign with the step during subsequent fall as seen in $t = 1499ms$. The control system was robust enough to compensate for the misalignment of the wheels with the edge of the step. The angle of the robot changes unpredictably when the robot rebounds. The prototype has only a single motor drive which makes it impossible to compensate for errors in heading. The heading errors could

Figure 6. Snapshot of stair traversing experiment. An alternate airborne/impact control scheme was used during step traversing. Post impact control scheme was activated once the robot had reached the final step and landed on flat ground surface.

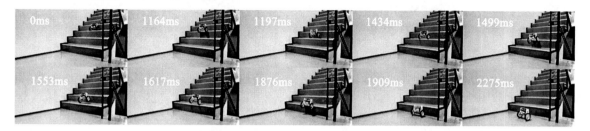

accumulate through successive steps, meaning that successful staircase traversal cannot be guaranteed. Differential drive could be used to correct the facing angle, and make it possible for the robot to traverse an unlimited number of steps.

The current prototype does not have actuators to mitigate disturbances in yaw and roll directions. Uneven weight distribution causes the robot to roll during rebound or when falling from extended height. Additional roll disturbances from uneven step surfaces will also reduce the stability of the robot. One possible solution to this problem is to introduce an additional flywheel actuator to actively control the roll attitude of the robot. Suspension system will also help reduced roll disturbances and increase the robustness of the robot in uneven terrains.

6. CONCLUSION

In this research, we have presented a unique solution enabling a two-wheeled pendulum robot to traverse through non-continuous ground terrain. During free fall, the drive wheels of the robot are used to generate attitude controlling torque to ensure a safe landing. On ground and airborne

dynamic models were presented along with a corresponding optimally designed control system. Additionally, a switching controller was used to vary the control effort of the controllers for smooth transition between phases. The experiment results show that the proposed method works well on single step and stair terrain. Landing angle were experimentally fine tuned due to the difficulty in measuring accurate linear velocities.

In our current prototype design no suspension system is used. The lack of suspension produces large impact forces and can cause the robot to rebound on impact. The rebound problem is mitigated by the robustness of the controller system design and implementation. The robot is designed to be durable and remains functional even after continuous fall experiment from various heights. For future work, we intend to introduce suspension system to the robot in order to tackle the rebound problem from increased falling height. We will also introduce differential drive to the system to reduce accumulative heading errors during successive steps traversal. Finally we will also investigate the possibility of introducing an active hopping mechanism so that the robot can also traverse step in both up and down directions.

REFERENCES

Astrom, K., Block, D. J., & Spong, M. W. (2007). *The reaction wheel pendulum.* New York: Morgan and Claypool.

Brown, H. J., & Xu, Y. (1997). A single wheel, gyroscopically stabilized robot. *IEEE Robotics & Automation Magazine*, *4*(3), 39–44. doi:10.1109/100.618022.

Cross, R. (2006). The fall and bounce of pencils and other elongated objects. *American Journal of Physics*, *74*(1), 26–30. doi:10.1119/1.2121752.

Grasser, F., D'Arrigo, A., Colombi, S., & Rufer, A. (2002). Joe: A mobile, inverted pendulum. *IEEE Transactions on Industrial Electronics*, *49*(1), 107–114. doi:10.1109/41.982254.

Guarnieri, M., Takao, I., Fukushima, E., & Hirose, S. (2007). Helios VIII search and rescue robot: Design of an adaptive gripper and system improvements. In *Proceedings of Intelligent Robots and Systems, 2007* (pp. 1775–1780). IEEE. doi:10.1109/IROS.2007.4399372.

Hayati, S., Volpe, R., Backes, P., Balaram, J., Welch, R., Ivlev, R., & Laubach, S. (1997). The rocky 7 rover: A mars sciencecraft prototype. In Proceedings of Robotics and Automation, 1997 (vol. 3, pp. 2458 2464). IEEE. doi: doi:10.1109/ROBOT.1997.619330.

Kikuchi, K., Sakaguchi, K., Sudo, T., Bushida, N., Chiba, Y., & Asai, Y. (2008). A study on a wheel-based stair-climbing robot with a hopping mechanism. *Mechanical Systems and Signal Processing*, *22*(6), 1316–1326. doi:10.1016/j.ymssp.2008.03.002.

Kumagai, M., & Ochiai, T. (2008). Development of a robot balancing on a ball. In Proceedings of Control, Automation and Systems, 2008 (pp. 433-438). ICCAS. doi: doi:10.1109/ICCAS.2008.4694680.

Lamon, P., Krebs, A., Lauria, M., Siegwart, R., & Shooter, S. (2004). Wheel torque control for a rough terrain rover. In *Proceedings of Robotics and Automation, 2004* (*Vol. 5*, pp. 4682–4687). IEEE. doi:10.1109/ROBOT.2004.1302456.

Lauwers, T., Kantor, G. A., & Hollis, R. (2006). A dynamically stable single-wheeled mobile robot with inverse mouse-ball drive. In *Proceedings of the 2006 IEEE International Conference on Robotics and Automation (* (pp. 2884 - 2889). IEEE.

Saranli, U., Buehler, M., & Koditschek, D. E. (2001). Rhex: A simple and highly mobile hexapod robot. *The International Journal of Robotics Research*, *20*(1), 616–631. doi:10.1177/02783640122067570.

Schmidt-Wetekam, C., & Bewley, T. (2011). An arm suspension mechanism for an underactuated single legged hopping robot. In *Proceedings of Robotics and Automation* (pp. 5529–5534). IEEE. doi:10.1109/ICRA.2011.5980339.

Schmidt-Wetekam, C., Zhang, D., Hughes, R., & Bewley, T. (2007). Design, optimization, and control of a new class of reconfigurable hopping rovers. In *Proceedings of Decision and Control* (pp. 5150–5155). IEEE. doi:10.1109/CDC.2007.4434975.

Spong, M. W., Corke, P., & Lozano, R. (2001). Nonlinear control of the reaction wheel pendulum. *Automatica*, *37*(11), 1845–1851. doi:10.1016/S0005-1098(01)00145-5.

Stilman, M., Olson, J., & Gloss, W. (2010). Golem Krang: Dynamically stable humanoid robot for mobile manipulation. In Proceedings of Robotics and Automation (pp. 3304-3309). IEEE. doi: doi:10.1109/ROBOT.2010.5509593.

Teeyapan, K., Wang, J., Kunz, T., & Stilman, M. (2010). Robot limbo: Optimized planning and control for dynamically stable robots under vertical obstacles. In *Proceedings of Robotics and Automation* (pp. 4519–4524). IEEE. doi:10.1109/ROBOT.2010.5509334.

Volpe, R., Balaram, J., Ohm, T., & Ivlev, R. (1996). The rocky 7 mars rover prototype. In *Proceedings of Intelligent Robots and Systems '96 (Vol. 3*, pp. 1558–1564). IEEE.

Xu, Y., & Au, S.-W. (2004). Stabilization and path following of a single wheel robot. *IEEE/ASME Transactions on Mechatronics, 9*(2), 407–419. doi:10.1109/TMECH.2004.828642.

Yap, H. E., & Hashimoto, S. (2012). Attitude control of an airborne two-wheeled robot. In *Proceedings of Artificial Life and Robotics*. AROB.

KEY TERMS AND DEFINITIONS

Airborne Attitude Control: The control of the orientation of the robot when in the robot is falling.

Ground Balance Control: The balance control algorithm of the robot when the robot is on the ground.

Inertia Measurement Unit: An electronic device equipped with accelerometers and gyroscopes used to detect the linear and rotational accelerations of an attached body. Raw data from the accelerometers and gyroscopes can be filtered to provide an accurate estimate of the angular orientation in space of an attached object.

Landing Angle: The angle of the robot body at the moment the robot hit the ground after falling a stepped terrain.

Reaction Wheel Pendulum: A simple pendulum with a rotating wheel attached at the end, which provides rotational torque.

State Feedback Control: A feedback controller that uses the current states of a system as subsequent input to the system to form a close loop control system.

Step-Traversing: The action of going down a stepped terrain without losing balance or falling.

Switching Controller: A deterministic controller that switches its states based on a set predefined criteria.

Two-Wheeled Robot: A two-wheeled robot is a dynamically stable mobile robot designed to balance on two wheels only. A two-wheeled robot requires active control to maintain balance but offers robust response to external disturbances.

Chapter 8
Ellipse Detection–Based Bin–Picking Visual Servoing System

Kai Liu
Tsinghua University, China

Hongbo Li
Tsinghua University, China

Zengqi Sun
Tsinghua University, China

ABSTRACT

In this chapter, the authors tackle the task of picking parts from a bin (bin-picking task), employing a 6-DOF manipulator on which a single hand-eye camera is mounted. The parts are some cylinders randomly stacked in the bin. A Quasi-Random Sample Consensus (Quasi-RANSAC) ellipse detection algorithm is developed to recognize the target objects. Then the detected targets' position and posture are estimated utilizing camera's pin-hole model in conjunction with target's geometric model. After that, the target, which is the easiest one to pick for the manipulator, is selected from multi-detected results and tracked while the manipulator approaches it along a collision-free path, which is calculated in work space. At last, the detection accuracy and run-time performance of the Quasi-RANSAC algorithm is presented, and the final position of the end-effecter is measured to describe the accuracy of the proposed bin-picking visual servoing system.

INTRODUCTION

Most of the manipulators practically used in industry operate in an open-loop mode. Their motions need to be programmed off-line or taught by operators, and the environment should be organized to suit them. In order to increase the flexibility and the accuracy of robot system, visual feedback control or visual servoing was introduced and widely researched. Typically, the visual feedback control system is classified into two groups: Position-Based Visual Servoing (PBVS) and Image-Based Visual Servoing (IBVS) (Seth, Gregory, & Peter, 1996; Danica & Henrik, 2002;

DOI: 10.4018/978-1-4666-4225-6.ch008

Deng, 2004). Both in position-based and image-based visual servoing process, the vision system extracts features of the target object in the image to provide input to the robot controller. The features could be points, lines, planes, regions, etc. Then in position-based visual servoing, three-dimensional information of the target object is estimated with respect to the camera (robot, world) coordinate system (Danica & Henrik, 2002). The aim is to steer the manipulator's end-effecter towards the target object in the task of picking, assembling, etc. While in image-based visual servoing, the main objective is to keep the features at a desired position in the 2D image plane (Bernard, Francois, & Patrick, 1992; Jenelle & Harvey, 2003).

The task of our bin-picking system is to pick parts (cylinders) from a bin. Hence, it is a position-based system. Our work is a further development of (Hao, Sun, & Fujii, 2007), where the detection and measurement method and visual servoing control algorithm to approach the cylinder were researched, and the experiments with an industrial 6-DOF manipulator with a hand-eye camera validated the proposed method. However, condition was simplified to only one horizontally placed target. In real industry, the most likely condition is that many objects of the same shape are stacked randomly in a container or a bin. Under this condition, the manipulator has to recognize and select one target for handling and avoid obstacles, e.g. the wall of the bin and other objects.

To recognize the cylinder targets, ellipse feature is used. There are mainly two types of ellipse detection methods: Hough transform based algorithms (Nair & Saunder, 1997; Guil & Zapata, 1997; Robert, 1998), and geometric methods (Hao, Sun, & Fujii, 2007; Ho, & Chen, 1995; Song & Wang, 2007). The Hough transform methods are computationally expensive and unsuitable for real-time control, so we choose geometric method. To improve the detection result's stability of the method in (Hao, Sun, & Fujii, 2007), a Quasi-Random Sample Consensus (Quasi-RANSAC) method is developed in this chapter.

QUASI-RANSAC ELLIPSE DETECTION

Since we choose cylinder as target object, and the projection of cylinders in image are some ellipses, so the target detection task is boiled down to ellipse detection task. The detection result is organized as

$$e_i = [x_i, y_i, a_i, b_i, \theta_i] \qquad (1)$$

where (x_i, y_i) denote the coordinates of ellipse center in image, (a_i, b_i) are its semi-major axis and semi-minor axis, and θ_i denotes its orientation angle. They are all the information about the target that the robot controller can utilize. The ellipse detection algorithm is based on the continuous edge feature which is called a contour (Cai, Yu, & Wang, 2004). In this way, we only regard the contour as a candidate of ellipse rather than all the edges in the edge image which cost lots of computation and memory.

The RANSAC algorithm is an effective method for model fitting (Fischler & Bolles 1981). In (Hao, Sun, & Fujii, 2007), the method was used to detect ellipse, but the detection results were not that stable. This is mainly because that the five points used to fit the ellipse are randomly sampled from the contour. The unstable ellipse will later lead to robot's motion error. In order to overcome this shortage, we change the sample strategy from random to quasi-random. Namely, we select the first point randomly while the other four points are uniformly distributed on the contour with respect to the first point, and ten groups of 5-point are sampled. For each group, an ellipse is fitted if could, using the Analytic Geometry Theory Based Ellipse Fitting method (Hao, Sun, & Fujii, 2007), and an average ellipse is computed from the fitted ellipses. Then the Mean Square Error (MSE) between the average ellipse and the candidate contour is computed. If the error is less than a threshold which changes corresponding to the mean square of the perimeter of the average ellipse, then the average ellipse is the final detection result.

The flow of the Quasi-RANSAC ellipse detection algori- thm is summarized as:

- Extract a contour from the image using OpenCV tools.
- Sample ten groups of 5-point from the contour using our Quasi-Random Sample strategy.
- Fit a conic for each group of 5 points using the Analytic Geometry Theory Based Ellipse Fitting method, and compute its geometric parameters in Formula (1) if the conic is an ellipse.
- Compute the average ellipse from the fitted ellipses.
- Compute the mean square error between the average ellipse and the candidate contour and compare it with a threshold. If the MSE is less than the threshold, the ellipse is detected.

POSITION AND POSTURE ESTIMATION

To estimate the position and posture of the target is a key step in PBVS. Since only a single hand-eye camera is utilized in the system, the camera is calibrated in order to use its pinhole model in conjuncture with the target's geometric model to construct the 3D information. The camera is mounted on the manipulator's end-effector and parallels to it. Figure 1 shows the structure of the system.

In order to describe the position and posture of the target, we create a coordinate system on the target, which we call the target coordinate system. Then we can use the position and posture of the target coordinate system in world coordinate system to represent that of the target. We suppose

Figure 1. System structure

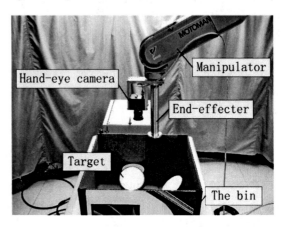

the position of the target is described as $\left(X_t, Y_t, Z_t\right)^T$, and its posture is described as a 3x3 rotational matrix R_t. Let $\left(x_t, y_t, z_t\right)^T$ denote the target's coordinates in camera coordinate system. According to camera pin-hole model and perspective projection, we get

$$\begin{cases} x_c = r_t(x_i - x_{ic}) / a_i \\ y_c = r_t(y_i - y_{ic}) / a_i \\ z_c = k / a_i \end{cases} \quad (2)$$

where $\left(x_{ic}, y_{ic}\right)$ is the image center, k is the camera pin-hole model parameter, r_t is the radius of the target cylinder and a_i is the length of the ellipse's semi-major axis.

To convert $\left(x_c, y_c, y_c\right)^T$ to $\left(X_t, Y_t, Z_t\right)^T$, the camera's position $\left(X_c, Y_c, Z_c\right)^T$ and rotational matrix R_c with respect to world coordinate system should be known. Since the camera is fixed to the end-effecter, the camera's position and posture in end-effecter coordinate system are constant. They are $\left(X_{ce}, Y_{ce}, Z_{ce}\right)^T$ and R_{ce}. So

$$R_c = R_e R_{ce} \quad (3)$$

$$\begin{pmatrix} X_c \\ Y_c \\ Z_c \end{pmatrix} = \begin{pmatrix} X_e \\ Y_e \\ Z_e \end{pmatrix} + R_c \begin{pmatrix} X_{ce} \\ Y_{ce} \\ Z_{ce} \end{pmatrix} \qquad (4)$$

where $(X_e, Y_e, Z_e)^T$ and R_e denote the position and posture of end-effecter in world coordinate system, which can be gotten from the manipulator. Then target's position $(X_t, Y_t, Z_t)^T$ can be calculated from

$$\begin{pmatrix} X_t \\ Y_t \\ Z_t \end{pmatrix} = \begin{pmatrix} X_c \\ Y_c \\ Z_c \end{pmatrix} + R_c \begin{pmatrix} x_c \\ y_c \\ z_c \end{pmatrix} \qquad (5)$$

The posture of the target is described by the rotational matrix R_t or the RPY-angle (γ, β, α) of the target coordinate system. According to (Sun, 1995), the relationship between the rotational matrix and the RPY-angle is

$$R_t = R_z(\gamma) R_{y'}(\beta) R_{x''}(\alpha) \qquad (6)$$

where

$$R_z(\gamma) = \begin{vmatrix} \cos\gamma & -\sin\gamma & 0 \\ \sin\gamma & \cos\gamma & 0 \\ 0 & 0 & 1 \end{vmatrix} \qquad (7)$$

$$R_{y'}(\beta) = \begin{bmatrix} \cos\beta & 0 & \sin\beta \\ 0 & 1 & 0 \\ -\sin\beta & 0 & \cos\beta \end{bmatrix} \qquad (8)$$

$$R_{x''}(\alpha) = \begin{bmatrix} 1 & 0 & 0 \\ 0 & \cos\alpha & -\sin\alpha \\ 0 & \sin\alpha & \cos\alpha \end{bmatrix} \qquad (9)$$

We can also get the RPY angle from the rotational matrix R_t according to Formula (6) ~ (9). If

$$R_t = \begin{bmatrix} n_x & t_x & b_x \\ n_y & t_y & b_y \\ n_z & t_z & b_z \end{bmatrix} \qquad (10)$$

then

$$\beta = \arcsin(-n_z) \qquad (11)$$

$$\alpha = \begin{cases} \arccos(\dfrac{b_z}{\cos\beta}), & \dfrac{t_z}{\cos\beta} \geq 0 \\ -\arccos(\dfrac{b_z}{\cos\beta}), & \dfrac{t_z}{\cos\beta} < 0 \end{cases} \qquad (12)$$

$$\gamma = \begin{cases} \arccos(\dfrac{n_x}{\cos\beta}), & \dfrac{n_y}{\cos\beta} \geq 0 \\ -\arccos(\dfrac{n_x}{\cos\beta}), & \dfrac{n_y}{\cos\beta} < 0 \end{cases} \qquad (13)$$

In order to estimate the posture of the target, we move the end-effecter to let the ellipse be at the center of image, where the target is at the optical axis of the camera, and also let the camera be perpendicular to horizontal plane, and the camera coordinate system's x, y axis should parallel to that of the world coordinate system respectively. This can be easily done since the position of the target is estimated. Figure 2 shows the camera's position and posture while estimating a target's posture, where $o_c - x_c y_c$ is the camera image coordinate system, $o_{c\_init} - x_{c\_init} y_{c\_init}$ is camera's initial position before end-effecter moves, $o_t - x_t y_t z_t$ is the target coordinate system, and $O - XYZ$ is the world coordinate system. Then the camera's optical axis coincides with $o_c o_t$ which collects the camera and the target center.

Figure 2. Camera position and posture while estimating target posture

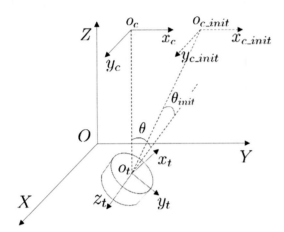

Figure 3. Symmetrical condition

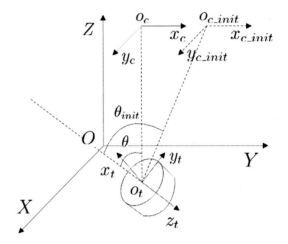

We should notice that the cosine value of the angle between $o_t o_c$ and target's rotational axis equals to b / a, where b is ellipse's semi-minor axis and a is the semi-major axis.

$$\cos \theta = b / a,$$
$$\theta = \arccos(b / a). \tag{14}$$

So when the camera is at position and posture shown in Figure 2, and noticing that $o_t x_t$ parallels to horizontal plane, some of the Direction Angles of the target coordinate system axes are

$$\theta_{zZ} = \pi - \theta,$$
$$\theta_{xZ} = \pi / 2,$$
$$\theta_{xX} = \pi / 2 - \theta_i, \tag{15}$$
$$\theta_{xY} = \pi - \theta_i,$$
$$\theta_{yZ} = \pi / 2 + \theta.$$

where θ_i is the ellipse's orientation angle, see Formula (1). Figure 3 shows a symmetrical condition which has the same ellipse image with Figure 2. In this condition,

$$\theta_{yZ} = \pi / 2 - \theta. \tag{16}$$

and other Direction Angles are the same with Formula (15).

Then some elements of target coordinate system's rotational matrix R_t can be calculated.

$$R_t = \begin{bmatrix} \cos \theta_{xX} & & \\ \cos \theta_{xY} & & \\ \cos \theta_{xZ} & \cos \theta_{yZ} & \cos \theta_{zZ} \end{bmatrix} \tag{17}$$

Substituting (17) in Formula (10) ~ (13), we can get the RPY-angle of the target. And then, substituting the RPY-angle in (6) yields the rotational matrix R_t. To distinguish the two symmetrical conditions in Figure 3 and Figure 4, two images' information is used: the image when the camera is at its initial position and the image when the camera is moved to where the target posture is estimated. We rewrite the value b / a of the ellipse in former image as b_{init} / a_{init}. Suppose o_{c_init} is the camera's initial position and θ_{init} is the angle between $o_t o_{c\_init}$ and the target's rotational axis. Conditions in Figure 3 and Figure 4 have different θ_{init} values. The correct condition's cosine value of θ_{init} is b_{init} / a_{init}. Knowing this truth, we can distinguish the two conditions.

Figure 4. Overlapping of targets and conditions of target's relationship with the bin

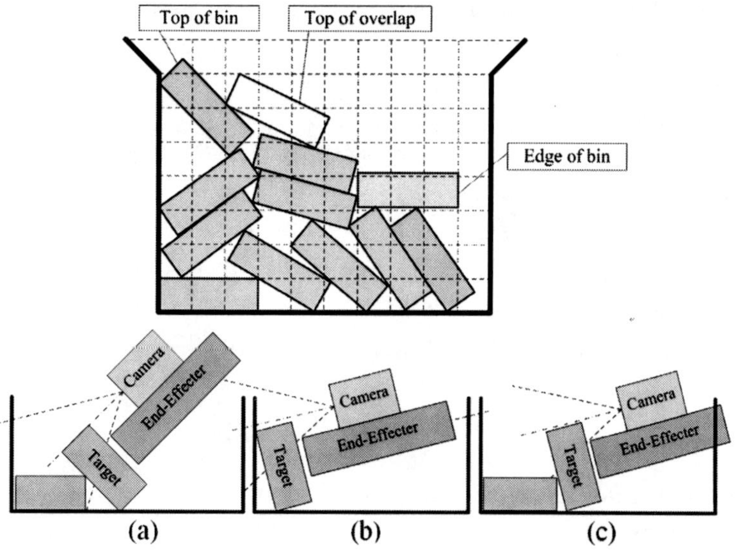

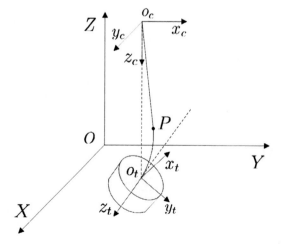

VISUAL SERVOING PROCESS

4.1. Target Selection

Since there are lots of objects overlapping together in the container shown in Figure 4, and in most cases more than one target are detected, so the first task of the bin-picking visual servoing process is to select a target which is the easiest one to pick for the manipulator. In order to avoid collision of the end-effecter with other objects, we first select the top one of overlap. This rule can be easily implemented when the position of the targets are estimated out. Then we should calculate whether the top one can be picked or not in consideration of the existence of the walls of the bin. The conditions are classified into three cases shown in Figure 4. Apparently, if the top one is in case (c), it can't be picked by the manipulator, then the second top one is considered, and so on.

4.2. Path Planning

The task of the path planning algorithm is to find a path towards the target without collision. According to the target selecting rules, there must be such a path for the selected target. The start point of the path is directly above the target where the target's posture is estimated, and the end point is the target center. The shortest path is the straight line $o_c o_t$, which also is free of collision. But this path is not good for the end-effecter's turning. So we add a little circular arc at the end of the path to let the end-effecter turn, shown in Figure 5. The path is restricted to plane $y_t o_t z_t$ of the target coordinate system.

Figure 5. End-effecter's path

The path consists of a line and a little circular arc. We can select several points on the path as the path points. During the manipulator's move, the end-effecter coordinate system, not the camera coordinate system, is at the path point, but except the start point which the camera is at to estimate the target's posture. At each path point, we should make sure that the posture of the end-effecter points to the target so the camera can see it. The final state of the end-effecter should be at the target's center and perpendicular to it.

4.3. Flow of Visual Servoing

The process of the bin-picking visual servoing includes the following steps:

1. Reset the manipulator to where the end-effecter is directly above the bin and perpendicular to the horizontal plane, and the x, y axis of image coordinate system respectively parallel to that of the world coordinate system.
2. Estimate position of multi-detected targets and select one according to the target selecting rules and start to track it.
3. Move the camera directly above the selected target and estimate its posture. If the target can be picked, go to step 4); otherwise, return to step 1).
4. Calculate a path according to the path planning rules.
5. Move the end-effecter close to the target along the path. At each path point, if the camera can see the whole target, estimate its position and replace the path end point with the newly estimated target position.

ACKNOWLEDGMENT

This work was also supported in part by the National Basic Research Program of China (973 Program) under Grant 2012CB821206, by the National Natural Science Foundation of China under Grants 61004021, 61174069, 61174103, and by the Beijing Natural Science Foundation under Grant 4122037.

REFERENCES

Bernard, E., Francois, C., & Patrick, R. (1992). A new approach to visual servoing in robotics. *IEEE Transactions on Robotics and Automation*, 8(3), 313–326. doi:10.1109/70.143350.

Cai, W. C., Yu, Q., & Wang, H. (2004). A fast contour-based approach to circle and ellipse detection. In *Proceedings of IEEE 5th World Conference on Intelligence Control and Automation*, (vol. 5, pp. 4686-4690). IEEE. doi: 10.1109/WCICA.2004.1342408

Danica, K., & Henrik, I. C. (2002). Survey on visual servoing for manipulation. *Technical Report ISRN KTH/NA/P-02/01-SE, CVAP259*. Retrieved from http://citeseerx.ist.psu.edu/

Deng, L. F. (2004). *Comparison of image-based and position-based robot visual servoing methods and improvements*. Waterloo, Canada: Waterloo University.

Fischler, M. A., & Bolles, R. C. (1981). Random sample consensus: a paradigm for model fitting with applications to image analysis and automated cartography. *Communications of the ACM, 24*, 381–395. doi:10.1145/358669.358692.

Guil, N., & Zapata, E. L. (1997). Lower order circle and ellipse hough transform. *Pattern Recognition*, 30(10), 1729–1744. doi:10.1016/S0031-3203(96)00191-4.

Hao, M., Sun, Z. Q., & Fujii, M. (2007). Ellipse detection based long range robust visual servoing. *Journal of Central South University, 38*, 432–439.

Ho, C. T., & Chen, L. H. (1995). A fast ellipse/circle detector using geometric symmetry. *Pattern Recognition*, 28(1), 117–124. doi:10.1016/0031-3203(94)00077-Y.

Jenelle, A. P., & Harvey, L. (2003). Uncalibrated eye-in-hand visual servoing. *The International Journal of Robotics Research*, *22*, 805–819. doi:10.1177/027836490302210002.

Nair, P. S., & Saunder, A. T. (1997). Hough transform based ellipse detection algorithm. *Pattern Recognition Letters*, *17*, 777–784. doi:10.1016/0167-8655(96)00014-1.

Robert, A. M. (1998). Randomized hough transform: Improved ellipse detection with comparison. *Pattern Recognition Letters*, *19*, 299–305. doi:10.1016/S0167-8655(98)00010-5.

Seth, H., Gregory, D. H., & Peter, I. C. (1996). A tutorial on visual servo control. *IEEE Transactions on Robotics and Automation*, *12*(5), 651–670. doi:10.1109/70.538972.

Song, G., & Wang, H. (2007). A fast and robust ellipse detection algorithm based on pseudo-random sample consensus. *Lecture Notes in Computer Science*, *4673*, 669–676. doi:10.1007/978-3-540-74272-2_83.

Sun, Z. Q. (1995). *Robot intelligence control.* Beijing, China: Beijing Education Press.

Chapter 9
Inferring Intention through State Representations in Cooperative Human– Robot Environments

Craig Schlenoff
NIST, USA & University of Burgundy, France

Zeid Kootbally
NIST, USA

Anthony Pietromartire
NIST, USA & University of Burgundy, France

Stephen Balakirsky
NIST, USA

Sebti Foufou
University of Burgundy, France & Qatar University, Qatar

ABSTRACT

In this chapter, the authors describe a novel approach for inferring intention during cooperative human-robot activities through the representation and ordering of state information. State relationships are represented by a combination of spatial relationships in a Cartesian frame along with cardinal direction information. The combination of all relevant state relationships at a given point in time constitutes a state. A template matching approach is used to match state relations to known intentions. This approach is applied to a manufacturing kitting operation[1], where humans and robots are working together to develop kits. Based upon the sequences of a set of predefined high-level state relationships that must be true for future actions to occur, a robot can use the detailed state information presented in this chapter to infer the probability of subsequent actions. This would enable the robot to better help the human with the operation or, at a minimum, better stay out of his or her way.

DOI: 10.4018/978-1-4666-4225-6.ch009

1. INTRODUCTION

Humans and robots working safely and seamlessly together in a cooperative environment is one of the future goals of the robotics community (CCC, 2009). When humans and robots can work together in the same space, a whole class of tasks can be automated, ranging from collaborative assembly to parts and material handling to delivery. Keeping humans, who are within a robotic work cell safe, requires the ability of the robot to monitor the work area, infer human intention, and be aware of potential dangers to avoid them. Robots are under development throughout the world that will revolutionize manufacturing by allowing humans and robots to operate in close proximity while performing a variety of tasks (Szabo, 2011).

Proposed standards exist for robot-human safety (such as work performed by International Organization for Standardization (ISO) / Robotics Industries Association (RIA) 10218-2 Safety Requirements – Part 2: Industrial robot systems and integration), but these standards focus on robots adjusting their speed based on the separation distance between the human and the robot (Chabrol, 1987). In essence, as the robot gets closer to a detected human, the robot gradually decreases its speed to ensure that if a collision between the human and robot occurs, minimal damage will be caused. These standards focus on where the human is at a given point in time. It does not focus on where they are anticipated to be in the future.

A key enabler for human-robot safety in cooperative environments involves the field of intention recognition. Intention recognition involves the robot attempting to understand the intention of an agent (the human) by recognizing some or all of his/her actions (Sadri, 2011) to help predict the human's future actions. Knowing these future actions will allow a robot to plan in such a way as to either help the human perform his/her activities or, at a minimum, not put itself in a position to cause an unsafe situation.

In this chapter, we present an approach to representing state information in an ontology for the purpose of ontology-based intention recognition. An overview of the intention recognition approach can be found in (Schlenoff, 2012a). In this context, we adopt the (Tomasello, 2005) definition of intention as "a plan of action the organism chooses and commits itself to in pursuit of a goal – an intention thus includes both an action plan as well as a goal." We also distinguish states from state relationships. In this context, we define a state as a set of properties of one or more objects in an area of interest that consist of specific recognizable configurations and or characteristics. A state relationship is a specific relation between two objects (e.g., Object 1 is on top of Object 2). A set of all relevant state relationships in an environment composes a state. This approach to intention recognition is different than many ontology-based intention recognition approaches in the literature (as described in the next section) as they primarily focus on activity (as opposed to state) recognition and then use a form of abduction to provide explanations for observations. We infer detailed state relationships using observations based on Region Connection Calculus 8 (RCC8) (Randell, 1992) and then combine these observations using the Semantic Web Rule Language (SWRL) (W3C_Member_Submission, 2004) to infer the overall state relationships that are true at a given time. Once a sequence of state relationships has been determined, we will use probabilistic procedures to associate those states with likely overall intentions to determine the next possible action (and associated state) that is likely to occur. This chapter focuses on the way that states are represented in the ontology and how intentions can be inferred from them.

We start by providing an overview of intention recognition efforts in the literature as well as various approaches for ontology-based state representation. We then present two evaluations, one in intention recognition and one in activity recognition, and show the difference in performance between the two. After that, we show a novel approach to state relation representation and ordering to describe intentions using RCC-8 and SWRL. We then describe the manufacturing kitting domain and provide a detailed scenario showing how the approaches can be used to represent state information in this domain. We then show how we can use the ordering of these states and state relationships to help to identify intentions that are occurring in the environment. We conclude the chapter by showing advantages of the state-based recognition approach as compared to other approaches in the literature.

2. INTENTION RECOGNITION AND STATE REPRESENTATION RELATED WORK

Intention recognition traditionally involves recognizing the intent of an agent by analyzing some of, or all of, the actions that the agent performs. Many of the recognition efforts in the literature are composed of at least three components: (1) identification and representation of a set of intentions that are relevant to the domain of interest, (2) representation of a set of actions that are expected to be performed in the domain of interest and the association of these actions with the intentions, (3) recognition of a sequence of observed actions executed by the agent and matching them to the actions in the representation. (Sadri, 2011)

There have been many techniques in the literature applied to intention recognition that follow the three steps listed before, including an ontology-based approach (Jeon, 2008), multiple probabilistic frameworks such as Hidden Markov Models (Kelley, 2008) and Dynamic Bayesian Networks (Schrempf, 2005), utility-based intention recognition (Mao, 2004), and graph-based intention recognition (Youn, 2007). In this chapter, we focus on ontology/logic-based approaches.

In many of these efforts, abduction has been used as the underlying reasoning mechanism in providing hypotheses about intentions. In abduction, the system "guesses" that a certain intention could be true based on the existence of a series of observed actions. For example, one could guess that someone may have watered the lawn if the lawn is wet. There may be other possible explanations for why the lawn is wet (e.g., it rained) but the fact that someone watered the lawn could be the most probable explanation given the circumstances. As more information is learned and activities are performed, probabilities of certain intentions can be refined to be consistent with the observations.

Intention recognition is a rich and challenging field. As shown previously, often multiple competing hypotheses are possible regarding the intentions of an observed agent or to describe the situation in an environment. Choosing between these hypotheses is one challenge, but there are many others. One challenge is that circumstances, including the adversarial nature of the observed agent, may afford only partial observability of the actions. In other words, the scene may be occluded by other objects so that it can not be seen in its entirety, or the agent performing the action may purposefully hide their actions or intentions from the observer. Furthermore, would-be-intruders and would-be attackers may even deliberatively execute misleading actions to throw off the intention recognition system.

Another challenge is the case where the agent may have multiple intentions that they are performing at the same time and may interleave the execution of their actions, or the case where the actor is concurrently trying alternative plans for achieving the same intention. As another example, intention recognition becomes more difficult when attempting to interpret the actions of cognitively impaired individuals who may be executing actions in error and with confusion, for example in the case of Alzheimer patients (Roy, 2007).

Additional complications arise when attempting to analyze the actions and intentions of multiple, cooperating agents (Sukthanker, 2001).

Much research in the intention recognition field focuses on pruning the space of hypotheses. In a given domain, there could be many possible intentions. Based on the observed actions, various techniques have been used to eliminate improbable intentions and assign appropriate probabilities to intentions that are consistent with the actions performed. Some of the ways that efforts have done this is by assigning weights to conditions of the rules used for intention recognition as a function of the likelihood that those conditions are true (Pereira, 2009). For example, it may be unlikely that a person is using an umbrella outside unless it is raining or very sunny, thus the cost associated with this condition would be very high since the probability of this happening would be low.

Once observations of actions have been made, different approaches exist to match those observations to an overall intention or goal. For example, in (Mulder, 2003), the authors use existentially quantified observations (not fully grounded observations) to match actions to plan libraries. Mulder can handle situations when they see an action occur (e.g., opening a door), without seeing or knowing who performed that action. Other approaches have focused on building plans with frequency information, to represent how often an activity occurs (Jarvis, 2005). The rationale behind these approaches is that there are some activities that occur very frequently and are often not relevant to the recognition process (e.g., a person cleaning his/her hands). These frequently-occurring activities can be mostly ignored, and only activities that are less commonly performed can be considered. In (Demolombe, 2006), the authors combine probabilities and situation calculus-like formalization of actions. In particular, Demolombe not only defines the actions and sequences of actions that constitute an intention, they also state which activities cannot occur for the intention to be valid.

For example, if the intention was to drive a car, the activity may be to open the door, get into the car, turn on the engine, release the emergency brake, and take the car out of park. Demolombe may also include that an activity cannot be to turn the car off after it is turned on and before the car is taken out of park.

All of these approaches have focused on the activity being performed as the primary basis for observation and the building block for intention recognition. However, as noted in (Sadri, 2011), activity recognition is a very hard problem and one that is far from being solved. There has been limited success in using Radio Frequency Identification (RFID) readers and tags attached to objects of interest to track their movement with the goal of associating their movement with known activities. For example, in (Philipose, 2005), the authors describe the process of making tea as a three step process involving using a kettle, getting a box of tea bags, and adding some combination of milk, sugar, or lemon. Each of these activities is identified by a user wearing a special set of gloves that can read RFID tags on objects of interest. However, this additional hardware can be inhibiting and unnatural. Recognizing and representing states as opposed to actions can help to address some of the issues involved in activity recognition (e.g., the location of the tea bag box and the milk carton with respect to the tea cup) will be the focus of the rest of this paper.

State representation is documented in the literature, although it has not been used (to the authors knowledge) for ontology-based intention recognition. An important aspect of an object's state is its spatial relationships to other objects. In (Bateman, 2006), an overview is given that describes the way that spatial information is represented in various upper ontologies including the Descriptive Ontology for Linguistics and Cognitive Engineering (DOLCE), Cyc, the Standard Upper Merged Ontology (SUMO), and Basic Formal Ontology (BFO). The conclusion of this work is the identification of a list of high-level requirements that were necessary for any spatial ontology, including:

1. A selection of an appropriate granular partition of the world that picks out the entity that we wish to locate with respect to other entities.
2. A selection of an appropriate space region formalization that brings out or makes available relevant spatial relationships.
3. A selection of an appropriate partition over the space region (e.g., RCC8, qualitative distance, cardinal direction).
4. The identification of the location of the entity with respect to the selected space region description.

Bateman ended up using a variation of the DOLCE ontology, but there is no mention in the literature about the detailed spatial relations that were developed as part of this effort.

Region Connection Calculus 8 (RCC-8) (Wolter, 2000) is a well-known and cited approach for representing the relationship between two regions in Euclidean space or in a topological space. There are eight possible relations, including disconnected, externally connected, partially overlapping, etc. However, RCC8 only addresses these relationships in two-dimensional space. There have been other approaches that have tried to extend this into a region connected calculus in three-dimensional space while addressing occlusions (Albath, 2010). There have also been other approaches to develop calculi for spatial relations. FlipFlop calculus (Ligozat, 1993) describes the position of one point (the referent) in a plane with respect to two other points (the origin and the relatum). Single Cross Calculus (SCC) (Freksa, 1992) is a ternary calculus that describes the direction of a point (C - the referent) with respect to a second point (B - the relatum) as seen from a third point (A - the origin) in a plane. Double Cross Calculus (DCC) (Freksa, 1992) extends SCC by allowing one to also determine the rela-

tive location of point A with respect to point B (in addition to point B with respect to point A as in SCC). Coarse-grained Dipole Relation Algebra (Schlieder, 1995) describes the orientation relation between two dipoles (an oriented line segment as determined by a start and end point). Oriented Point Relation Algebra (OPRA) (Moratz, 2005) relates two oriented points (a point in a plane with an additional direction parameter) and describes their relative orientation towards each other. All of these approaches, apart from RCC8, focus on points and lines as opposed to regions. Also, despite the large variety of qualitative spatial calculi, the amount of applications employing qualitative spatial reasoning techniques is comparatively small (Wallgrun, 2006).

Throughout the rest of this paper, we will describe an approach for ontology-based state representation within the context of a typical manufacturing scenario.

3. STATE RECOGNITION VS. ACTIVITY RECOGNITION

One of the core assumptions of this work is that it is easier to perform state recognition in an environment as opposed to activity recognition. While there have been many efforts that have attempted to do each, they have primarily been self-evaluated in configurations and environments that have been most conducive to their approaches. There have, however, been some open, impartial competitions/evaluations in these areas (DARPA, 2012; Marvel, 2012). Next we describe a state recognition competition and an activity recognition evaluation performed by external, impartial parties. We will use the results of these competitions as benchmarks to characterize the state of the art in these two fields.

State recognition can encompass many things, including recognizing objects' color, size, shape, location, pose, as well of the identification of the object itself. Later in this chapter, we describe an approach that relies on the identification of an object as well as determining its location and pose to characterize its spatial relationships with other objects. As such, we will use the competition described in Section 3.1 as the basis for the performance of state recognition technology since it is well aligned with the focus of this paper.

3.1. State of the Art in State Recognition

In 2011, as part of the International Conference on Robotics and Automation (ICRA) in Shanghai, China, a Solutions in Perception Challenge (SPC) was held (Newman, 2011). The purpose of establishing the SPC was to determine the current state of maturity for robotic perception algorithms. There are many algorithms that currently exist world-wide for identifying objects and determining their pose (location and orientation with respect to a coordinate frame), yet it is difficult to ascertain whether an algorithm could be applied to a given task or to know with confidence its robustness. In addition, efforts to develop these algorithms are being duplicated, but there is no convenient way of readily knowing what algorithms have already solved a particular aspect of the perception problem (Marvel et al., 2012). The SPC seeks to identify the best available perception algorithms that will be documented in the form of open source software to prevent duplication of development efforts and, in turn, accelerate the development of the next generation of perception algorithms.

The topic of the 2011 Challenge was single and multiple rigid object identification and 6 degree of freedom (6DOF) pose estimation in structured scenes. Competing teams were required to develop algorithms that could "learn" an arbitrary number of objects from the provided 3D point cloud data that had been augmented with the corresponding red-green-blue point color, and then to correctly identify and locate the same objects in a presented scene.

The data sets composing the training and evaluation sets were assembled using 16 machined aluminum artifacts representing commonly-encountered features of manufactured parts (Figure 1). Each artifact was created from a unique computer-aided design model, and was augmented with semi-Lambertian, optically-textured decals. The artifacts were first categorized into three classification groups based on perceived physical features. Group 1 consisted of objects with maximal heights greater than 2 inches. Group 2 objects were shorter than 2 inches, but had raised features that gave them non-level surfaces. And Group 3 objects were shorter than 2 inches, and had level surfaces. The artifacts were designed to be congruent with industrial assembly parts like automotive or aircraft components, and could be rigidly fixtured to a low-cost ground truth system (Figure 2).

Evaluation of the submitted algorithms was broken into two distinct rounds. Round 1 consisted of image frames featuring only one artifact at a time, while the frames in Round 2 contained three artifacts each. Both rounds were composed of several sub-runs based on variations in object translation and rotation. Run 1 consisted of only object translations; Run 2 had only object rotations

Figure 1. Sample machined NIST artifacts used in the 2011 SPC, arranged randomly

Figure 2. The ground truth fixture with an attached sensor and an artifact in the rotation plate

using the four fixture-based alignment holes; and Run 3 had a combination of translations and rotations. The teams were not aware of the composition of the data sets prior to the competition.

For each run in Round 1, a sample artifact was randomly selected from the three classification groups. Poses compliant with the Run-based transformation restrictions were then randomly selected for each run, and the chosen artifacts were each applied to the same subset of transformations. For Round 2, one object from each classification group was randomly selected to form the test group. Random poses were generated for each run, with each artifact being assigned a different location on the base plate. There were a total of 399 frames for the contestants to assess.

For each frame of data, the ground truth consisted of a set of one or more objects. A true positive count (*hits*, c_h) reflects an algorithm's ability to correctly identify when an object is in the scene. A non-zero false positive count (*noise*, c_n) indicates that an algorithm identified objects that were not actually present in the ground truth, and a non-zero false negative count (*misses*, c_m) implies that the algorithm could not correctly identify objects that were in the scene.

The true positive counts were tallied over all 399 frames. The competing teams correctly identified over 80% of all objects over all frames (665 artifacts or more of the 831 present over all 399 frames).

Of the seven competing teams, two teams scored above 80% on the translation test (i.e., the estimated translation was within tolerance greater than 80% of the time). In addition, three of the seven teams scored greater than 72% on the rotation test (See Figure 3). The baseline score in the figures was the performance obtained by a Willow Garage test system used to develop the data sets and evaluation metrics.

3.2. State of the Art in Activity Recognition

Activity recognition can be roughly divided into two camps. The first focuses on sensor-based activity recognition where the person in the environment wears some type of sensor, which could be a smartphone, accelerometer, or other tracking device, that provides data as to where the person is and the motion s/he is taking. Examples of this include (Choudhury, 2008) and (Ravi, 2005). The second is vision-based activity recognition, where

Figure 3. Translation scores over all 399 frames

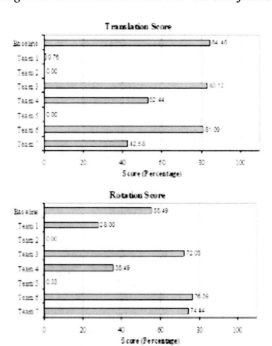

the behaviors of agents are characterized using videos taken by various cameras. Examples of this type of work can be found in (Bodor, 2003) and (Hoogs, 2008).

For the work described in this chapter, the focus is on approaches that do not require the agent to wear any external or internal device, as this is not reasonable or practical in an industrial setting. Therefore we will only consider vision-based activity recognition. In addition, as mentioned earlier in this section, we will limit our analysis to evaluations that were performed by an external evaluator (i.e., not the developers of the systems) to remove any conflict of interest and ensure an unbiased evaluation. In the next paragraph, we will describe one such evaluation and the subsequent results.

The Defense Advanced Research Projects Agency (DARPA) Mind's Eye Program[3] seeks to develop machine-based visual intelligence by automating the ability to learn generally applicable and generative representations of actions between objects in a scene directly from visual inputs, and then reason over those learned representations. The focus of this project is on the military domain, where Army scouts are commonly tasked with covertly entering uncontrolled areas, setting up a temporary observation post, and then performing persistent surveillance for 24 hours or longer. A truly "smart" camera would describe with words everything it sees and reason about what it cannot see. These devices could be instructed to report only on activities of interest, which would increase the relevancy of incoming data to users. Thus, smart cameras could permit a single scout to monitor multiple observation posts from a safe location.

To evaluate the activity recognition, the DARPA program identified 2,588 short vignettes (approximately 15 s to 20 s each) of 48 different activities. Each activity was represented by approximately 54 of the short vignettes. The activities that were explored are shown in Table 1.

The eight systems under evaluation had to determine if one or more of the verbs listed in Table 1 were present in the vignettes. There were two metrics used. The first was precision which is defined as:

$$Precision = \frac{VerbMatches}{VerbMatches + FalsePositives} \quad (1)$$

The precision results from each team are shown in Figure 4. Team scores ranged from 21% to 60%.

For the second metric, the program used the Matthews Correlation Coefficient (MCC), which is a balanced measure of correlation which can be used even if the classes are of different sizes (in lieu of accuracy). The formula is:

$$MCC = \frac{TP * TN - FP * FN}{\sqrt{(TP + FP)(TP + FN)(TN + FP)(TN + FN)}} \quad (2)$$

Table 1. Verbs used on the DARPA mind's eye program

Approach	Close	Flee	Have	Open	Run
Arrive	Collide	Fly	Hit	Pass	Snatch
Attach	Dig	Follow	Hold	Pick up	Stop
Bounce	Drop	Get	Kick	Push	Take
Bury	Enter	Give	Jump	Put down	Throw
Carry	Exchange	Go	Leave	Raise	Touch
Catch	Exit	Hand	Lift	Receive	Turn
Chase	Fall	Haul	Move	Replace	Walk

Figure 4. Precision results from mind's eye evaluation

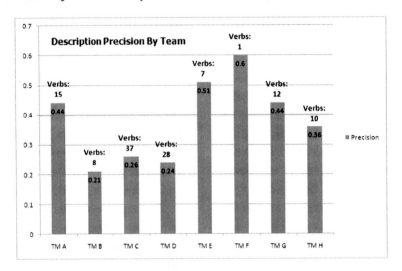

where TP are true positives, TN are true negatives, FP are false positives, and FN are false negatives. The results of applying this metric are very similar to the precision metrics, where systems scored between 12% - 63% when using this metric, with an average of approximately 35%.

In summary, in both cases, the system was given a scene or video and asked to characterize it by stating either the type and pose/orientation of an object or the existence of an activity. While it is impossible to perform a direct comparison between these two approaches, these examples provide a sampling of the type of performance that can be expected by each type of technology. Based on this analysis, it appears that state recognition systems are approximately 20% (80% vs. 60%) more accurate than the activity recognition systems. This shows the benefit in using state recognition systems in performing intention recognition as opposed to using activity recognition systems.

4. REPRESENTING STATES AND SPATIAL RELATIONS

In this section, we describe an approach that uses RCC8 to model state relationships based on the relative position of objects in the environment.

However, we will extend RCC8, which was initially developed for a two-dimensional space, to a three-dimension space by applying it along all three planes (x-y, x-z, y-z). The frame of reference will be with respect to the fixed object (e.g., a worktable), with the z-dimension pointing straight upwards and the tabletop extending in the x- and y- dimension (with detailed orientation specific to the application). Each of the high-level state relationships (to be discussed later in the chapter) will have a set of logical rules that associate these RCC8 relations to them. These RCC8 relations, in theory, should easily allow a sensor system to characterize the corresponding state relations.

As mentioned earlier, RCC8 abstractly describes regions in Euclidean or topological space by their relations to each other. RCC8 consists of eight basic relations that are possible between any two regions:

- Disconnected (DC)
- Externally Connected (EC)
- Equal (EQ)
- Partially Overlapping (PO)
- Tangential Proper Part (TPP)
- Non-Tangential Proper Part (NTPP)
- Tangential Proper Part Inverse (TPPi)
- Non-Tangential Proper Part Inverse (NTPPi)

These are shown pictorially in Figure 5.

RCC8 was created to model the relationships between two regions in two dimensions. In many domains, these relations need to be modeled in all three dimensions. As such, every pair of objects has a RCC8 relation in all three dimensions. To address this, we are prepending an x-, y- or z- before each of the RCC8 relations. For example, to represent the RCC8 relations in the x-dimension, the nomenclature would be:

- x-DC
- x-EC
- x-EQ
- x-PO
- x-TPP
- x-NTPP
- x-TPPi
- x-NTPPi

Similar nomenclature would be used in the y- and z- dimensions. The combination of all 24 RCC relations starts to describe the spatial relations between any two objects in the scene. However, more information is needed to represent the cardinal direction between any two objects. For example, to state that a worktable is empty (worktable-empty(wtable)), one needs to state that there is nothing on top of it. If we assume that the vertical dimension is the z-dimension, then saying that:

$$z\text{-}EC(wtable, obj1) \qquad (3)$$

Figure 5. RCC8 relations (credit: http:// en.wikipedia.org/wiki/RCC8)

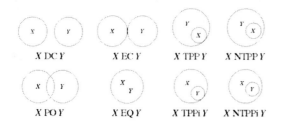

which intuitively means *obj1* is externally connected to the worktable in the z-dimension - is not sufficient because *obj1* could be either on top of or below the worktable. In other words, we need to represent directionality. We do this using the following Boolean operators:

$$greater\text{-}x(A,B) \qquad (4)$$

$$smaller\text{-}x(A,B) \qquad (5)$$

$$greater\text{-}y(A,B) \qquad (6)$$

$$smaller\text{-}y(A,B) \qquad (7)$$

$$greater\text{-}z(A,B) \qquad (8)$$

$$smaller\text{-}z(A,B) \qquad (9)$$

which intuitively means, in Equation 4, that the edge of the bounding plane (discussed later in Section 8) of object A is greater than (in the x-dimension in the defined frame of reference) the edge of the bounding plane of object B.

There are undoubtedly many other relationships that may be needed in the future to describe a scene of interest. These could include absolute locations and orientations of objects (x, y, z, roll, pitch, yaw) and relative distance (closer, farther, etc.). However, these spatial relations are sufficient for describing the manufacturing kitting example later in this chapter.

From these RCC8 spatial relations, we can define more complex spatial relations such as the ones next:

- **Contained-In:** An object is enclosed in a second object from all sides
- **Not-Contained-In:** An object is not enclosed in a second object from all sides
- **Partially-In:** An object is inside of another object in at least two dimensions but not fully contained in

- **In–Contact-With:** Touching at least one side and not contained with (i.e., toughing outer edges)
- **On-Top-Of:** The edge of the bounding plane of the one object is greater (in the z-dimension) than that of a second object
- **Under:** The edge of the bounding plane of the one object is less (in the z-dimension) than a second object

And from these, we can define composite spatial relationships such as:

- **Under and in Contact With:** An object is both under and in contact with a second object
- **Partially in and in Contact With:** An object is inside of another object in at least two dimensions and touching the object in at least one dimension

Next, we formalize these spatial relationships by defining them using the RCC8 state representation. All of the formulas described are represented by SWRL rules. Each atom in the SWRL rules references a predicate, which is represented by object properties in the ontology.

In natural language, Equation 10 states that object 1 (obj1) is contained in object 2 (obj2) if obj1 is tangentially or non-tangentially a proper part of object2 in the x, y, and z-dimension. One can logically envision this by drawing two convex figures, and the first convex hull is completely inside of the second convex hull in all three dimensions, with it touching or not touching the second convex hull in all of the three dimensions.

Contained-In(*obj1, obj2*) →

(x-TPP(*obj1, obj2*) ∨ x-NTPP(*obj1, obj2*)) ∧

(y-TPP(*obj1, obj2*) ∨ y-NTPP(*obj1, obj2*)) ∧

(z-TPP(*obj1, obj2*) ∨ z-NTPP(*obj1, obj2*))   (10)

Not-Contained-In(*obj1, obj2*) →

¬Contained-In(*obj1, obj2*)   (11)

Partially-In(*obj1, obj2*) →

Not-Contained-In(*obj1, obj2*) ∧

((x-TPP(*obj1*, obj2) ∨ x-NTPP(*obj1, obj2*)) ∧

(y-TPP(*obj1, obj2*) ∨ y-NTPP(*obj1, obj2*))) ∨

((x-TPP(*obj1, obj2*) ∨ x-NTPP(*obj1, obj2*)) ∧

(z-TPP(*obj1, obj2*) ∨ z-NTPP(*obj1, obj2*))) ∨

((y-TPP(*obj1, obj2*) ∨ y-NTPP(*obj1, obj2*)) ∧ (z-TPP(*obj1, obj2*) ∨ z-NTPP(*obj1, obj2*)))   (12)

In-Contact-With(*obj1, obj2*) →

x-EC(*obj1, obj2*) ∨ y-EC(*obj1, obj2*) ∨ z-EC(*obj1, obj2*)   (13)

On-Top-Of(*obj1, obj2*) →

greater-z(*obj1, obj2*) ∧ (x-EQ(*obj1, obj2*) ∨ x-NTPP(*obj1, obj2*) ∨

x-TPP(*obj1, obj2*) ∨ x-PO(*obj1, obj2*) ∨ x-NTPPi(*obj1, obj2*) ∨

x-TPPi(*obj1, obj2*)) ∧ (y-EQ(*obj1, obj2*) ∨ y-NTPP(*obj1, obj2*) ∨

y-TPP(*obj1, obj2*) ∨ y-PO(*obj1, obj2*) ∨ y-NTPPi(*obj1, obj2*) ∨

y-TPPi(*obj1, obj2*))   (14)

Under($obj1$, $obj2$) $\rightarrow$

smaller-z($obj1$, $obj2$) $\wedge$ (x-EC($obj1$, $obj2$) $\vee$ x-NTPP($obj1$, $obj2$) $\vee$

x-TPP($obj1$, $obj2$) $\vee$ x-PO($obj1$, $obj2$) $\vee$ x-NTPPi($obj1$, $obj2$) $\vee$

x-TPPi($obj1$, $obj2$)) $\wedge$ (y-EC($obj1$, $obj2$) $\vee$ y-NTPP($obj1$, $obj2$) $\vee$

y-TPP($obj1$, $obj2$) $\vee$ y-PO($obj1$, $obj2$) $\vee$ y-NTPPi($obj1$, $obj2$) $\vee$

y-TPPi($obj1$, $obj2$)) $\qquad$ (15)

Under-And-In-Contact-With($obj1$, $obj2$) $\rightarrow$

Under($obj1$, $obj2$) $\wedge$ In-Contact-With ($obj1$, $obj2$) $\qquad$ (16)

Partially-In-And-In-Contact-With($obj1$, $obj2$) $\rightarrow$

Partially-In($obj1$, $obj2$) $\wedge$ In-With-Contact($obj1$, $obj2$) $\qquad$ (17)

These spatial relationships will be used later in the chapter to define two manufacturing kitting intentions.

5. ORDERING STATES TO FORM INTENTIONS

In this work, an ordering of state relationships represents an intention. As such, we need a formal mechanism to allow for this ordering. To do this, we borrow some concepts that are described in OWL-S (Web Ontology Language – Services) (Martin, 2004). OWL-S is described on the Website (http://www.w3.org/Submission/OWL-S/) as an ontology of services enabling a user and software agents to discover, invoke, compose, and monitor Web resources offering particular services and having particular properties. Though intended for Web-based services, many of the same ordering constructs are equally applicable to the representation of the sequencing of states. OWL-S defines eight control constructs:

- **Perform:** Execution of an action.
- **OrderedList:** A list of control constructs to be done in order.
- **Split:** A "bag" of process components to be executed concurrently. The Split completes when all of its component processes have been scheduled for execution.
- **Split+Join:** Concurrent execution of a bunch of process components with synchronization. Split+Join completes when all of its components processes have completed.
- **Any-Order:** Process components (specified as a bag) to be executed in some unspecified order but not concurrently. All components must be executed.
- **Choice:** The execution of a single control construct from a given bag of control constructs.
- **If-Then-Else:** Intended as "Test If-condition; if True do Then, if False do Else"
- **Iterate:** Makes no assumption about how many iterations are made or when to initiate, terminate, or resume. The initiation, termination, or maintenance condition could be specified with a whileCondition or an untilCondition.
- **Repeat-While/Repeat-Until:** Both of these iterate until a condition becomes false or true. Repeat-While tests for the condition, exits if it is false and does the operation if the condition is true, then loops. Repeat-Until does the operation, tests for the condition, exits if it is true, and otherwise loops.

We adapt some of the these control constructs to represent the ordering of states by changing their name and definition as shown in Table 2. Column 2 in Table 2 shows the state ordering constructs that we define in this work. This is not an exhaustive list, but the constructs shown should be sufficient for representing the kit assembly shown later.

Another interesting aspect not represented in OWL-S is the representation of state relationships that cannot occur for an intention to be true. For example, in the assembly of Kit 2 (shown later), we state that Part D is not a part of that kit. Thus, if we see a state in which Part D is placed within the Kit Tray, we can logically assume that Kit 2 is not the kit being assembled. Therefore, in addition to needing an "Exists" construct, we also need a "NotExists" construct, as shown in Table 2.

Table 2. Initial state representation ordering constructs

OWL-S Control Construct	Adapted State Representation Ordering Construct	State Representation Definition
Perform	Exists	A state relationship must exist
Sequence	OrderedList	A set of state relationships that must occur in a specific order
Any-Order	Any-Order	A set of state relationships that must all occur in any order
Iterate	Count	A state relationship that must be present multiple times.
n/a	NotExists	State relationship that can't exist.
Choice	Choice	A set of possible state relationships that can occur after a given state relationship
Join	Co-Exist	Two or more state relationships that must be true

In the ontology, we represent the ordering constructs previously as subclasses of the overall "OrderingConstructs" class. In addition, the state representations listed previously are modeled as subclasses of the overall StateRepresentation class. States refer to an unordered list (a bag) of state relationships which completely describe the state. A state contains one to many state relationships. StateRelationships are defined using the RCC8 relationships described earlier. Table 3 shows what each construct points to and the cardinality restrictions.

6. THE MANUFACTURING KITTING DOMAIN

Though we expect the approaches described in this chapter to be generic, we are initially applying them to a specific manufacturing domain to show their feasibility. In this domain, we focus on manufacturing kitting operations as described

Table 3. Construct details

Construct	Points to:	Cardinality
State	StateRelationshipBag	1:n StateRelationships
StateRelationships	SolidObject	Exactly 2 SolidObjects
Ordering Constructs		
AnyOrder	OrderingConstructsBag	2:n OrderingConstructs
Choice	OrderingConstructsBag	2:n OrderingConstructs
CoExists	OrderingConstructsBag	2:n OrderingConstructs
Count	OrderingConstruct	Exactly 1 OrderingConstruct
Exists	StateRelationship	Exactly 1 StateRelationship
NotExists	StateRelationship	Exactly 1 StateRelationship
OrderedList	OrderingConstructsList	2:n OrderingConstructs

in (Balakirsky, 2012). Kitting is the process in which several different, but related items are placed into a container and supplied together as a single unit (kit) as shown in Figure 6. Kitting is often performed prior to final assembly in industrial assembly of manufactured products so all of the necessary parts are gathered in one location. Manufacturers utilize kitting due to its ability to provide cost savings, including saving manufacturing or assembly space, reducing assembly workers' walking and searching times, and increasing line flexibility and balance.

In batch kitting, the kit's component parts may be staged in containers positioned in the workstation or may arrive on a conveyor. Component parts may be fixtured, for example, placed in compartments on trays, or may be in random orientations, for example placed in a large bin. In addition to the kit's component parts, the workstation usually contains a storage area for empty kit boxes as well as completed kits.

Kitting has not yet been automated in many industries where automation may be feasible. Consequently, the cost of building kits is higher than it could be (Balakirsky et al., 2012). We are

addressing this problem by building models of the knowledge that will be required to operate an automated kitting workstation in an agile manufacturing environment. For our automated kitting workstation, we assume that a robot performs a series of pick-and-place operations in order to construct the kit. These operations include:

1. Pick up empty kit and place on work table.
2. Pick up multiple component parts and place in kit.
3. Pick up completed kit and place in full kit storage area.

Each of these actions may be a compound action that includes other actions such as end-of-arm tool changes, path planning, and obstacle avoidance. Finished kits are moved to the assembly floor where components are picked from the kit for use in the assembly procedure. The kits are normally designed to facilitate component picking in the correct sequence for assembly. Component orientation may be constrained by the kit design in order to ease the pick-to-assembly process. Empty kits are returned to the kit building area for reuse.

Figure 6. Example kit (courtesy of littlemachineshop.com)

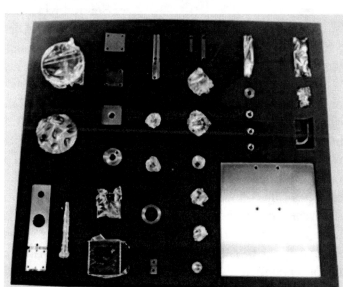

7. A KITTING ONTOLOGY

In late 2011, IEEE formed a working group entitled Ontologies for Robotics and Automation (ORA) (Schlenoff, 2012b). The goal of the working group is to develop a standard ontology and associated methodology for knowledge representation and reasoning in robotics and automation, together with the representation of concepts in an initial set of application domains. The working group understood that it would be extremely difficult to develop an ontology that could cover the entire space of robotics and automation. As such, the working group is structured in such a way as to take a bottom-up and top-down approach to addressing this broad domain. This group is comprised of four sub-groups entitled: Upper Ontology/Methodology (UpOM), Autonomous Robots (AuR), Service Robots (SeR), and Industrial Robots (InR). The InR, AuR, and SeR sub-groups are producing sub-domain ontologies that will serve as a test case to validate the upper ontology and the methodology developed by UpOM.

The industrial robots group is focusing on manufacturing kitting operations as a test case. This kitting ontology is focusing on activities that are expected to be performed in a sample kitting operation along with pertinent objects that are expected to be present.

7.1. Representing Kitting Objects in the Ontology

The industrial kitting objects ontology is written in Web Ontology Language (OWL) (Harmelen, 2004). Conceptually, the model is an object model in which there are classes with attributes (but no functions, constructors, etc.). OWL has classes but does not have attributes; it has ObjectProperties and DataProperties instead. They may be used to model attributes. OWL Properties are global, not local to a class, so localizing each attribute to a class is done by a naming convention that includes the class name and the attribute name. OWL sup-

ports multiple inheritance, but that has not been used in the kitting ontology. Except by subclass relationship, no object is in more than one class.

The model has two top-level classes, **SolidObject** and **DataThing**, from which all other classes are derived. **SolidObject** models solid objects, things made of matter. **DataThing** models data. Subclasses of **SolidObject** and **DataThing** are defined as shown in Table 4. The level of indentation indicates subclassing. Items in *italics* following class are names of class attributes. Derived types inherit the attributes of the parent. An attribute with a solid underline must occur at least once and can also occur many times. An attribute with a dashed underline may occur zero, one, or many times. An attribute with a dotted underline may occur once or not at all. An attribute with no underline must occur exactly once. Each attribute has a specific type not shown in Table 4. If an attribute type has derived types, any of the derived types may be used.

Each **SolidObject** has a native coordinate system conceptually fixed to the object. The native coordinate system of a **BoxyObject**, for example, has its origin at the middle of the bottom of the object, its Z axis perpendicular to the bottom, and the X axis parallel to the longer horizontal edges of the object.

Each **SolidObject A** has at least one **PhysicalLocation** (the *PrimaryLocation*). A **PhysicalLocation** is defined by giving a reference **SolidObject B** and information saying how the position of **A** is related to **B**. Two types of location are required for operation of the kitting workstation. Relative locations, specifically the knowledge that one **SolidObject** is in or on another, are needed to support making logical plans for building kits. Mathematically precise locations are needed to support robot motion. The mathematical location, **PoseLocation**, gives the pose of the coordinate system of **A** in the coordinate system of **B**. The mathematical information consists of the location of the origin of **A**'s coordinate system and the directions of its Z and X axes. The mathemati-

Table 4. Subset of the kitting object ontology

SolidObject *PrimaryLocation SecondaryLocation*
BoxyObject *Length Width Height*
KitTray *SkuRef SerialNumber*
LargeContainer *SkuRef SerialNumber*
PartsBin *PartQuantity PartSkuRef*
PartsTray *SkuRef SerialNumber*
Kit *DesignRef Tray Parts*
KittingWorkstation *WorkTable Robot ChangingStation OtherObstacles AngleUnit LengthUnit WeightUnit Skus KitDesigns*
LargeBoxWithEmptyKitTrays *LargeContainer Trays*
LargeBoxWithKits *LargeContainer Kits Capacity*
Part *SkuRef SerialNumber*
PartsTray *SkuRef SerialNumber*
PartsTrayWithParts *Tray Parts*
Robot *Description Id WorkVolume EndEffector Maximum-LoadWeight*
WorkTable *SolidObjects*
DataThing
PhysicalLocation *RefObject*
PoseLocation *Point ZAxis XAxis*
PoseLocationIn
PoseLocationOn
PoseOnlyLocation
RelativeLocation
RelativeLocationIn
RelativeLocationOn

cal location variety has subclasses representing that, in addition, **A** is in **B** (**PoseLocationIn**) or on **B** (**PoseLocationOn**). The subclasses of **RelativeLocation** are needed not only for logical planning, but also for cases when the relative location is known, but the mathematical information is not available. This occurs, for example when a **PartsBin** is being used, since by definition, the **Parts** in a **PartsBin** are located randomly.

All chains of location from **SolidObject**s to reference **SolidObject**s must end at a **Kitting-Workstation** (which is the only class of **Soli-dObject** allowed to be located relative to itself).

The kitting ontology includes several sub-classes of **SolidObject** that are formed from components that are **SolidObject**s. These are: **Kit, PartsTrayWithParts, LargeBoxWithEmp-tyKitTrays,** and **LargeBoxWithKits**. Combined objects may come into existence or go out of existence dynamically when a kitting workstation is operating. For example, when all the parts in a **PartsTrayWithParts** have been removed and put into kits, the **PartsTrayWithParts** should go out of existence and the **PartsTray** that was holding **Part**s should have its location switched from its location relative to the **PartsTrayWithParts** to the former location of the **PartsTrayWithParts**.

In the current version of the kitting ontology, the only way a shape can be described is by using a **BoxyObject**. This is adequate for making kits of boxes, but is not adequate for most industrial forms of kitting. The kitting ontology includes **ShapeDesign** as a stub, but its only attribute is a string giving a description of the shape. For manipulating non-boxy **SolidObject**s some mathematically usable representation of shape will be required.

7.2. Representing Activities in the Ontology

To represent activities in the manufacturing kitting ontology, both the actions and the pre- and post-conditions of those actions need to be represented. OWL-S (Martin et al., 2004) is used to represent the actions that need to be performed and SWRL atoms are used to represent the preconditions and effects of the actions. Preconditions and effects are represented as state relationships. An example of the take-kt (take kit tray) action is shown in Equation 18. In natural language, the take-kt action involves a robot (r) equipped with an end effector (*eff*) picking up a kit tray (*kt*) from within a large box with empty kit trays (*lbwekt*). The action is formally defined in the State Variable Representation (Nau, 2004) as:

$$\text{take-kt}(r,\ kt,\ lbwekt,\ eff,\ wtable) \qquad (18)$$

Figure 7 shows the preconditions and effects that are associated with this action.

Each of the state relationships in Table 4 is described:

1. **rhold-empty(*r*):** TRUE iff robot (*r*) is not holding anything.
2. **lbwekt-non-empty(*lbwekt*):** TRUE iff Large Box With Empty Kit Trays (*lbwekt*) is not empty.
3. **r-with-eff(*r, eff*):** TRUE iff Robot (*r*) is equipped with an EndEffector (*eff*).
4. **ktlocation(*kt, lbwekt*):** TRUE iff the Kit Tray (*kt*) is in the Large Box With Empty Kit Trays (*lbwekt*).
5. **efflocation(*eff, r*):** TRUE iff the EndEffector (*eff*) is being held by the robot (*r*).
6. **worktable-empty(*wtable*):** TRUE iff there is nothing on the Worktable (*wtable*).
7. **efftype(*eff, kt*):** TRUE iff the EndEffector (*eff*) is designed to handle the Kit Tray (*kt*).
8. **rhold-empty(*r*):** TRUE iff the robot (*r*) is holding something.
9. **ktlocation(*kt, r*):** TRUE iff the Kit Tray (kt) is being held by the Robot (*r*).
10. **rhold(*r, kt*):** TRUE iff the Robot (*r*) is holding the Kit Tray (*kt*).
11. **ktlocation(*kt, lbwekt*):** TRUE iff the Kit Tray (*kt*) is not in the Large Box With Empty Kit Trays (*lbwekt*).

There are many other actions that can be performed during the kitting operation, including putting down a kit tray, picking up and putting

Figure 7. Preconditions and effects for the take kit tray action

precond	effects
rhold-empty(*r*), lbwekt-non-empty(*lbwekt*), r-with-eff(*r,eff*), ktlocation(*kt,lbwekt*), efflocation(*eff,r*), worktable-empty(*wtable*), efftype(*eff,kt*)	¬rhold-empty(*r*), ktlocation(*kt,r*), rhold(*r,kt*), ¬ktlocation(*kt,lbwekt*)

down a part, attaching/removing an end effector, etc. Each of these actions has associated preconditions and effects. Sections 10, 11, and 12 will show a novel approach to how the truth-value of these pre- and post-conditions (state relationships) can be inferred using the output of object recognition and pose recognition algorithms from a sensor system.

8. STATE REPRESENTATION IN THE MANUFACTURING KITTING DOMAIN

When modeling the state relationships (preconditions and effects) shown in Section 7.2, the first step is to precisely define the state relationships in such a way as to determine if there were similar spatial relations that could be leveraged. We can start to formalize the previous definition of the state relationships as follows:

The rholds state relationships (1, 8, and 10) depend on the type of effector that is being used to define the state relationship. We will assume there are two types of end effectors: a vacuum end effector and a parallel gripper end effector. The vacuum end effector picks objects up by positioning itself on top of the object and uses air to create a vacuum to adhere to the object. The parallel gripper end effector picks objects up by squeezing them from both sides. Because the vacuum end effector would not reasonably be used to pick up the kit tray, the vacuum-rhold(*r, kt*) state is not included next. In the case of the vacuum end effector, the relevant state relationships would be:

- **vacuum-rhold-empty(*r*):** There is no object **under and in contact with** the robot vacuum effector
- **vacuum-rhold-empty(*r*):** There is an object **under and in contact with** the robot vacuum effector

In the case of the parallel gripper end effector, the state relationships would be

- **gripper-rhold-empty(*r*):** There is no object **partially in and in contact with** the robot gripper
- **gripper-rhold-empty(*r*):** There is an object **partially in and in contact with** the robot gripper
- **gripper-rhold(*r, kt*):** The Kit Tray is **partially in and in contact with** the robot gripper (See Table 5)

There is also one state relationship that does not rely on spatial relations. The definition of efftype(*eff, kt*) states that a specific effector must be able to be used on a kit tray. It is envisioned that this information will be included in the ontology class to describe the kit tray and therefore is out of scope of this document.

Also, in the manufacturing kitting domain, not all objects can be represented as convex regions, as is required by the RCC8 formalism. For example, the robot gripper in Figure 8 is not convex and thus does not neatly fit into the RCC8 approach. To address this, we develop a convex hull along each relevant plane (as shown in Figure 8) around objects of this sort and use that convex hull to represent the region of the object in that plane.

Based on the manufacturing kitting ontology described in Section 7.1. and the spatial relations described in Section 4, we can formally define the 11 manufacturing kitting state relationships:

Figure 8. Convex hull around robot gripper

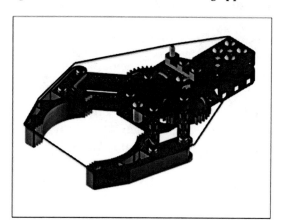

Table 5. Revised definitions of spatial relationships

# from Above	State Relationship	Previous Definition	Revised Definition
(2)	lbwekt-non-empty(*lbwekt*)	TRUE iff an object is in the Large Box With Empty Kit Trays (*lbwekt*)	there is an object that is **contained in** the Large Box With Empty Kit Tray
(3)	r-with-eff(*r, eff*)	TRUE iff Robot (*r*) is equipped with an EndEffector (*eff*)	the effector is **in contact with** the robot
(4)	ktlocation(*kt, lbwekt*)	TRUE iff the Kit Tray (*kt*) is in the Large Box With Empty Kit Trays (*lbwekt*)	the Kit Tray is **contained in** the Large Box With Empty Kit Trays
(5)	efflocation(*eff, r*)	TRUE iff the EndEffector (*eff*) is being held by the robot (*r*)	the effector is **in contact with** the robot (Note: Same as (3) above)
(6)	worktable-empty(*wtable*)	TRUE iff there is nothing on the Worktable (*wtable*)	there is no object that is **on top of** and **in contact with** the Worktable
(7)	efftype(*eff, kt*)	TRUE iff the EndEffector (*eff*) is designed to handle the Kit Tray (*kt*)	the Effector can handle the Kit Tray
(9)	ktlocation(*kt, r*)	TRUE iff the Kit Tray (kt) is being held by the Robot (*r*)	the Kit Tray is **in contact with** the Robot and there is nothing **under and in contact with** the Kit Tray
(11)	¬ktlocation(*kt, lbwekt*)	TRUE iff the Kit Tray (*kt*) is not in the Large Box With Empty Kit Trays (*lbwekt*)	the Kit Tray is **not contained in** the Large Box With Empty Kit Trays

lbwekt-non-empty(*lbwekt*) →

SolidObject(*obj1*) ∧ Contained-In(*obj1*, *lbwekt*)
$$(19)$$

r-with-eff(*r, eff*) → In-Contact-With(*r, eff*) (20)

ktlocation(*kt, lbwekt*) → Contained-In(*kt, lbwekt*) (21)

worktable-empty(*wtable*) →

SolidObject(*obj1*) ∧ ¬On-Top-Of(*obj1*, *wtable*) ∧

¬In-Contact-With(*obj1*, *wtable*) (22)

gripper-rhold(*r, kt*) →

GripperEffector(*eff*) ∧ r-with-eff(*r, eff*)

∧ Partially-In-And-In-Contact-With(*kt, eff*) (23)

¬ktlocation(*kt, lbwekt*) → ¬Contained-In(*kt, lbwekt*) (24)

vacuum-rhold-empty(*r*) →

SolidObject(*obj1*) ∧ SolidObject(*obj2*) ∧ VacuumEffector(*eff*) ∧

r-with-eff(*r, eff*) ∧ ¬(Under-And-In-Contact-With(*obj1, eff*) ∧

¬Under-And-In-Contact-With(*obj2, obj1*)) (25)

¬vacuum-rhold-empty(*r*) →

SolidObject(*obj1*) ∧ SolidObject(*obj2*) ∧ VacuumEffector(*eff*) ∧

r-with-eff(*r, eff*) ∧ Under-And-In-Contact-With(*obj1, eff*) ∧

¬Under-And-In-Contact-With(*obj2, obj1*) (26)

gripper-rhold-empty(*r*) →

SolidObject(*obj1*) ∧ GripperEffector(*eff*) ∧

r-with-eff(*r, eff*) ∧

¬Partially-In-And-In-Contact-With(*obj1, eff*)
$$(27)$$

¬gripper-rhold-empty(*r*) →

SolidObject(*obj1*) ∧ GripperEffector(*eff*) ∧

r-with-eff(*r, eff*) ∧

Partially-In-And-In-Contact-With(*obj1, eff*) (28)

ktlocation(*kt, r*) → gripper-rhold(*r, kt*) (29)

The formal definitions of these high-level state relationships will allow their existence to be recognized in a manufacturing environment, which in turn can be used by a state-based intention recognition system. The presence of states in certain predefined orders can help a robot recognize the intention of a human in the environment, which would allow the robot to better assist the human in performing upcoming activities. In the following detailed, yet simple, example, we will show how these state relationships can be represented, sequenced, and used to infer the intention of the performer (human).

9. KITTING EXAMPLE

In this section, we will use the detailed scenario shown in Figure 9. In this scenario, a person (not shown) is constructing one of two possible kits. Both kits use the same kit tray (Kit Tray) and contain a series of parts placed in the kit in any order. Kit 1 contains two Part A's, two part B's, one Part C, and one Part D. Kit 2 contains three part A's, one part B, and one part C. This

Figure 9. Sample kitting scenario

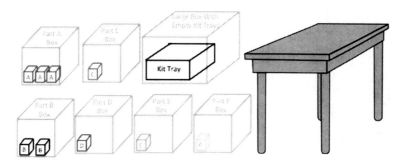

is shown in Figure 10. In this scenario, all parts with the same letter are identical (e.g., all A's are the same, all B's are the same). Other parts are available at the workstation and could be used for other kit assemblies that are not of interest for this example. A table is provided on which all work is performed. Parts and part trays are available from a set of boxes that can be refilled as needed. When a kit is completed, it is placed in a completed kit box (not shown).

A person is tasked with creating these kits. The person may choose to create either of these kits at any given time. A robot is available to assist the person by inferring which kit the person is intending to create at a given time and providing support where needed. The goal of the robot is to infer the intention of the human based on what it perceives (i.e., which kit tray the human is assembling).

10. REPRESENTING STATES, STATE RELATIONSHIPS, AND INTENTIONS IN AN ONTOLOGY

For Kit 1, we can infer from the previous description that the process for assembling this kit would initially involve placing the kit tray (Kit Tray) on the table and then adding, in any order, two Part A's, two Part B's, one Part C, and one Part D. Similarly, for Kit 2, we can deduce from the previous description that for the process for

Figure 10. Completed kits 1 and 2

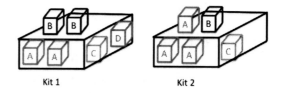

assembling this kit would initially involve placing the kit tray (Kit Tray 1) on the table and then adding, in any order, three Part A's, one Part B, and two Part C's. We can now use the state relations described in Section 4 and the ordering relationships in Section 5 to build intentions.

For Kit 1, we start by requiring that the Kit Tray be on the table (where SR stands for state relation and OC stands for Ordering Construct):

$$SR1 = \text{On-Top-Of(KitTray, Table} \qquad (30)$$

$$OC1 = \text{Exists(SR1)} \qquad (31)$$

Next, we represent all of the states that can happen in any order:

$$SR2 = \text{Contained-In(PartA, KitTray)} \qquad (32)$$

$$SR3 = \text{Contained-In(PartB, KitTray)} \qquad (33)$$

$$SR4 = \text{Contained-In(PartC, KitTray)} \qquad (34)$$

$$SR5 = \text{Contained-In(PartD, KitTray)} \qquad (35)$$

$$OC2 = Count(SR2, 2) \qquad (36)$$

$$OC3 = Count(SR3, 2) \qquad (37)$$

$$OC4 = Count(SR4, 1) \qquad (38)$$

$$OC5 = Count(SR5, 1) \qquad (39)$$

$$OC6 = AnyOrder(OC2, OC3, OC4, OC5) \quad (40)$$

$$SR6 = Contained\text{-}In(KitTray, CompletedKit\text{-}Box) \qquad (41)$$

$$OC7 = Exists(SR6) \qquad (42)$$

We also need to represent the states that cannot be true if this kit assembly is indeed the intention. In reality, there are likely an infinite number of states that could make this intention be deemed false. However, for this work, we will specifically focus on the states that are possible in other intentions but are not possible here. Some of these states may include:

$$SR7 = Contained\text{-}In(PartE, KitTray) \qquad (43)$$

$$OC8 = \neg Exists(SR7) \qquad (44)$$

Based on the prior, we can represent the intention of the Kit 1 assembly as follows:

$$OC9 = OrderedList(OC1, OC6, OC7) \& OC8 \qquad (45)$$

The & symbol in Equation 45 represents that the ordering constraint after the & symbol (OC8 in this case) must be true throughout the entire process (OC9).

Using OC9 , and applying the same approach to the Kit 2 intention (not shown), we can use a template-based approach to represent the sequence associated with Kit 1 and Kit 2 intention, which may look something like what is shown in Table 6 and Table 7.

All state relationships are listed, and the ones that must be true more than once (as represented by the Count construct in the ontology) are rep-

Table 6. Kit 1 template-based intention sequences

State Relation	1	2	2	3	3	4	5	6	7
Kit 1	On_Top_Of (Kit Tray, Table)	Contained-In (PartA, KitTray)	Contained-In (PartA, KitTray)	Contained-In (PartB, KitTray)	Contained-In (PartB, KitTray)	Contained-In (PartC, KitTray)	Contained-In (PartD, KitTray)	Contained-In (KitTray, Completed KitBox)	Not(Contained-In(PartE, KitTray))
Previous State Relation	n/a	1	1	1	1	1	1	2, 3, 4, 5	n/a

Table 7. Kit 2 template-based intention sequence

State Relation	1	2	2	2	3	4	5	6	7
Kit 2	On_Top_Of (Kit Tray, Table)	Contained-In (PartA, KitTray)	Contained-In (PartA, KitTray)	Contained-In (PartA, KitTray)	Contained-In (PartB, KitTray)	Contained-In (PartC, KitTray)	Contained-In (KitTray, Complet-edKitBox)	Not(Contained-In(PartD, KitTray	Not(Contained-In(PartE, KitTray))
Previous State Relation	n/a	1	1	1	1	1	2, 3, 4	n/a	n/a

resented in separate columns. The state relationships that cannot be true for a specific intention are represented at the end of the table. Under each state relationship is a pointer to the state relationship that must occur before this one is relevant. For example, before the Contained-In(PartA, KitTray) state relationship can be evaluated as being true, the On-Top_Of(KitTray, Table) (as indicated by the number 1 at the top of the table) must have occurred. This table will be used in subsequent sections to show how intentions can be recognized based upon observations.

11. TRACKING STATES AND STATE TRANSITIONS AS ACTIONS OCCUR

Now that we have specified how the states, state relationships, and intentions are represented in the ontology, we will show how state relationships are tracked by observations and associated with the intentions. Based on the detailed kitting example described in Section 9, the first step is to determine which objects (and spatial relations) are relevant to be tracked. In this domain, the relevant objects are:

- Part A
- Part B
- Part C
- Part D
- Part E
- Kit Tray
- Table
- Part A Box
- Part B Box
- Part C Box
- Part D Box
- Completed Kit Box

Note that Part E is of interest not because it is a part if the kit assembly, but because it is explicitly prohibited from being in one of the kit assemblies. Even though Part F is available within the environment, it is not of interest to either

intention and is therefore not tracked. Using the approaches described previously, the robot first extracts the state relationships that are relevant to the various intentions it is trying to perceive. These state relations are described in the previous section. In the cases of the two kit assemblies, the relevant state relationships include:

- Part A is in Part A Box
- Part B is in Part B Box
- Part C is in Part C Box
- Part D is in Part D Box
- KitTray is in Large Box With Empty Kit Trays
- KitTray is on Table
- Part A is in the KitTray
- Part B is in the KitTray
- Part C is in the KitTray
- Part D is in the KitTray
- Part D is not in the KitTray
- Part E is not in the KitTray
- Kit Tray is in Completed Kit Box

The truth-value of these state relationships can be evaluated at a given point in time by applying the state representation approach based on RCC8 as described in the previous section. In tabular format, the state relationships can be represented as shown in Table 8. Actions occur between each state to cause the truth-value of the relevant state relations to change. However, for this approach, all we care about is the truth-value of the state relations without needing to track the actions that cause them to be true or false.

State 1 shows the state of the environment as described in Figure 9. Numbers in the cells represent how many instances of the state relationship are true. In this example, there are three instances of Part A in Part A Box, as shown in the figure. If, at the next state, the Kit Tray was removed from the Kit Tray Holder and placed on the Table, the next state would be represented as in State 2. If one of the Part A's is then removed from the Part A Box and placed into the Kit Tray, the state representation would look like State 3.

Table 8. Tabular state and state relationship representation

State Relation	State 1 (Initial State)	State 2 (Kit Tray on Table)	State 3 (Part A in Kit Tray)	...
Part A is in Part A Box	3	3	2	
Part B is in Part B Box	2	2	2	
Part C is in Part C Box	1	1	1	
Part D is in Part D Box	1	1	1	
Kit Tray is in Large Box With Empty Kit Trays	1	0	0	
Kit Tray is on Table	0	1	1	
Part A is in Kit Tray	0	0	1	
Part B is in Kit Tray	0	0	0	
Part C is in Kit Tray	0	0	0	
Part D is in Kit Tray	0	0	0	
Part E is in Kit Tray	0	0	0	
Kit Tray in Completed Kit Box	0	0	0	

Understanding the states that are relevant also plays a significant role in determining which sensors to use and where to place the sensors that are meant to track objects in the environment. For example, if we assume that Part Box A is not a clear box and there is an opening at the top, a sensor would need to be placed above the box so that it is evident if (and how many) objects are present within the box (addressing the state relationship "Part A is in Part A Box").

In this approach, object locations are tracked in real-time by sensors. However, the state relationships shown in Table 8 are only evaluated when a new state is reached. States are defined as a period of time in which an object of interest (as described before) has moved greater than a predefined distance from its previous position and has stopped moving for greater than a predefined period of time. The distance and time values are up to the user, and are often a function of the domain. For an operation where objects are relatively close together (such as a scenario where all objects are already on a table), the distance metric will likely be very small. In other domains where objects are moved greater distances (such as a large assembly operation), this distance metric will likely be much greater. States can happen over various periods of

time. In Table 8, State 1 is meant to represent the initial state of the system. State 2 may be achieved a few seconds after the observations begin, while State 3 may not be achieved until minutes or hours afterwards. Many objects may have changed state between State 1 and State 2, but in this case, those objects of interest and their corresponding state relationships of interest have not changed.

12. MATCHING PERCEIVED STATES TRANSITIONS TO ONTOLOGY INTENTION REPRESENTATIONS

As mentioned previously, a new state becomes true when a state relationship that is being tracked changes its truth value. As an example, if the Kit Tray goes from not being on the table to being on the table, a new state is formed. During each state, the state relationships that are true are compared to the intention templates shown in Table 6. However, only the state relationships in Table 6 that are eligible are available for matching.

A state relationship becomes eligible if all required previous state relationships have occurred. For example, in Table 6 for Kit 1, the state relationship Contained-In(PartA, KitTray) can not occur until On_Top_Of (Kit Tray, Table) has occurred. Therefore, even if Contained-In(PartA, KitTray) is true in the observed scene, it is not active unless On_Top_Of (Kit Tray, Table) has previously occurred and thus should not be "checked off" unless that previous condition is met.

As each observed state occurs, the corresponding state relationships are compared with those associated with the possible intentions. When the observed state relationship(s) match an active state relationship in an intention, it is "checked off" indicating that the state relationship has occurred. This process occurs with every new observed state that occurs. In addition, after each observed state and associated matching with the intentions, the number of state relations that are true for that intention are summed. Continuing with our example, if the state table was updated to look like what is shown in Table 9, the corresponding intention table would look like what is shown in Table 10 and Table 11, with green cells representing those state relationships which have been observed and whose precondition state relations have been met.

In this simple example, at a given state, the Kit 1 intention has five state relations that are matched and the Kit 2 intention also has five state relations that are matched. To determine which intention is more probable, we use the following equation:

$$P_{i,t} = \frac{S_{i,t}}{\sum_{i=1}^{n} S_{i,t}} \tag{46}$$

where $P_{i,t}$ represents the probability of intention i at time t, $S_{i,t}$ represents the sum of the state relationships that are true for intention i at time t and n represents the number of intentions that are being evaluated. So in this case, for the Kit 1 and Kit 2 intentions, the ratio would be 5/10 = 0.5 (or 50%).

In addition, we need to handle state relationships that cannot occur in an intention and adjust the probabilities accordingly. We modify Equation (46) by adding a coefficient (F_t for forbidden at time t) in front of the equation, so it becomes:

Table 9. Tabular state and state relationship representation (version 2)

State Relation	State 1 (Initial State)	State 2 (Kit Tray On Table)	State 3 (Part A In Kit Tray)	State 3 (2nd Part A In Kit Tray)	State 4 (Part B in Kit Tray)	State 5 (Part C In Kit Tray)
Part A is in Part A Box	3	3	2	1	1	1
Part B is in Part B Box	2	2	2	2	1	1
Part C is in Part C Box	1	1	1	1	1	0
Part D is in Part D Box	1	1	1	1	1	1
Kit Tray is in Large Box With Empty Kit Trays	1	0	0	0	0	0
Kit Tray is on Table	0	1	1	1	1	1
Part A is in Kit Tray	0	0	1	2	2	2
Part B is in Kit Tray	0	0	0	0	1	1
Part C is in Kit Tray	0	0	0	0	0	1
Part D is in Kit Tray	0	0	0	0	0	0
Part E is in Kit Tray	0	0	0	0	0	0
Kit Tray in Completed Kit Box	0	0	0	0	0	0

Table 10. Matching observed states to the kit 1 intention

State Relation	1	2	2	3	3	4	5	6	7	Sum
Kit 1	On_Top_ Of (Kit Tray, Table)	Contained-In (PartA, KitTray)	Contained-In (PartA, KitTray)	Contained-In (PartB, KitTray)	Contained-In (PartB, KitTray)	Contained-In (PartC, KitTray)	Contained-In (PartD, KitTray)	Contained-In (KitTray, Completed KitBox)	Not (Contained-In(PartE, KitTray	5
Previous State Relations	n/a	1	1	1	1	1	1	2, 3, 4, 5	n/a	

Table 11. Matching observations to the kit 2 intention

State Relation	1	2	2	2	3	4	5	6	7	Sum
Kit 2	On_Top_ Of (Kit Tray, Table)	Contained-In (PartA, KitTray)	Contained-In (PartA, KitTray)	Contained-In (PartA, KitTray)	Contained-In (PartB, KitTray)	Contained-In (PartC, KitTray)	Contained-In (KitTray, Completed KitBox)	Not (Contained-In(PartD, KitTray	Not (Contained-In(PartE, KitTray	5
Previous State Relations	n/a	1	1	1	1	1	2, 3, 4	n/a	n/a	

$$P_{i,t} = F_{i,t} \frac{S_{i,t}}{\sum_{i=1}^{n} S_{i,t}} \tag{47}$$

F_t has a value of 0 or 1 depending if any observed state relationships are explicitly prohibited in the intention at the given time. F_t is initially set to 1. If a prohibited state relationship is observed, the value of F_t is set to 0 and the overall probably of the intention becomes 0. If no such state relationships exist, the value of F_t remains at 1.

So in Table 10, if we have a subsequent observation where Part D is placed into the kit tray, the Kit 1 intention would have six state relationships which are true and the Kit 2 intention would have five true relationship and one prohibit relationship. The number of true state relationships in the Kit 2 intention would drop to zero because of the prohibited state relationship. Thus, the probability of the Kit 1 intention would be 6/6 = 1 (100%) and the probability of Kit 2 would be 0/6 = 0 (0%).

It is important to note that prohibited state relations are treated differently than state relations that occur but do not advance the intention. For example, if an additional Part A is placed inside the kit, it would "advance" the Kit 2 intention since three Part A's are required for it. However, a third Part A is not explicitly prohibited for Kit 1 as it is for Part D. Therefore, the probability that Kit 2 is the intention would increase but the probability for Kit 1 would not drop to zero since the addition of an third Part A is not explicitly forbidden.

One of the assumptions in many intention recognition systems is that there is a closed world. In other words, only the intentions that are being compared can happen and no other intentions can occur. In this work, we relax that assumption by ensuring that the intentions are progressing with each state observation. If we find that no tracked intention has increased the number of state relationships that are associated with it after a state observation, we lower the probability of all intentions and correspondingly increase the probability of an unknown intention category. In essence, we are saying that there is some intention being observed, but it does not match the intentions that we know about.

We capture this using a progress factor (PF), where the equation is shown:

$$\begin{cases} if \ (S_{i,t} - S_{i,t-1}) > 0 \ then \ PF_{i,t} = 1 \\ if \ (S_{i,t} - S_{i,t-1}) = 0 \ then \ PF_{i,t} = 0.8 * PF_{i,t-1} \quad (48) \\ if \ (S_{i,t} - S_{i,t-1}) < 0 \ then \ PF_{i,t} = 0.6 * PF_{i,t-1} \end{cases}$$

where $S_{i,t}$ represents the number of state relationships that are true at time t for intention i and $S_{i,t-1}$ represents the number of state relationships that are true at time $t-1$ for intention i. If at any time $PF_{i,t} > 1$, then it is set to 1.

The values of 0.6 and 0.8 allow for degradation in the confidence that the intention is occurring. These values can be set by the user to be any values that are deemed appropriate for the domain of interest. At this point, the probability of the ith intention would be:

$$(P_{i,t})_{temp} = PF_{i,t} F_{i,t} \frac{S_{i,t}}{\sum_{i=1}^{n} S_{i,t}} \quad (49)$$

The reason for the *temp* after the $P_{i,t}$ is because of the probability adjustment; the sum of the probabilities does not equal 100% and needs to be normalized. This can happen in one of two fashions. In the case where at least one intention retains a PF of 1 (in other words, at least one intention progresses since the last state), all probabilities are normalized to result in a total probability of 100%, as shown next:

$$P_{i,t} = \frac{(P_{i,t})_{temp}}{\sum_{1}^{n} (P_{i,t})_{temp}} \quad (50)$$

where n is the total number of intentions being matched.

In the case where no intentions retain a PF of 1 (in other words, no intention progresses since the last state), we assume that something is occurring in the environment that does not correspond to one of the predefined intentions that we are exploring. In this case, we continue to lower the probability of each known intention as shown in Equation 49, and assign the remaining probability that was deducted from the known intentions to a new "unknown" intention. As more time passes and no intention continues to be matched, the known intentions continue to get less probable and the unknown intention becomes more probable.

In Table 12, a new state has occurred in which a Part A was removed from the Part A Box, but not placed in either of the existing kit trays. Because this is a relevant state relationship, it is recorded in the state table but does not directly impact the two intentions we are tracking. As such, the intention table (Table 10) does not change. What this implies is that a new state occurred, but did not advance either of the two intentions. In this example, the probability of the Kit 1 intention becomes 0.9 ((progress factor) * 1 (forbidden state relations) * 5 (true state relationships)) / 10 (total true state relationships for all intentions) = 0.45 or 45%. Similarly for the Kit 2 intention, the probability becomes 45%. The remaining 10% is assigned to the unknown intention, since no intention progressed, thus implying that there is an intention going on in the environment that we are unaware of.

13. CONCLUSION

In this chapter, we present an approach to representing state information as the basis for intention recognition and then show how these states and their sequences can be used to associate probabilities with various intentions. The state representation is based on RCC8 and various cardinal directions along the x-, y-, and z- axes. Intention recognition is based upon a form of template matching that incorporates sequence information and state relationships that cannot occur. A progress metric is also introduced to decrease the probability that an intention is occurring if no progress has been made on it in recent states, which also allows for an open world assumption that intentions could be occurring in the environment that are not previously known. A small set of abstract state relationships is defined (e.g., under, on-top-of, contained-in) as the basis for defining domain-specific state relationships (e.g., worktable-empty, ktlocation, etc.). A simple example of two kit intentions is included to explain each step in the process.

State-based intention recognition offers some interesting advantages over activity-based recognition (which has traditionally been performed in the literature), including:

- States are often more easily recognizable by sensor systems than actions, as shown in Section 3.
- Using activities, intention recognition is often limited to inferring the intention of a single person. State-based intention recognition eliminates this shortfall, in that the state relationship is independent of who created it.
- State information is often more ubiquitous than activity information, thus allowing for reusability of the ontology.

Because of the similarity of state representation with activity representation, many of the same approaches that were described in the "Intention Recognition and State Representation Related Work" section can also be applied to this approach, which will also be the subject of future work. Additional future work will explore the association of weights with various state relationships. For example, if a state relationship exists in the environment that is only relevant to one intention, that state relationship should receive a higher weight since it is a strong indicator that the associated intention is occurring.

Table 12. Tabular state and state relationship representation (version 2)

State Relation	State 1 (Initial State)	State 2 (Kit Tray On Table)	State 3 (Part A In Kit Tray)	State 3 (2nd Part A In Kit Tray)	State 4 (Part B in Kit Tray)	State 5 (Part C In Kit Tray)	State 6 (Part A Removed From Part A Box)
Part A is in Part A Box	3	3	2	1	1	1	0
Part B is in Part B Box	2	2	2	2	1	1	1
Part C is in Part C Box	1	1	1	1	1	0	0
Part D is in Part D Box	1	1	1	1	1	1	1
Kit Tray is in Large Box With Empty Kit Trays	1	0	0	0	0	0	0
Kit Tray is on Table	0	1	1	1	1	1	1
Part A is in Kit Tray	0	0	1	2	2	2	2
Part B is in Kit Tray	0	0	0	0	1	1	1
Part C is in Kit Tray	0	0	0	0	0	1	1
Part D is in Kit Tray	0	0	0	0	0	0	0
Part E is in Kit Tray	0	0	0	0	0	0	0
Kit Tray in Completed Kit Box	0	0	0	0	0	0	0

ACKNOWLEDGMENT

Certain commercial software and tools are identified in this chapter in order to explain our research. Such identification does not imply recommendation or endorsement by the National Institute of Standards and Technology, nor does it imply that the software tools identified are necessarily the best available for the purpose.

REFERENCES

Albath, J., Leopold, J., Sabharwal, C., & Maglia, A. (2010). *RCC-3D: Qualitative spatial reasoning in 3D*. Paper presented at the 23rd International Conference on Computer Applications in Industry and Engineering (CAINE). Las Vegas, NV.

Balakirsky, S., Kootbally, Z., Schlenoff, C., Kramer, T., & Gupta, S. (2012). *An industrial robotic knowledge representation for kit building applications*. Paper presented at the International Robots and Systems (IROS) Conference Vilamoura. Algarve, Portugal.

Bateman, J., & Farrar, S. (2006). *Spatial ontology baseline version 2.0. OntoSpace Project Report - Spatial Cognition SFB/TR 8: I1* [OntoSpace]. University of Bremen.

Bodor, R., Jackson, B., & Papanikolopoulos, N. (2003). *Vision-based human tracking and activity recognition*. Paper presented at the 11th Mediterranean Conference on Control and Automation. Athens, Greece.

CCC. (2009). *A roadmap for US robotics: From internet to robotics*. Retrieved from http://www.us-robotics.us/reports/CCC Report.pdf

Chabrol, J. (1987). Industrial robot standardization at ISO. *Robotics*, *3*(2). doi:10.1016/0167-8493(87)90012-X.

Choudhury, T., & Borriello, G. (2008). *The mobile sensing platform: An embedded system for activity recognition*. IEEE Pervasive Magazine. doi:10.1109/MPRV.2008.39.

DARPA. (2012). *Information innovation office: Mind's eye program*. Retrieved from http://www.darpa.mil/Our_Work/I2O/Programs/Minds_Eye.aspx

Demolombe, R., Mara, A., & Fern, O. (2006). *Intention recognition in the situation calculus and probability theory frameworks*. Paper presented at the Computational Logic in Multi-Agent Systems (CLIMA) Conference. New York, NY.

Freksa, C. (1992). Using orientation information for qualitative spatial reasoning. In Frank, A. U., Campari, I., & Formentini, U. (Eds.), *Theories and methods of spatio-temporal reasoning in geographic space* (pp. 162–178). Heidelberg, Germany: Springer. doi:10.1007/3-540-55966-3_10.

Harmelen, F., & McGuiness, D. (2004). *OWL web ontology language overview*. Retrieved from http://www.w3.org/TR/2004/REC-owl-features-20040210/

Hoogs, A., & Perera, A. G. A. (2008). *Video activity recognition in the real world*. Paper presented at the American Association of Artificial Intelligence (AAAI) Conference. New York, NY.

Jarvis, P. A., Lunt, T. F., & Myers, K. L. (2005). Identifying terrorist activity with AI plan-recognition technology. *AI Magazine*, *26*(3), 9.

Jeon, H., Kim, T., & Choi, J. (2008). *Ontology-based user intention recognition for proactive planning of intelligent robot behavior*. Paper presented at the International Conference on Multimedia and Ubiquitous Engineering. Busan, Korea.

Kelley, R., Tavakkoli, A., King, C., Nicolescu, M., Nicolescu, M., & Bebis, G. (2008). *Understanding human intentions via hidden Markov models in autonomous mobile robots*. Paper presented at the 3rd ACM/IEEE International Conference on Human Robot Interaction. Amsterdam, The Netherlands.

Ligozat, G. (1993). Qualitative triangulation for spatial reasoning. In Campari, I., & Frank, A. U. (Eds.), *CPSIT 1993 (Vol. 716*, pp. 54–68). Heidelberg, Germany: Springer.

Mao, W., & Gratch, J. (2004). *A utility-based approach to intention recognition*. Paper presented at the AAMAS Workshop on Agent Tracking: Modeling Other Agents from Observations. New York, NY.

Martin, D., Burstein, M., Hobbs, J., Lassila, O., McDermott, D., McIlrath, S., et al. (2004). *OWL-S: Semantic markup of web services*. Retrieved from http://www.w3.org/Submission/OWL-S/

Marvel, J., Hong, T.-H., & Messina, E. (2012). *2011 solutions in perception challenge performance metrics and results*. Paper presented at the Performance Metrics for Intelligent Systems (PerMIS) Conference. College Park, Maryland.

Moratz, R., Dylla, F., & Frommberger, J. (2005). *A relative orientation algebra with adjustable granularity*. Paper presented at the Workshop on Agents in Real-Time and Dynamic Environments. Edinburgh, UK.

Mulder, F., & Voorbraak, F. (2003). A formal description of tactical plan recognition. *Information Fusion, 4*(1). doi:10.1016/S1566-2535(02)00102-1.

Nau, D., Ghallab, M., & Traverso, P. (2004). *Automated planning: Theory and practice*. San Francisco, CA: Morgan Kaufmann Publishers Inc..

Newman, M., & Balakirsky, S. (2011). Contests in China put next-generation robot technology to the test. *IEEE Robotics & Automation Magazine*. doi:10.1109/MRA.2011.942540 PMID:23028210.

Pereira, L. M., & Ahn, H. T. (2009). *Elder care via intention recognition and evolution prospection*. Paper presented at the 18th International Conference on Applications of Declarative Programming and Knowledge Management (INAP'09). Evora, Portugal.

Philipose, M., Fishkin, K., Perkowitz, M., Patterson, D., Hahnel, D., Fox, D., & Kautz, H. (2005). Inferring ADLs from interactions with objects. *IEEE Pervasive Computing / IEEE Computer Society [and] IEEE Communications Society*.

Randell, D., & Cui, Z. (1992). *A spatial logic based on regions and connection*. Paper presented at the 3rd International Conference on Representation and Reasoning. San Mateo, CA.

Ravi, N., Dandekar, N., Mysore, P., & Littman, M. (2005). *Activity recognition from accelerometer data*. Paper presented at the Seventeenth Conference on Innovative Applications of Artificial Intelligence (IAAI/AAAI). New York, NY.

Roy, P., Bouchard, B., Bouzouane, A., & Giroux, S. (2007). *A hybrid plan recognition model for Alzheimer's patients: Interleaved-erroneous dilemma*. Paper presented at the IEEE/WIC/ACM International Conference on Intelligent Agent Technology. New York, NY.

Sadri, F. (2011). Logic-based approaches to intention recognition. In Chong, N.-Y., & Mastrogiovanni, F. (Eds.), *Handbook of Research on Ambient Intelligence and Smart Environments: Trends and Perspectives* (pp. 346–375). Academic Press. doi:10.4018/978-1-61692-857-5.ch018.

Schlenoff, C. (2012a). *An approach to ontology-based intention recognition using state representations.* Paper presented at the Fourth International Conference on Knowledge Engineering and Ontology Development. Barcelona, Spain.

Schlenoff, C. (2012b). *An IEEE standard ontology for robotics and automation.* Paper presented at the International Conference on Intelligent Robots and Systems (IROS). Algarve, Portugal.

Schlieder, C. (1995). Reasoning about ordering. In Kuhn, W., & Frank, A. U. (Eds.), *COSIT* (Vol. 988, pp. 341–349). Heidelberg, Germany: Springer.

Schrempf, O., & Hanebeck, U. (2005). *A generic model for estimating user-intentions in human-robot cooperation.* Paper presented at the 2nd International Conference on Informatics in Control, Automation, and Robotics ICINCO 05. Barcelona, Spain.

Sukthanker, G., & Sycara, K. (2001). *Team-aware robotic demining agents for military simulation.* Paper presented at the Innovative Applications of Artificial Intelligence (IAAI). New York, NY.

Szabo, S., Norcross, R., & Shackleford, W. (2011). *Safety of human-robot collaboration systems project.* Retrieved from http://www.nist.gov/el/isd/ps/safhumrobcollsys.cfm

Tomasello, M., Carpenter, M., Call, K., Behne, T., & Moll, H. (2005). Understanding and sharing intentions: The origins of cultural cognition. *The Behavioral and Brain Sciences, 28,* 675–735. doi:10.1017/S0140525X05000129 PMID:16262930.

W3C_Member_Submission. (2004). *SWRL: A semantic web rule language combining OWL and RuleML.* Retrieved from http://www.w3.org/Submission/SWRL/

Wallgrun, J. O., Frommberger, L., Wolter, D., Dylla, F., & Freksa, C. (2006). Qualitative spatial representation and reasoning in the SparQ-toolbox. In Barkowsky, T., Knauff, M., Ligozat, G., & Montello, D. R. (Eds.), *Spatial Cognition V* (pp. 39–58). Springer.

Wolter, F., & Zakharyaschev, M. (2000). *Spatio-temporal representation and reasoning based on RCC-8.* Paper presented at the 7th Conference on Principles of Knowledge Representation and Reasoning (KR2000). Breckenridge, CO.

Youn, S.-J., & Oh, K.-W. (2007). Intention recognition using a graph representation. World Academy of Science, Engineering and Technology, 25.

ENDNOTES

[1] Kitting is the process in which several different, but related items are placed into a container and supplied together as a single unit (kit).

[2] Kitting is the process in which several different, but related items are placed into a container and supplied together as a single unit (kit).

[3] http://www.darpa.mil/Our_Work/I2O/Programs/Minds_Eye.aspx

Chapter 10
Smart Sensor Systems

Hiroo Wakaumi
Tokyo Metropolitan College of Industrial Technology, Japan

ABSTRACT

This chapter addresses smart sensor systems. In recent years, goods identification technology using a soft magnetic barcode, radio frequency identification, and automated wheelchair guidance technology using a magnetic field usable in dirty environments as part of Robotics and Mechatronics are becoming important in many areas, such as factories, physical distribution, office, security, etc. These identification and guidance technologies are based on sensing of magnetic field. Therefore, smart magnetic sensing technologies suitable for these identification and guidance techniques are described in this chapter.

1. INTRODUCTION

Optical paper-based barcodes are widely used in Point Of Sale (POS) systems for the management of foods, daily necessities, and goods management in clean factories. These commercial available systems are not usable for goods in dirty factories or in outdoor environments, where it is easy to be contaminated by oil, dust, and or mud. In 1990, researchers proposed a magnetic grooved barcode detection system using a magneto-resistive sensor usable in these dirty environments (Okabe & Wakaumi, 1990). This was an experimental system, in which the detection of a grooved barcode

on an iron plate was confirmed in a low detection height of 0.1 mm. Afterward, its detection performance was improved to practical levels by developing a new scanner head structure using the Tape-Automated-Bonding (TAB) technology and by developing a new MR sensor with a slant-element sensor structure for realizing highly sensitive scanners (Wakaumi, Komaoka, & Hankui, 2000; Wakaumi, Ajiki, Hankui, & Nagasawa, 2000). These systems have an ability of detecting a barcode pattern engraved on a soft magnetic substance. So, it enables us to manage by easily identifying goods contaminated by oil, dust, and/ or mud. In 2000, a highly sensitive GMR spin

DOI: 10.4018/978-1-4666-4225-6.ch010

valve sensor being currently used in hard disks has been also developed (Lenz & Edelstein, 2006). This device has an ability to realize highly stable magnetic scanners due to its high sensitivity in low magnetic fields.

Magnetic sensing technology using a magnetic field is now being used for ID tag detection. This magnetic sensing technology had been already developed for use in detection of a magnetic ferrite marker. Ferrite can be cheaply made as a sub-product of Mn-Zn core and a product generated at refinement of Fe/Ti. Ferrite marker is a solid block or soft tile of ferrite sub-product bound in place with asphalt, concrete, and resin etc. This ferrite marker is detected by a magnetic sensor consisting of a magnetic field generating exciting coil and detecting coils. As concrete examples of this technology, an automated wheelchair guidance and control system has been developed (Wakaumi, Nakamura, Matsumura, & Yamauchi, 1989). This system can be constructed at low cost because of using the ferrite sub-product. Since it is also extremely useful for protection of global environment, this system is worth to be described. In the era of 2000 after that, computer controlled wheelchairs using several kinds of sensors such as a panoramic camera, a sonar proximity sensor, and a laser range finder have been developed instead of the magnetically guided wheelchair, in which their representative is the Bremen Autonomous Wheelchair (Rofer & Lankenau, 2000). It is thought that these technologies are very important in the process of realizing fully automation control wheelchairs required in advancing aging society.

As a Radio Frequency Identification (RFID) technology, an electromagnetic induction technology using a magnetic wave of 13.56 MHz is known and is widely used. This technology is basically the same as an operation principle in magnetic sensing technology for detection of ferrite markers using the magnetic field. However, each sensor structure is different from each other. The former uses one detection unit unified an oscillating coil with its sensing coil for detecting magnetic distortion. Basically, transmitted magnetic waves

to ferrite sub-product are not modulated. The latter uses reader and tag receiver antenna coils for transmitting and receiving modulated electromagnetic waves. Transmitting of waves from the reader and receiver is performed in separate time periods using a half duplex mode or in the form of frequency-modulated signals within the same time period using a full duplex mode. Mainly, an amplitude modulation signal in a transmitter and receiver consisting of antenna coils is used in this system. This RFID identification system was developed to identify concrete-mixer vehicles to monitor the running status in 2003 (Nikkei Computer, 2003; RFID Technology Editorial Department, 2004). Presently, this technology is used for identification of personal information and goods such as products.

Thus, sensing technologies using magnetic field sensors of MR or magnetic coil and identification technologies using magnetic waves are very useful for identification or automated vehicle guidance in dirty environments and so these are ranked to important technologies. In recent years, goods identification using a barcode and automated wheelchair guidance technologies in dirty environments as part of Robotics and Mechatronics are becoming important in many areas such as factories, physical distribution, office, security etc. So, smart magnetic sensing technologies suitable for these identification and guidance are mainly described in this chapter.

In this chapter, barcode pattern sensing techniques engraved on a soft magnetic substance using an MR magnetic sensor are presented. A GMR spin valve sensor which has an ability to achieve highly sensitive scanners is also introduced. Ferrite sensing techniques using ferrite markers and magnetic waves are further described. As relatively new rising control technologies of wheelchairs, the trendy technologies of computer controlled autonomous wheelchairs are additionally introduced. Finally, the principle and utilized status of electromagnetic induction technology using magnetic waves of 13.56 MHz in current use are described. Its application feasibility is also added.

2. MR MAGNETIC SENSING

A magnetic grooved barcode detection system consists of a grooved barcode engraved on a soft magnetic substance such as an iron plate and a pen type scanner with a Magneto-Resistive (MR) sensor (made up of NiFe thin film) and a magnet. It features that a magnetic barcode covered with a non-magnetic metal (e.g. Al, Cu) and dirty substances such as oil, dust and /or mud, can be detected magnetically by a sensor. This sensor system was proposed in 1990 and the basic operation of its prototype was examined (Okabe & Wakaumi, 1990). When a magnet is placed above a groove on the iron plate, magnetic flux density distribution near groove is shown in Figure 1. The Bx component in the magnetic flux density parallel to the iron plate shows a positive or negative peak near the edge of the groove. This means that a groove pattern is reconstructed by detecting the points where the component shows the peak level. Figure 2 shows the calculated groove depth dependence of dBx at a height of 1 mm (simulated

Figure 1. Magnetic flux density distribution near groove

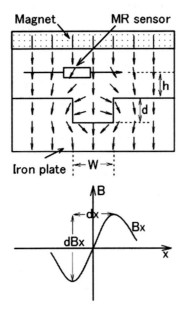

Figure 2. Magnetic flux density change vs. groove depth (W=1 mm, h=1 mm, magnet magnetization is 10 kG)

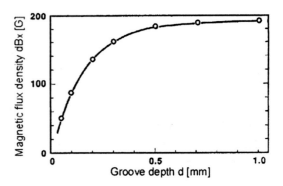

by the FEM program). Because the magnetic flux density change is saturated at a depth of 0.5 mm, this shows that one-half of the groove width is enough for the groove depth. Though the magnetic flux density change decreases linearly with height near the groove, the peak distance dx keeps constant within a height of 0.5 mm above the groove (Figure 3). As a test result of experimental apparatus, a barcode consisting of different groove pattern of 0.3mm wide–0.3mm deep and 0.8mm wide–0.1mm deep grooves has been recognized up to a 0.1 mm detection height.

This experimental apparatus has a problem of short detection height. In order to increase the detection height, a highly-sensitive magnetic barcode detection system with a TAB head scan-

Figure 3. Magnetic flux density peak distance vs. height above iron plate (W=1 mm, d=1 mm)

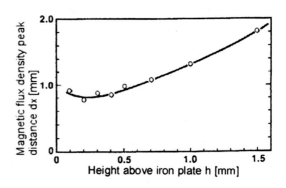

ner was developed (Wakaumi, Komaoka, & Hankui, 2000). In this system, a tape automated bonding technique was used to connect an MR sensor (installed at the scanner head) and inner leads through bumps (Figure 4). By this technique, the gap between the scanner head tip and the MR sensor is set narrow to enable highly sensitive detection. Because the inner lead consisting of a flat sheet covers the bumps completely, the connective sections are hard to be destroyed by external pressure. Thus, this TAB head scanner enables highly sensitive barcode pattern recognition, while still retaining high endurance strength.

The performance of the TAB head scanner prototype was tested using two different groove depth structure barcodes. Figure 5 shows a model of magnetization of MR sensor. By using the TAB head structure, the surface magnetic field at the sensor's scanning over the iron plate becomes larger due to the effectively lowering the height of MR sensor. Test results showed that the TAB

head structure is effective for recognizing a high-density barcode at a detection height of 0.15 mm higher than in a conventional scanner. Figure 6 shows a recognition rate per 50 times scanning versus the detection height. Barcodes used are a barcode with two different groove depths 2GD of wide shallow grooves and narrow deep ones (d_1=0.3 mm and d_2=0.1 mm) and one with a single groove depth SGD (d_1=d_2=0.3 mm). The detection height for the barcode with two different groove depths increases over an amount of nearly 0.05 mm higher than for barcode with the single groove depth. Thus, we can see that the barcode with two-different groove depths is effective in recognizing high-density barcode patterns. Also, the effect of contaminating substances was investigated using cutting oil (uniway 68 with color of union 7) and grinding oil (with color of nearly black). When these oils were attached on the barcode with a SGD consisting of widths of 0.5 mm and 1.3 mm, and a depth of 0.3 mm, the recognition rate is shown compared with that for an optical barcode detection system in Table 1. When the cutting oil was attached on each barcode surface, the recognition rates for both systems were almost the same. However, when black grinding oil (dot type 0.8 mm φ) was attached to the optical barcode, the system could not recognize the barcode. The grooved barcode detection system with the TAB head scanner was able to recognize this barcode with the oil. Thus,

Figure 4. Head configuration in TAB head scanner

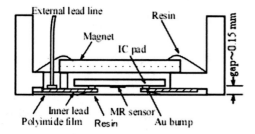

Figure 5. Magnetization change of MR sensor

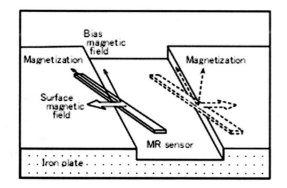

Figure 6. Recognition rate per 50 times scanning versus detection height. A barcode with SGD of d_1=d_2=0.3 mm and one with 2GD of d_1=0.3 mm and d_2=0.1 mm were used.

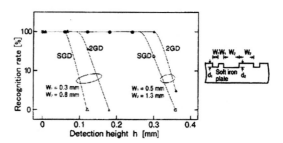

Table 1. Recognition rate for optical scanner and groove barcode detection system with TAB head scanner

Attached material on barcode pattern	Optical scanner	TAB head scanner
Cutting oil	84 %	100 %
Grinding oil	0 %	100 %

Figure 8. Photograph of fabricated slant element MR sensor

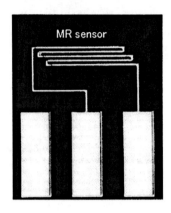

this magnetic barcode detection system is immune to stain. Therefore, it is expected that this system will be available for goods management in dirty environments.

To increase the detection height important in realizing a practical barcode scanner, another approach for an increase of MR sensor sensitivity was considered. That is, a Slant-Element MR Sensor (SEMS) consisting of some slant MR sensor elements with a slant-angle of 45 degrees arranged in parallel was proposed to achieve highly sensitive detection (Wakaumi, Ajiki, Hankui, & Nagasawa, 2000). Figure 7 shows the SEMS structure. Figure 8 also shows a fabricated slant MR sensor. Table 2 shows the comparison of barcode recognition characteristics. The scanner

Table 2. Comparison of barcode recognition characteristics. Scanner: TAB head scanner. Barcode: NW7 patterns of W1=0.5 mm, W2=1.3 mm. h: distance between film and barcode plate.

Barcode pattern	Single MR sensor	Slant element MR sensor
2GD with $d_1=0.3$ mm, $d_2=0.1$ mm	$h \leqq 0.3$ mm	$h \leqq 0.52$ mm
SGD with $d_1=d_2=0.3$ mm,	$h \leqq 0.25$ mm	$h \leqq 0.42$ mm

can detect at a detection height h of 0.42 mm, which is nearly 0.15 mm higher than that in the conventional single MR sensor. The SEMS enables the increase in detection height by nearly 0.2 mm when using the 2GD barcode. Thus, the SEMS is effective to achieve highly sensitive detection due to detection sensitivity increase.

In 2000, a Giant Magneto-Resistance (GMR) sensor which enables us to apply for highly sensitive read heads in magnetic recording was developed and has become commercially available in 2006 (Lenz & Edelstein, 2006; Noetzel, Meisenberg, & Bartos, 2006). Especially, GMR spin-valves are used in hard disk read heads. The GMR spin valve sensor consists of two ferromagnetic layers separated by a conducting layer (Cu) whereas one of the ferromagnetic layers is mag-

Figure 7. Slant element MR sensor structure

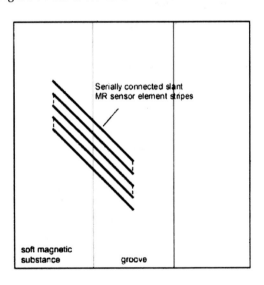

netically coupled to an anti-ferromagnet (such as PtMn or IrMn) (Figure 9). The coupled ferromagnet is magnetically pinned whereas the other layer can follow an external magnetic field freely. Low magnetic fields may turn only the free-layer magnetization and thus result in a GMR resistance. The minimum resistance occurs when the magnetizations of pinned and free layer are parallel. On the other hand, the maximum resistance occurs when they are anti-parallel. This sensor has a sensitivity even in 0.4-4 kA/m (5 – 50 Gauss) of a 1/40-1/20 lower magnetic field than conventional MR sensors. Therefore, if the GMR senor is used for the TAB head scanner, it is predicted that its detection height will be increased up to about 6 mm considering the characteristic of magnetic flux density change in proportion to the square of distance. That is, highly stable barcode detection scanners should be realized by means of application of such technology.

3. ELECTROMAGNETIC FERRITE SENSING

A unique magnetic sensing technology using a ferrite marker is presented. Ferrite can be cheaply made as a sub-product of Mn-Zn core and a product generated at refinement of Fe/Ti. Ferrite marker is a solid block or soft tile of ferrite sub-product bound in place with asphalt, concrete, and resin etc. This ferrite marker is detected by a magnetic sensor consisting of a magnetic field generating exciting coil and detecting coils. As applications of this sensing technology, the following wheelchair

Figure 9. GMR spin valve sensor structure

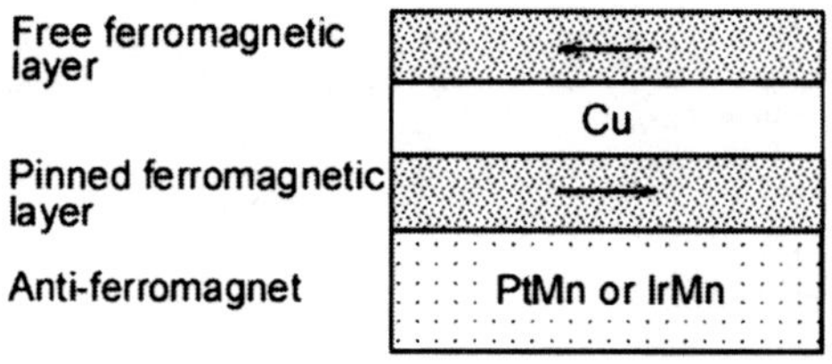

Free ferromagnetic layer	
	Cu
Pinned ferromagnetic layer	
Anti-ferromagnet	PtMn or IrMn

guidance system is presented here. This system can be constructed at low cost because of using the ferrite sub-product.

3.1. Automated Wheelchair Guidance using a Ferrite Marker

Until now, many kinds of hand-propelled wheelchairs and electrically controlled wheelchairs have been developed. Hand-propelled wheelchairs do not permit aged riders an extensive range of travel, because they must be moved by the occupant's arms and are tiring to use. Though electrically controlled wheelchairs are usually guided by manually operated joysticks, driving the wheelchairs require the operator to be skilled in turning and in direction change operations in places such as curved roads. Therefore, it is difficult for severely handicapped and very old persons to operate them skillfully. Under such situations, a unique guiding technology with a magnetic sensor was developed. Here, an automated wheelchair, whose movement is guided by a magnetic ferrite marker, is described (Wakaumi, Nakamura, Matsumura, & Yamauchi 1989; Wakaumi, Nakamura, & Matsumura, 1989; Wakaumi, Nakamura, & Matsumura, 1992). This wheelchair permits easy use by severely handicapped persons and old people, because of its simple operation. Figure 10 shows a configuration of an automated guided wheelchair

Figure 10. Automated guided wheelchair configuration (Wakaumi, Nakamura, & Matsumura, 1989)

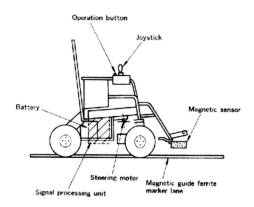

with a magnetic exciting unit and sensor guided by ferrite marker lanes, which are laid in/on the sidewalk or floor. The wheelchair is partially modified by installing the magnetic exciting and sensor unit, signal processing circuit and operation button. The magnetic exciting and sensor unit is installed at the front of wheelchair to control the steering wheels. The sensor, which is 7 cm away from the road surface or floor, picks up guidance signals from ferrite markers. The sensor unit is 25 cm wide, 15 cm deep, and 10 cm high. The ferrite marker lane, using soft ferrite material bound in place with resin, is 10 cm wide and 5 mm thick (Magnetization μ=5~8).

Figure 11 shows a magnetic ferrite marker sensing system configuration. The magnetic sensor consists of an exciting coil L at the center of

the sensor unit, detecting coils L_1 and L_2, placed on its left and right sides, and signal processing circuit (Figure 12). The distance between the centers of two detecting coils is 128 mm. The distance between the centers of exciting coil and detecting coils is 61 mm. The exciting coil generates a magnetic field by driving at an exciting frequency of 40 kHz, which is lower than those in RFID systems to be described later. The ferrite marker is magnetized by this field and sets up a different resonant magnetic field. As a result, the original magnetic field is deviated. The detecting coils detect this deviated magnetic field deviation caused by the magnetic ferrite markers. The detected output signals obtained by detecting coils, v_{L1} and v_{L2}, reach a peak level and then reduce to a small value as the sensor deviates further (Figure 13). The output signals are subtracted from each other to obtain an S-shaped characteristic, suitable for use as a steering control signal. Because the difference between these detection signals is proportional to the wheelchair deviation from the center of the marker, this difference signal permits it controlling the wheelchair's steering. For example, when the sensor output signal voltage is increasing to a higher level, a controller for controlling the steering motor rotation direction permits it to rotate the forward wheels in a direction to bring the sensor unit position back to the marker lane center. On the other hand, when the sensor output voltage is

Figure 11. Magnetic ferrite marker sensing system configuration (Wakaumi, Nakamura, & Matsumura, 1989)

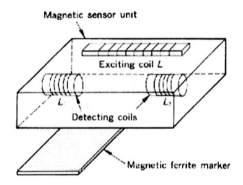

Figure 12. Signal processing circuit in the exciting and detecting coils (Wakaumi, Nakamura, & Matsumura, 1989)

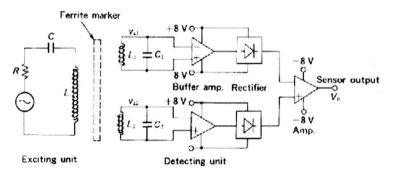

Figure 13. Detecting coil and sensor output voltage versus sensor position. Ferrite marker width: 10 cm, ferrite marker thickness: 5 mm (Wakaumi, Nakamura, & Matsumura, 1989).

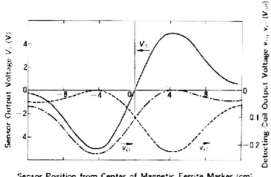

decreasing to a lower negative level, the controller permits the forward wheels to rotate in the opposite direction. Thus, the wheelchair can be controlled nearly along the center of the marker lane.

3.1.1. Techniques for High-Level Performance

Several considerations for safety, comfort, and convenience are required to realize a high-level performance wheelchair system.

3.1.1.1. Smooth Running Operation for Comfort

To realize smooth running operation, partially steering-free operation by a nonlinear circuit and a pulse steering drive method have been developed. A non-linear signal processing circuit as shown in Figure 14 has been realized to obtain steering-free operation within a small deviation range, preventing its oscillation operation. The small deviation range means a small area along the center of the ferrite marker lane. This circuit consists of an amplifier and several diodes to shift nearly 1-2 V of the sensor output voltage point when a steering control voltage rises.

Figure 14. Nonlinear signal processing circuit configuration (Wakaumi, Nakamura, & Matsumura, 1989)

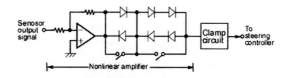

Also, a pulse drive method has been developed as an approach of controlling steering to minimize a zigzag running to the small extent, which tends to occur when the inertia movement of the wheel movement is too large in its movement direction after the steering angle rises sharply at the starting point of steering control (Wakaumi, Nakamura, & Matsumura, 1989; Wakaumi, Nakamura, & Matsumura, 1992). The pulse drive method was used within a small range of steering wheel deviation. In this drive method, by providing intermittent drive pulses of 10 Hz (with from 25% to 10% duty range) added to the sensor output signal as shown in Figure 15, steering is gradually bent from a little lower sensor output voltage than that required for the steering-free limit (Wakaumi, Nakamura, & Matsumura,1989; Wakaumi, Nakamura, & Matsumura, 1992). As gradual steering control is achieved through this drive, a delicate steering wheel angle control is possible (does not cause zigzag running).

3.1.1.2. Automatic Stop Operation for Convenience

To permit the wheelchair to stop automatically at the desired destinations, an infrared position sensor system has been installed (Figure 16). The

Figure 15. Pulse steering drive method, compared with conventional drive method (Wakaumi, Nakamura, & Matsumura, 1989)

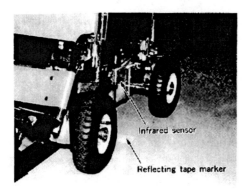

sensor detects a reflecting tape previously set up at the destination. Road sign reflecting material was used for the reflecting tape as providing a different reflection coefficient, compared with that for the surrounding floor or pavement surface. When the position sensor detects this reflecting tape, it automatically stops the wheelchair movement.

3.1.1.3. Emergency Stop Operation for Safety

Maintaining safety is important for every user of automated wheelchairs. Two infrared obstacle detection sensors have been constructed on the front of the wheelchair to maintain safety (preventing collision with people, chairs, animals, etc.). When an obstacle appears in front of the wheelchair, it stops temporarily, after detecting the obstacle, and starts running after the obstacle is moved out of the wheelchair's path.

3.1.2. Experiment Results

Running experiments of automated wheelchair with a riding adult man (of normal weight) have been carried out along a corridor with a soft ferrite marker on the floor. The infrared position

detection sensor was installed on the left side of the wheelchair about 25 cm above the floor. The reflecting tape marker was laid on the floor. Figure 17 shows steering angle versus sensor output voltage. Circuit 1 and 2 having different gains and diodes with non-acting zones in the nonlinear circuit have given ±0.9 V and ±1.4 V steering-free operation voltage area as a sensor output voltage. The wheelchair did not oscillate in either case because steering is not controlled within these areas. However, it showed zigzag running when the gradual steering control was not used.

The gradual steering control scheme gave steering angle change, gradually changed from nearly ±0.5 V. The wheelchair with this control did not cause overrun, that is, zigzag running. It successfully ran along the ferrite marker lane at less than 2.2 km/h (walking speed), even when carrying a man weighing 80 kg. At higher speeds, slight zigzag running phenomena have been observed.

When an obstacle (a man) appeared nearly 1 m in front of the wheelchair (in this case, the detection range of the infrared sensor has been set up at about 1 m), it came to a stop safely. In this way, the safety function has been confirmed.

Figure 17. Steering angle versus sensor output voltage, determined by a nonlinear circuit and a pulse steering drive method (Wakaumi, Nakamura, & Matsumura, 1989)

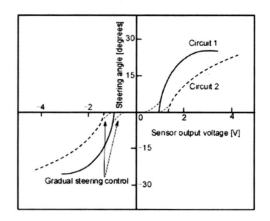

Furthermore, the wheelchair stopped at a predetermined position by detecting the reflection tape marker. When there are some destinations, users can move to the desired destination by counting the number of reflection tape markers, when the wheelchair passes.

Thus, the wheelchair could move without zigzag at less than 2.2 km/h walking speed, using simple operation, and automatically stopped at the desired destination. Furthermore, it accurately detected obstacles and prevented safely collision. Therefore, this wheelchair allows handicapped and aged persons to move about merely by pushing a button easily, while keeping safety, comfort and convenience. Since the ferrite marker can be used without any influence from pollution or small non-magnetic materials, this kind of marker lane allows us to use both inside and outside. This simply operated wheelchair, using such a marker, should be a useful tool for increasing the activity range for both the physically handicapped and the elderly.

3.2. Burst Excitation Method

Increasing the magnetic field intensity is important for realizing highly sensitive sensing in electromagnetic ferrite sensing systems. The magnetic field intensity is measured by an exciting voltage V_E in the AC exciting type system. The exciting voltage V_E depends on the resonant frequency f_E in the LCR serial resonant exciting system as shown in Figure 18. This shows that a resonance frequency of 100 kHz is better than the conven-

tional frequency of 40 kHz to realize the higher exciting voltage, because it is near the resonant frequency. It is also seen that doubling the resonance frequency is better to obtain a high detecting coil output voltage for the system (Figure 19). However, the power dissipation of the resonance power supply is doubled as the f_E is doubled. As a method to eliminate this power loss, the burst resonance method was proposed (Wakaumi & Yokoyama, 1988). As an experimental result, when the duty ratio of the burst excitation wave was settled to 1/8, the power dissipation was reduced to 60% than the conventional continuous value. Thus, optimum control of the resonance frequency with the burst resonance method is effective for realizing a highly sensitive low-power sensing system.

3.3. Application for Golf Cart

As an application example of magnetic ferrite sensing, a golf cart guidance system was developed. An automated guided golf cart with a magnetic exciting unit and detection sensors is guided along ferrite marker lanes which are laid in/on the golf course. This cart substitutes for a caddy. Basic operation principle is the same as the automated wheelchair mentioned previously.

Figure 19. Detecting coil output voltage vs. resonance frequency. Sensor height against ferrite marker is 7 cm. Ferrite marker width is 10 cm. Ferrite marker thickness is 5 mm.

Figure 18. Exciting voltage vs. resonance frequency. Coils are enclosed on Al box. L=4.94 mH.

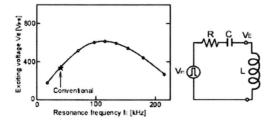

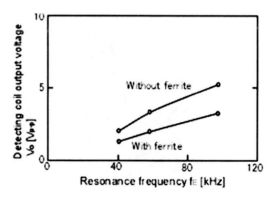

4. AUTONOMOUS WHEELCHAIR GUIDANCE BY NON-MAGNETIC SENSING TECHNOLOGIES

Since 1989, there has not been any development of autonomous wheelchairs with magnetic sensing technology. Though a human-oriented operation wheelchair with functions of an obstacle avoidance using fuzzy control based on skilled operator's operation record and supervision of environment using the CCD video camera placed on a building's ceilings was proposed to achieve safe operation in 1995 (Sakai & Kitazawa, 1995), the system is just in a simulation phase.

Afterward, computer controlled wheelchairs using various kinds of sensors appeared instead of the magnetically guided wheelchair. As one of these ones, an autonomous wheelchair equipped with a control PC, 27 sonar proximity sensors, and a laser range finder behind the seat (called as the Bremen Autonomous Wheelchair "Rolland") (Figure 20), has been developed for use as a rehabilitation service robot in 2000 (Rofer & Lankenau, 2000; Lankenau, 2001). This is controlled by its human operator and an automated safety module. This wheelchair's algorithm works in unchanged environments and provides a sufficient precision to its wheelchair navigation in large building complexes and outdoor scenarios. Several modules with functions of obstacle avoid-

Figure 20. Bremen autonomous wheelchair "Rolland"

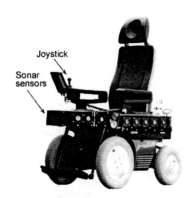

ance (smoothy detouring around objects in the pass of the wheelchair), assistance for passing the doorway, behavior-based travelling (wall following, turning on the spot, etc), provide the driver with various levels of support for speed control and for steering. These modules permit the wheelchair to avert collision with obstacles and facilitate difficult maneuvers. If the system carries out a basic automation behavior, such as obstacle avoidance or wall following or turning on the spot, it ignores the position of the joystick completely. After the obstacle avoidance in the automation mode, a case is apt to occur that the driver does not react immediately. At this time, the driver does not adapt to the new situation of the operator control mode. As a result, the wheelchair would follow the wrong track. Thus, there is a case that the human operator cannot track the behavior of the automation after the obstacle avoidance. Therefore, to increase the safety, a speech module that the system indicates changes will be required as the human-machine interface.

In addition to the previously one, an autonomous robot wheelchair with a PC controller and proximity ultrasonic and orientation encoder sensors, has been developed for indoor navigation use (Sgouros, 2002). This wheelchair uses for guidance the Qmap consisting of grid cells such as the floor plan of a building made off-line. The navigation system guides the wheelchair according to the planning course on the process map. The wheelchair is equipped with a ring of eight proximity sensors, which is used for Qmap construction and positioning. In the fully autonomous test mode, the wheelchair was instructed to navigate automatically to a specific Qmap point from a known starting position and orientation with an average speed of 0.2 m/sec. Sensors had a minimum sensing distance of 32 cm. As a test result in home environment with average complexity (Figure 21a), the wheelchair successfully navigated between points occluded by static obstacles. In environments involving points in adjacent rooms (Figure 21b), a 60% success in a total of 15 trials

Figure 21. Wheelchair trajectories in the arenas with average complexity (a) and with above average complexity (b)

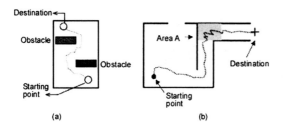

was achieved. Most of the time, failure was caused in area A. This is because the spatial resolution of the chair was limited in the front direction during closed turns due to the limited accuracy of the scanning sonars. In all of the failures in area A, the wheelchair followed a turning trajectory that led to a collision with one of the side walls without attaining a minimum distance of 32 cm from these walls (the sonars could not detect the obstacle).

In 2002, a semi-autonomous wheelchair, which is navigated with a panoramic camera for following a moving target and sonars to measure its distance to target for avoiding obstacles, has been developed (Argyros, Georgiadis, Trahanias, & Tsakiris, 2002). The camera can process the color images using the color information and the orientation of the target. In the person-following behavior, the orientation of the moving person target with respect to the wheelchair based on color information from the panoramic images is computed. Sonar data provide the distance between the wheelchair and the moving target for the controller. This algorithm identifies an object by comparing person color features to the color features of objects in a database. The wheelchair with 6 sonars and the panoramic vision camera with a 360° fields of view robustly operated avoiding obstacles at a specified direction under moderate lightning conditions. However, it has still a problem that when significant variations in lightning occur during the movement, the color-based visual tracking may lose the target or confuse it with another one.

5. IDENTIFICATION TECHNOLOGY USING RADIO-FREQUENCY ELECTROMAGNETIC INDUCTION

RFID is an identification system consisting of a reader and an IC tag for communicating each other using the radio frequency electromagnetic wave. 13.56 MHz radio frequency systems are widely used for applications on electronic car keys, electronic tickets, and in/out management office cards, etc., due to the properties of the relatively high-immunity against metals and water. Basically this system uses an electromagnetic induction method with loop coils in the reader/tag. The IC tag configuration is standardized by ISO and IEC (Table 3) (Usami & Yamada, 2005). Proximity coupling IC cards with a detection distance shorter than 10 cm and vicinity coupling IC cards with a detection distance shorter than 70 cm are provided by ISO/IEC 14443 and ISO/IEC15693, respectively. On the contrary, IC tag itself is standardized by ISO/IEC 18000-3. Table 4 shows its major technical specifications. Mode

Table 3. IC tag standards of ISO/IEC

Classification	Standard	Specifications
Card type	ISO/IEC 14443	13.56 MHz Proximity type (Communication distance: 10 cm)
	ISO/IEC 15693	13.56 MHz Vicinity type (Communication distance: 70 cm)
Tag type	ISO/IEC 18000-3	13.56 MHz

Table 4. Major standard specifications of ISO/IEC18000-3

Specification		Mode1 (ISO/IEC15693-2)	Mode2
RF tag		Passive	
Communication from reader to RF tag	Carrier frequency	13.56 MHz±7 kHz	
	Communication speed	26.48kbit/sec or 1.65kbit/sec	423.75kbit/sec
	Encoding method	PPM	Double Frequency Modified Frequency Modulation
	Modulation method	ASK100% or 10%	Phase Jitter Modulation
Communication from RF tag to reader	Subcarrier frequency	423.75kHz or 423.75kHz・484.28kHz	969/1233/1507/1808/2086/2465/2712/3013kHz
	Communication speed	26.48kbit/sec・6.62kbit/sec or 26.69kbit/sec・6.67kbit/sec	106kbit/sec × 8channel (848kbit/sec)
	Modulation method	ASK・FSK	BPSK
	Encoding method	Manchester	Modified Frequency Modulation
Anticollision		Time-slot method	Frequency and Time Division Multiple Access method

1 is almost the same as ISO/IEC 15693. Mode 2 has the properties of high communication speed of maximum 848 kbs. Here, IC card's basic system configuration and operation principles are mainly introduced.

5.1. Radio Wave Transmission

The transmission characteristic of electromagnetic wave greatly varies depending on the distance from the generating source. Within the vicinity of the source, that wave is strongly connected by the magnetic field. On the contrary, away from the source, that wave is strongly connected by the electric field. The distances within $\lambda/2\pi$ (λ represents a wave length) from the source is called as a Near filed area. Electromagnetic wave transmission in RFID tags using 13.56 MHz is done in this area. The magnetic field H in the distance d from the generating source is described as follows (Figure 22).

$$H = n\ r^2 I\ /\ [2\sqrt{(r^2+d^2)^3}] \quad (1)$$

Here, r, n, and I represent coil radius, winding number of coil, and flowing current, respectively. When the distance d is over r, H suddenly decreases in inversely proportional to d. This area is called as a far field region. Therefore, passive tags using the RF of 13.56 MHz can be used only

Figure 22. Magnetic field strength vs. distance

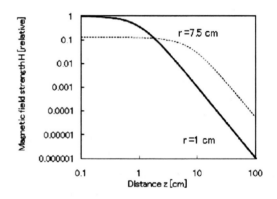

within the near field area. An operation area of IC cards is determined by the magnetic field H generating a minimum voltage in the card.

5.2. RFID System Configuration and Data Transmission Method

Figure 23 shows a passive tag's operation principle block diagram (Kishigami, 2005). Digital signals in the reader are modulated and unnecessary signals filtered out. After the signal modulated by a carrier is amplified, it is emitted as an electromagnetic wave including some commands from the reader antenna. An antenna of the reader sends its magnetic energy required for an IC tag's operation and modulated data to the IC tag. The

Figure 23. Passive tag's operation principle block diagram

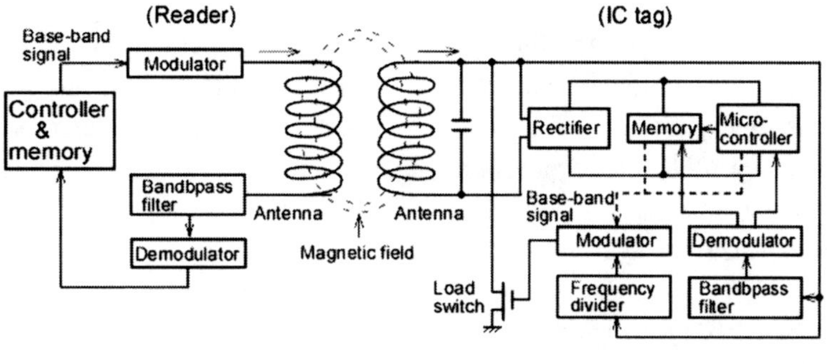

IC tag transforms its magnetic wave energy to a DC power supply voltage using a rectifier and provides its DC power supply voltage to demodulator, memory, micro-controller (including a congestion control circuit), modulator and frequency divider. At the same time, just signals within the specified bandwidth are extracted in the filter and then demodulated employing a demodulation circuit. These signals are sent to the micro-controller and re-encoded. According to some command signals, transmitted signals are written in the memory and subsequently emit the Unique ID (UID) written in the memory to the reader. This UID has usually 16 or 32 bit information. Base-band signals representing data saved in the tag, modulate the carrier signal sent from the reader. After the frequency of modulated signals is changed, those signals are emitted from the tag loop coil to the reader. In the reader, carrier signals are filtered out and only signals from the tag are extracted. These ones are changed to base-band signals and re-encoded.

There are two methods for transmitting the data from the IC tag shown as follows.

1. **Half Duplex Communication Mode:** After the electromagnetic energy from the reader was stored in the tag, the modulation signal is transmitted to the reader (Kaiser & Steinhagen, 1995).

2. **Full Duplex Communication Mode:** Data transmission and reception between the reader and the tag are simultaneously done.

In the half duplex communication, energy provision from the reader stops when the tag transmits signals to the reader. In this method, some measure storing an electric power in the tag is needed.

In the full duplex communication mode, usually, the frequency converted signal consisting of a frequency different from the reader carrier frequency is transmitted from the tag. Using the frequency divider, the carrier signal fc of 13.56 MHz from the reader is converted to the frequency signal fm =423.75 kHz of 1/32 as an example. When this divided signal is used for modulation signals, transmission signals of 13.13625 MHz and 13.98375 MHz from the tag can be used. In the transmission of signals, the load modulation method is used. A load switch FET (without or with a resistor connected in series depending on the modulation mode) is loaded parallel to the loop coil constructing an oscillating circuit in the tag (Figure 23). When this load is turned on, the antimagnetic field (magnetic flux density change) occurs in the coil of the tag due to this load current. The reader detects this antimagnetic field through coil current change flowing the monitoring resistor and can consequently receive tag data signals by turning FET's on and off.

5.3. RF Communication Scheme

Digital code data in the RFID system are modulated using the carrier and then transmitted through the antenna. Factors to be superposed on RF signals are three parameters of amplitude, frequency, and phase. Modulation methods using these factors are known as follows (Yoshioka, 2004).

5.3.1. Amplitude Shift Keying (ASK)

In this modulation method, a modulation signal of base-band (digital data signal) modulates the amplitude of carrier according to "0" or "1" (Figure 24). The modulated signal is described in the following.

$$X(t)=A(t)\cos(2\pi f_c \cdot t) \qquad (2)$$

Here, A(t) represents a modulation signal (0 or 1). In the load modulation method, 80-90% of the full amplitude is kept to continuously supply the electromagnetic energy to the tag. A modulation degree m is smaller than 100%. In this method, the ASK is often used because the system changes an antenna load using the load switch. When a 100% on/off switching of the carrier is made (ASK100%), the power saved on a large capacitor is used within its off period ("0").

5.3.2. Frequency Shift Keying (FSK)

In the FSK method, one of two frequency carriers is allocated to "0" and the other to "1". The modulation signal is described as follows.

$$X(t)=A(t)\cos(2\pi f_{c1} \cdot t)+A(t)\cos(2\pi f_{c2} \cdot t) \qquad (3)$$

This method is used for transmission from the card in the vicinity coupling IC card system because a small level of carrier energy is not limited by the radio wave regulation.

5.3.3. Phase Shift Keying (PSK)

In this method, different phase carrier signals are allocated to "0 or "1" of the modulation signal. The carrier with the phase of 0° and one with 180° correspond to "0" or "1" of base-band digital signals (It is called as a Binary Phase Shift Keying (BPSK). In this method, 2 bit (quaternary) data can be transmitted at once by using the phase shift carrier signals of 0°, 90°, 180°, and 270°. This is called as a Quadrature Phase Shift Keying (QPSK).

The BPSK method using a sub-carrier is widely used for the non-contact IC card and RFID. This sub-carrier is created by dividing the carrier by 1/16 or 1/32 and used for the data transmission from the card (Figure 25) (Karibe, 2008). In the non-contact IC Card, the sub-carrier of 847 kHz is used. At first, a BPSK coded signal (0°/180°) is created by modulating the sub-carrier using base

Figure 24. Modulation methods

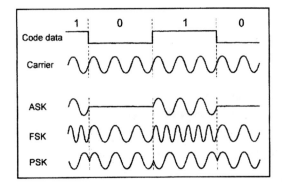

Figure 25. BPSK using the subcarrier

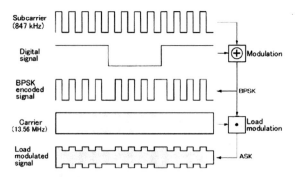

band digital data. Next, this BPSK coded signal modulates the carrier in the ASK method turning on/off the load FET. In this case, data information is transmitted on the sub-bands of the sub-carrier.

Digital data are encoded for transmission and store of information. Encoding methods of RZ, NRZ, Manchester, Mirror, Pulse interval encoding, Pulse position encoding are known (Figure 26) (Yoshioka, 2004). In the NRZ encoding, a previous state of high level ("1") or low level ("0") is kept until the next state changes. This encoding method is used for FSK and PSK.

In the RZ encoding, after the logic state transitions to a high level of "1", its level is returned to the previous low "0" level during a bit period. In the logic "0" state, pulses are not generated. Because the bandwidth of side waves is great, the modulation intensity is necessary to be reduced.

In the Manchester encoding, a logic "1" changes to negative level in the center of the bit period while a bit "0" changes to positive level "1" in the center of the bit period. In this encoding, the mean voltage within a bit period is 0 V and so any DC bias is not generated.

In the Mirror encoding, when the "1" appears, the state changes from "H" to "L" or from "L to "H" in the center of a bit period, while there is no change when "0" after the state of "1" continues. The state changes at the start of a bit period after the state of "0". It has a bandwidth narrower than the Manchester encoding. There is also a Modified

mirror encoding. In this method, negative pulses generated by differentiating the transition of the signal, coded by the mirror encoding, is used. This is useful for transmission from the reader, because the stored energy is saved even in the ASK100% operation, by using pulses of narrow width.

In the Pulse interval encoding, the pulse width of the initial start signal is used as a reference. When the time period between the subsequently appearing pulse and the initial start signal is longer than the reference pulse width, its logic state is acknowledged as "1". When the time period is shorter than the reference pulse width, its logic state is acknowledged as "0". In this encoding, it features that it is able to settle the transmission speed arbitrarily and continuously. However, it has a defect that the total transmission time period changes depending upon the number and distribution of bits.

In the Pulse position encoding, a pulse is generated in the nth position from the frame start signal (SOF). If one bit delay time is assumed to be 10 µs in the 1/256 position method using 8 bit data, a transmission time period will be 2.56 ms. Therefore, its transmission speed will be nearly 3 kb/sec. Though the transmission speed is slow like this, the spread of the sub-band is little. Therefore, it can cover a long transmission distance due to the possibility of strong magnetic field generation. This encoding method is used for the vicinity coupling IC card and RF ID tag.

5.4. Signal Transmission of IC Card

For the proximity coupling type IC card, there are two interfaces of type A and B (Figure 27) (Karibe, 2008). In the type A IC card, the modulation scheme from the reader to card is ASK100% with the modified mirror encoding. Its transmission speed is 106 kb/sec, or changeable to 212 kb/sec, 424 kb/sec, and 847 kb/sec. For the transmission from the card to the reader it uses a sub-carrier of 847 kHz. It transmits signals turning this sub-carrier on/off with the Manchester encoding.

Figure 26. Typical encoding methods

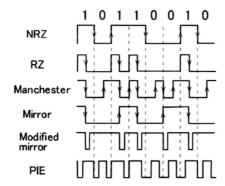

Figure 27. Modulation in the proximity coupling IC card

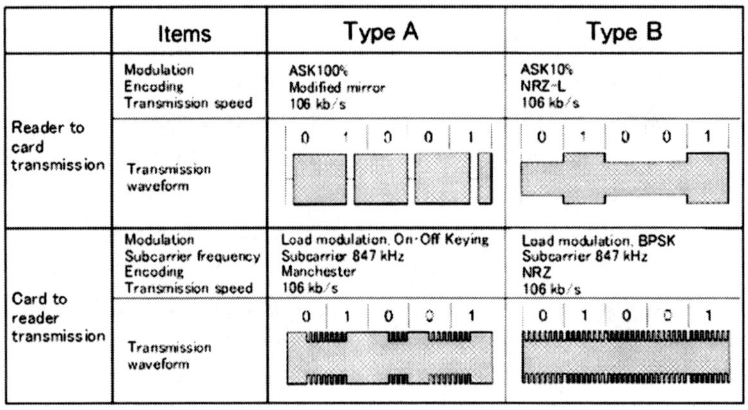

	Items	Type A	Type B
Reader to card transmission	Modulation Encoding Transmission speed	ASK100% Modified mirror 106 kb/s	ASK10% NRZ-L 106 kb/s
	Transmission waveform	0 1 0 0 1	0 1 0 0 1
Card to reader transmission	Modulation Subcarrier frequency Encoding Transmission speed	Load modulation, On-Off Keying Subcarrier 847 kHz Manchester 106 kb/s	Load modulation, BPSK Subcarrier 847 kHz NRZ 106 kb/s
	Transmission waveform	0 1 0 0 1	0 1 0 0 1

In the type B, transmission from the reader is done by the modulation scheme of ASK10% with the NRZ encoding. On the other hand, transmission from the card is done using the BPSK method combined the sub-carrier of 847 kHz with the NRZ encoding. Its transmission speed is 106 kb/sec, or changeable to 212 kb/sec, 424 kb/sec, and 847 kb/sec.

For the vicinity coupling type IC card, the transmission distance is 70 cm. The diameter of loop antenna is from 50 cm through 1 m. Operational magnetic field is 0.15-5 A/m. Transmission from the reader is done by ASK100% or ASK10% with the pulse position encoding. There are low speed and high speed transmission modes. In the low speed transmission mode, a pulse within 256 slots is generated corresponding to 8-bit data (Figure 28) (Karibe, 2008). The transmission time with a pulse period of 19 μs is about 4.8 ms. In the high-speed transmission mode, 8 bit data are divided per two bits. Pulses corresponding to two bit data are generated within each of 4 phase time slots. The transmission time is nearly 0.3 ms. In the transmission from the card, a sub-carrier generated by the load switch with the Manchester encoding is used in both modes.

Figure 28. Reader to tag modulation and encoding method for the vicinity coupling IC card

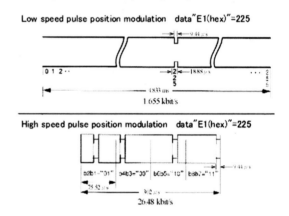

5.5. Operation Region

In the design of the system, consideration for an operation region from the reader is also needed. The operation region is determined by the distance (from the reader antenna), in which the minimum operation voltage V_2 for the card is kept. The induction voltage V_2 on the card antenna coil is described considering the united equivalent circuit of reader and card (Figure 29) as follows (Karibe, 2008).

$$V_2 = j\omega \cdot \mu_0 \cdot H \cdot A \cdot N / [(j\omega L_2 + R_2)(1/R_L + j\omega C_2)] \quad (4)$$

Figure 29. Equivalent circuit combining reader and card

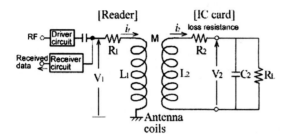

Here, A, N, H, ω means area of card antenna, wire's number wound up on card antenna coil, magnetic field strength, oscillating angle frequency, respectively. Therefore, it is seen that the magnetic field strength H determines the operation region. For the proximity coupling IC card, H is regulated as 1.5 – 7.5 A/m. When H is 7.5 A/m in the center of reader antenna coil, H reduces to 1.5 A/m at the distance of 10 cm away from the reader antenna coil (Figure 30). Therefore, the operation region in this case is nearly 10 cm. This distance is also influenced by the power dissipation of IC chip and directivity of antenna coils. When the flow current on the IC chip is larger, the greater magnetic field strength is needed.

Figure 30. Magnetic field strength change from the antenna. Radius of reader loop antenna =7.5 cm.

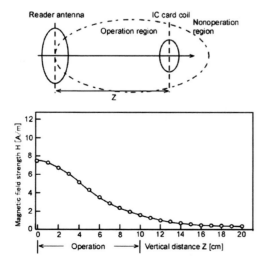

5.6. Feasibility of Application of RFID

At present, as an application on FA production lines, these IC tags are used for identification of goods. Its system is also being introduced partly into goods management in physical distribution and warehouses. In the future, it has a feasibility of using for management of a small capsule of medicine and a bottle of medicine. Its application possibility for articles of clothing, electrical appliances, books, and shoes etc. is now being investigated (Yoshioka, 2004). It is expected that the RFID system will be used as a basic technology supporting the connection of man and man, man and material, material and material.

6. CONCLUDING REMARKS

Goods identification using a soft magnetic substance barcode and RFID using a magnetic wave of 13.56 MHz usable in dirty environments were described. Automated wheelchair guidance technology was also described. Though a magnetic grooved barcode detection system with practically usable detection height using SEMS has been developed, decreasing production cost of iron soft magnetic material is a crucial and future issue to be available on the market. It was also introduced that a highly sensitive GMR spin valve sensor is hopeful to achieve highly stable scanners. Regarding the guidance technologies using the ferrite consisting of a sub-product of Mn-Zn core or a refinement product of Fe/Ti, it has a feature that markers can be made cheaply. This technology should be used for protection of environment by standardizing its specifications considering the current RFID standard. As relatively new rising technologies, some of the trendy computer control autonomous wheelchair technologies were introduced. In the future, it is expected that they will be improved for realizing fully automation control wheelchairs required in running aging society. The RFID technology is currently being used for various applications. Though the IC card system of 13.56 MHz was introduced, its application area with tags will be expanded resolving false detection problems in multiple tags and interference problems with irons. In practical use of this technology, its applications must be considered separating or living together with current barcode systems.

ACKNOWLEDGMENT

The author would like to thank the President Y. Aragane and the Head of School Affairs K. Tominaga of Tokyo Metropolitan College of Industrial Technology for their supports concerning the publication of this document. He would like to also thank Mr. K. Nakamura and Mr. T. Matsumura for their kind permissions for reusing figures from previously published materials (Wakaumi, Nakamura, & Matsumura, 1989).

REFERENCES

Argyros, A., Georgiadis, P., Trahanias, P., & Tsakiris, D. (2002). Semi-autonomous navigation of a robotic wheelchair. *Journal of Intelligent & Robotic Systems, 34,* 315–329. doi:10.1023/A:1016371922451.

Kaiser, U., & Steinhagen, W. (1995). A low-power transponder IC for high- performance identification systems. *IEEE Journal of Solid-State Circuits, 30,* 306–310. doi:10.1109/4.364446.

Karibe, H. (2008). *Easy book for non-contact IC card.* Tokyo, Japan: Nikkan Kogyo Shinbun Publishing.

Kishigami, J. (2005). *Textbook on RFID – Whole RF IC tags towards ubiquitous society.* Tokyo, Japan: ASKII.

Lankenau, A. (2001). Avoiding mode confusion in service robots – The Bremen autonomous wheelchair as an example. In *Proceedings of the 7th International Conference on Rehabilitation Robotics (ICORR 2001),* (pp. 162-167). Evry, France: ICORR.

Lenz, J., & Edelstein, S. A. (2006). Magnetic sensors and their applications. *IEEE Sensors Journal, 6,* 631–649. doi:10.1109/JSEN.2006.874493.

Nikkei Computer. (2003). *IC tag (RFID).* Tokyo, Japan: Nikkei BP.

Noetzel, R., Meisenberg, A., & Bartos, A. (2006). Customized GMR-spin valve sensors for low field applications. [Daegu, Korea: IEEE.]. *Proceedings of IEEE Sensors, 2006,* 1020–1023. doi:10.1109/ICSENS.2007.355798.

Okabe, H., & Wakaumi, H. (1990). Grooved barcode pattern recognition system with magnetoresistive sensor. *IEEE Transactions on Magnetics, 26,* 1575–1577. doi:10.1109/20.104451.

RFID Technology Editorial Department. (2004). *All radio-frequency IC tags.* Tokyo, Japan: Nikkei BP.

Rofer, T., & Lankenau, A. (2000). Architecture and applications of the Bremen autonomous wheelchair. *Science Direct.com–. Information Sciences, 126,* 1–20. doi:10.1016/S0020-0255(00)00020-7.

Sakai, Y., & Kitazawa, M. (1995). Human-centered wheelchair supervisory system utilizing human ideas for operation. *Biomedical Fuzzy Human Science, 1,* 57–70.

Sgouros, N. M. (2002). Qualitative navigation for autonomous wheelchair robots in indoor environments. *Autonomous Robots, 12*, 257–266. doi:10.1023/A:1015265514820.

Usami, M., & Yamada, J. (2005). *Ubiquitous technology IC tag*. Tokyo, Japan: Ohmusha Publishing.

Wakaumi, H., Ajiki, H., Hankui, E., & Nagasawa, C. (2000). Magnetic grooved bar-code recognition system with slant MR sensor. *IEE Proceedings. Science Measurement and Technology, 147*, 131–136. doi:10.1049/ip-smt:20000369.

Wakaumi, H., Komaoka, T., & Hankui, E. (2000). Grooved bar-code recognition system with tape-automated-bonding head detection scanner. *IEEE Transactions on Magnetics, 36*, 366–370. doi:10.1109/20.822548.

Wakaumi, H., Nakamura, K., & Matsumura, T. (1989). A new automated wheelchair guided by magnetic ferrite marker lane. *NEC Research & Development, 95*, 62–68.

Wakaumi, H., Nakamura, K., & Matsumura, T. (1992). Development of an automated wheelchair guided by a magnetic ferrite marker lane. *Journal of Rehabilitation Research and Development, 29*, 27–34. doi:10.1682/JRRD.1992.01.0027 PMID:1740776.

Wakaumi, H., Nakamura, K., Matsumura, T., & Yamauchi, F. (1989). *Automated wheelchair guided by magnetic ferrite marker*. Paper presented at the RESNA 12th Annual Conference. (pp. 47-48). New Orleans, LA.

Wakaumi, H., & Yokoyama, S. (1988). An exciting method for highly-sensitive magnetic sensors. In *Proceedings of the 12th Institute of Magnetic Engineers of Japan Conference*.(pp. 341). IMEJ.

Yoshioka, T. (2004). *Radio frequency IC tags illustrated with pictures – The world of radio frequency identification expanding*. Tokyo, Japan: Ohmusha Publishing.

Section 3

Chapter 11
Development and Simulation of an Adaptive Control System for the Teleoperation of Medical Robots

Vu Trieu Minh
Tallinn University of Technology, Estonia

ABSTRACT

This chapter presents the design and calculation procedure for a teleoperation and remote control of a medical robot that can help a doctor to use his hands/fingers to examine patients in remote areas. This teleoperation system is simple and low cost, connected to the global Internet system, and through the interaction with the master device, the medical doctor is able to communicate control signals for the slave device. This controller is robust to the time-variant delays and the environment uncertainties while assuring the stability and the high transparent performance. A novel theoretical framework and algorithms are developed with time forward observer-based adaptive controller and neural network-based multiple model. The system allows the medical doctor to feel the real sense of the remote environments.

1. INTRODUCTION

Teleoperation indicates the operation of a device at a distance. It is similar to remote control but normally associated with robotic systems in which a device operated by a human (master) is used to control a robot from a distance (master-slave control). Master-slave teleoperation has many applications including dangerous environments. The human operator interacts with the master device to generate control signals to send to the slave device. The slave device actually interacts with the remote environments while staying under the control of the human operator. Information collecting at the remote environments is then transmitted back to the human operator through the communication networks and the master device.

Varying time delays through the communication channel and the uncertainties of the remote environments are the most critical problems for

DOI: 10.4018/978-1-4666-4225-6.ch011

the stability and the transparent performance of a teleoperation system since they can cause bad performance and instability to the system. In this study, a novel framework for medical remote control is developed. It can assure the high level of transparent performance as the impedance felt by the medical doctor on the local site. This allows him feel the real sense of the remote environments to examine patients in remote locations.

The evolution of teleoperation has generated sophisticated systems in order to provide better solutions that the operator can feel as if he is present in the operation sites. Kikuchi et al. (1998) proposed a teleoperation system in dynamic environment with varying communication time delays. The proposed system consists of the stable bilateral teleoperation subsystems using the virtual time delay method. The visual information offers the prediction of the slave manipulator and the environment. Zhu and Salcudean (1999) introduced a novel stability guaranteed controller design for bilateral teleoperation under both position and rate control modes with arbitrary motion/force scaling. Boukhnifer and Ferreira (2006) showed that the application of wave variable transformation preserves the passivity of the teleoperation system in spite of communication delays and the varying scale factors.

Lawrence (1993) explored the trade-off transparency and stability in the presence of communication delays based on the concept of impedance. Impedance is a quantity that maps the input of a system to the output force. When a teleoperation system is ideal, the operator feels as if he is executing the task with his own hands on the scene. Slawinski and Mut (2008) proposed defining transparency in the time domain and established a quantitative measure of how the human operator feels the remote system. It allows analyzing the effect of the time-varying delays on the system transparency.

Unknown and highly changeable attributes of remote environments make the development of fully comprehensive models impractical. A method of creating a selection of different simple models and dynamically selecting an appropriate one for a given time is proposed. Various methods have been developed for model detection. One of the most effective methods of model detection and usage is using multiple model neural networks. Multiple model neural networks are useful because they are powerful but also run fast enough to operate in real time. Most neural networks that are presently employed for artificial intelligence are based in statistical estimation, optimization and control theory. Chen et al. (2007) proposed a neural network based multiple model adaptive control for teleoperation systems. Decision controllers are designed to adaptively switch among all predictive controllers according to the performance target. This method can ensure the stability and transparent performance of the system. Smith and Hashtrudi-Zaad (2005) used two neural networks at the master and slave devices to improve the transparency and to compensate the effect of the time delays.

The aim of the chapter is to develop a simple and low cost teleoperation system operated over the Internet using TCP/IP and without a camera in the slave device. The system can be supported by a telephone or voice chat over Internet Protocol (VoIP) for a medical doctor who can use the hands/fingers to examine the remote patients. The idea for the system is based on the design for a medical tele-analyzer in Suebsomran and Parnichkun (2005). The developed system was shown to be robust to internal and external disturbances, model uncertainty and friction. Hashtrudi-Zaad and Salcudean (2002) introduced the mathematics the defines the stability of a teleoperation system. Minh et al. (2007) performed further research and analysis environmental uncertainty and verification. Chen et al. (2010) introduced methods to maintain the stability of uncertain cellular neural networks with interval time-varying delays. Park and Kwon (2009) proposed a novel stability criterion for the stability based on the Lyapunov function in terms of Linear Matrix Inequalities

(LMIs), which can be solved easily by various optimization algorithms. And Chen *et al.* (2910) investigated the problem of stability of neural networks with time-varying delay in a given range.

The contents of the chapter are as follows: Section 2 introduces the system modeling; Section 3 develops the system time forward observer; Section 4 sets up stability and transparency conditions for the system controller; Section 5 illustrates simulation results; and finally conclusions are drawn in Section 6.

2. SYSTEM MODELING

A teleoperation system consists of five fundamental components: the human operator; the master device; the communication channel; the slave manipulator; and the environment. Commands and sensor information are exchanged in both directions through these components. The control channel enables the human operator to assign tasks to the slave and control the slave as desired. Information about the task execution is fed back to the human operator. Both channels are interconnected via the normal Internet system. To reduce costs and simplify the system, no camera is fitted to the teleoperation system. Communications such as cameras and voice can instead be fitted to a supplementary system separate from the teleoperation. Figure 1 shows the basic configuration for a bilateral teleoperation system as developed by Lawrence (1993): The operator moves the master manipulator and via the communication channel (e.g. Internet), the slave follows. Forces exerted by the environment on the slave are transmitted back to the master and hence felt by the operator.

The human force (medical doctor) on the master f_h and the master motion x_m should have the same relationship with the force on the environment f_e and the slave motion x_s, i.e., for the same forces, $f_e = f_h$, the motions should be the same, $x_s = x_m$. This requirement assures the system completely transparent. So the operator feels the real sense of the remote environments. Notations: u, u_{dm}, u_d and u_s are the signals transmitted among the system's components.

The master and slave are all modeled as mechanical devices with masses, dampers and springs, as shown in Figure 2.

The human force f_h is applied to the master manipulator and x_m is the position of the master; f_e is the contact force between the slave manipulator and the environment and x_s is the position of the slave; u_{dm} is the feedback signal from the slave and u_d is the control signal given by the master through the Internet. Notations, m_m, m_s, b_m, b_s, k_m, and k_s, are the inertia, damping and spring stiffness of the master and slave manipulator, respectively.

The dynamics of the master and the slave are given by the following equations:

$$f_h - u_{dm} = m_m \ddot{x}_m + b_m \dot{x}_m + k_m x_m \qquad (1)$$

$$u_d - f_e = m_s \ddot{x}_s + b_s \dot{x}_s + k_s x_s \qquad (2)$$

For the variable time delays and the environmental uncertainties, it is assumed that the dynamics of the environment are also regulated by a finite set of mechanical devices with masses,

Figure 1. Teleopration system configuration

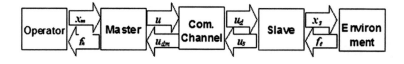

Figure 2. Physical model of master and slave manipulator

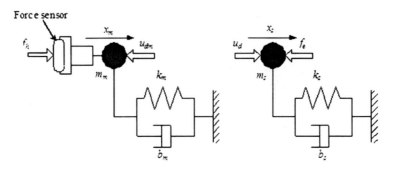

dampers, and springs, shown in Figure 3: x_{si} denotes the position of the slave device and f_{ei} denotes the force applied on the environment i. Similarly, m_e, k_e, and b_{ei} are mass, stiffness and damping parameters of the unknown and time-variant environment models.

The dynamics of the environment model are formulated by the following equation:

$$f_e = m_e \ddot{x}_s + b_e \dot{x}_s + k_e x_s$$
$$\text{or } f_{ei} = m_{ei} \ddot{x}_{si} + b_{ei} \dot{x}_{si} + k_{ei} x_{si}, \qquad (3)$$
$$\text{for } i = 1,...,n$$

Since the environment models are uncertain, a neural network with the radial basis function (RBF) is used to detect the environmental dynamics since RBF network is easy to approximate the

parameters and the training speed is fast. For n given environment models, the input vectors U is the movement of the slave device,

$$U = \begin{bmatrix} x_{s1} & \cdots & x_{sn} \\ \dot{x}_{s1} & \ddots & \dot{x}_{sn} \\ \ddot{x}_{s1} & \cdots & \ddot{x}_{sn} \end{bmatrix},$$

and the target vectors Y is the environment model parameters,

$$Y = \begin{bmatrix} k_{e1} & \cdots & k_{en} \\ b_{e1} & \ddots & b_{en} \\ m_{e1} & \cdots & m_{en} \end{bmatrix}.$$

Figure 3. Physical model of environment

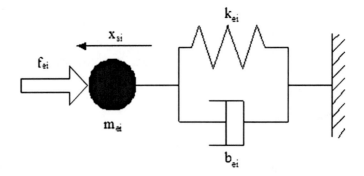

From the online archived slave movements, $\begin{bmatrix} x_{si} & \dot{x}_{si} & \ddot{x}_{si} \end{bmatrix}'$, RBF will online calculate the corresponding environment parameters, $\begin{bmatrix} \hat{k}_{ei} & \hat{b}_{ei} & \hat{m}_{ei} \end{bmatrix}'$, and then, the corresponding model errors $\in_i = \sqrt{\left[m_{ei} - \hat{m}_{ei} \right]^2 + \left[b_{ei} - \hat{b}_{ei} \right]^2 + \left[k_{ei} - \hat{k}_{ei} \right]^2}$. The environment model with the smallest error, $\in_m$, is selected. Environment modeling and verification are performed as described by Minh et al. (2007). It is assumed that the environmental uncertainties can be represented by mathematical models. To each model, the system behavior changes and should be estimated by a different modes. The system mode may jump up or vary continuously in a discrete mapping set.

The master and the slave device are connected via Internet. Using the Internet introduces significant and variable time delays. Thus, the equations for signal transmitted via communication channel must be included the forward time delay, $T_R(t)$, and the backward time delay, $T_L(t)$:

$$u_d = u(t - T_R(t)) \qquad (4)$$

$$u_{dm} = u_s(t - T_L(t)) \qquad (5)$$

Finally the overall schematic diagram of the proposed system shows in Figure 4.

The operator moves the master plant, which causes an transmitted signal, $u(t)$. In conventional methods proposed by Lawrence (1993) and Hashtrudi-Zaad and Salcudean (2002), the transmitted signals include position, velocity, and acceleration. But in this proposed system, the operator (medical doctor) wants to directly touch the patient; the acceleration is difficult to be realized and can be omitted from the dynamic equations in order to guarantee the high transparency for the operator. Thus,

$$u(t) = f_{11}x_m(t) + f_{12}\dot{x}_m(t) + c_{11}f_h(t) \qquad (6)$$

where f_{11}, f_{12}, and c_{11} are the feedback coefficients. Because the forward time delay, $T_R(t)$, causes $u_d(t) \neq u(t)$, we use $x_s(t)$, $\dot{x}_s(t)$, and $f_e(t)$ to adjust $u_d(t)$ as follows:

$$\begin{aligned} u_d(t) &= f_{11}x_m(t - T_R(t)) \\ &+ f_{12}\dot{x}_m(t - T_R(t)) + c_{11}f_h(t - T_R(t)) \\ &+ f_{13}x_s(t) + f_{14}\dot{x}_s(t) + c_{12}f_e(t) \end{aligned} \qquad (7)$$

Figure 4. Schematic diagram of the teleoperation system

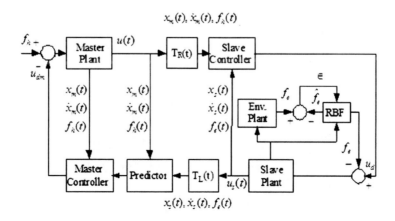

where f_{13}, f_{14}, and c_{12} are also feedback coefficients.

Similarly, the slave transmitted signal, $u_s(t)$ should include force, position, and velocity:

$$u_s(t) = f_{23}x_s(t) + f_{24}\dot{x}_s(t) + c_{22}f_e(t) \qquad (8)$$

where f_{23}, f_{24}, and c_{22} are the feedback coefficients. And since $u_{dm}(t) \neq u_s(t)$ caused by the backward time delay, $T_L(t)$, we $x_m(t - T(t))$, $\dot{x}_m(t - T(t))$, and $f_h(t - T(t))$ use to adjust $u_{dm}(t)$ as follows:

$$
\begin{aligned}
u_{dm}(t) &= f_{23}x_s(t - T_L(t)) \\
&+ f_{24}\dot{x}_s(t - T_L(t)) + c_{22}f_e(t - T_L(t)) \\
&+ f_{21}x_m(t - T(t)) + f_{22}\dot{x}_m \\
&(t - T(t)) + c_{21}f_h(t - T(t))
\end{aligned} \qquad (9)
$$

where $T(t) = T_R(t) + T_L(t)$ is the total time delay and f_{21}, f_{22}, and c_{21} are the feedback coefficients. In this equation, $u_{dm}(t)$ can response the control effect of $u(t - T(t))$. As a result, if we use the predicted value of $u_{dm}(t + T(t))$ as the feedback value from the slave, then the operator can feel that the time delay does not existed since the effect of time delay on transparency has been eliminated.

The predicted value of $u_{dm}(t + T(t))$ is labeled as $\hat{u}_{dm}(t)$ and:

$$
\begin{aligned}
\hat{u}_{dm}(t + T(t)) &= f_{21}x_m(t) + f_{22}\dot{x}_m(t) \\
&+ c_{21}f_h(t) + f_{23}\hat{x}_s(t + T_R(t)) \\
&+ f_{24}\dot{\hat{x}}_s(t + T_R(t)) + c_{22}\hat{f}_e(t + T_R(t))
\end{aligned} \qquad (10)
$$

where $\hat{x}_s(t + T_R(t))$, $\dot{\hat{x}}_m(t + T_R(t))$, and $\hat{f}_e(t + T_R(t))$ are the predicted values of $x_s(t + T_R(t))$, $\dot{x}_m(t + T_R(t))$, and $f_e(t + T_R(t))$.

According to the environment model in (10), $\hat{f}_e(t + T_R(t))$ can be estimated by the following equation:

$$
\begin{aligned}
\hat{f}_e(t + T_R(t)) &= m_e\ddot{\hat{x}}_s(t + T_R(t)) \\
&+ b_e\dot{\hat{x}}_s(t + T_R(t)) + k_e\hat{x}_s(t + T_R(t))
\end{aligned} \qquad (11)
$$

The predicted values of $\hat{x}_s(t + T_R(t))$ and $\dot{\hat{x}}_s(t + T_R(t))$ can be achieved through the time forward observer design in the next section.

3. OBSERVER DESIGN

The teleoperation observer is used to estimate the system parameters for the time delay compensation. The system can only achieve a high level of transparency with a good observer based on the predicted system dynamics. In order to predict $x_s(t + T_R(t))$ and $\dot{x}_s(t + T_R(t))$, Equation (7) is substituted into Equation (2):

$$
\begin{aligned}
&f_{11}x_m(t - T_R(t)) + f_{12}\dot{x}_m(t - T_R(t)) \\
&+ c_{11}f_h(t - T_R(t)) + f_{13}x_s(t) \\
&+ f_{14}\dot{x}_m(t) + c_{12}f_e(t) - f_e(t) \\
&= m_s\ddot{x}_s(t) + b_s\dot{x}_s(t) + k_sx_s(t)
\end{aligned} \qquad (12)
$$

We define:

$$
\begin{aligned}
\bar{u}_d(t) &= f_{11}x_m(t - T_R(t)) \\
&+ f_{12}\dot{x}_m(t - T_R(t)) + c_{11}f_h(t - T_R(t))
\end{aligned} \qquad (13)
$$

Equation (12) becomes:

$$
\begin{aligned}
\bar{u}_d(t) &= m_s\ddot{x}_s(t) + (b_s - f_{14})\dot{x}_s(t) \\
&+ (k_s - f_{13})x_s(t) + (1 - c_{12})f_e(t)
\end{aligned} \qquad (14)
$$

Substituting Equation (3) into Equation (14):

$$\bar{u}_d(t) = (m_s + (1 - c_{12})m_e)\ddot{x}_s(t)$$
$$+((b_s - f_{14}) + (1 - c_{12})b_e)\dot{x}_s(t) \quad (15)$$
$$+((k_s - f_{13}) + (1 - c_{12})k_e)x_s(t)$$

Converting Equation (15) into state-space form, we have:

$$\dot{\bar{x}}_s(t) = A_s\bar{x}_s(t) + B_s\bar{u}_d(t)$$
$$\bar{y}(t) = C_s\bar{x}_s(t) \quad (16)$$

where

$$\bar{x}_s(t) = \begin{bmatrix} x_s(t) \\ \dot{x}_s(t) \end{bmatrix},$$

$$C_s = \begin{bmatrix} 1 & 0 \end{bmatrix},$$

$$B_s = \begin{bmatrix} 0 \\ 1 \\ \hline m_s(1 - c_{12})m_e \end{bmatrix},$$

and

$$A_s =$$
$$\begin{bmatrix} 0 & 1 \\ -\dfrac{(k_s - f_{13}) + (1 - c_{12})k_e}{m_s + (1 - c_{12})m_e} & -\dfrac{(b_s - f_{14}) + (1 - c_{12})b_e}{m_s + (1 - c_{12})m_e} \end{bmatrix}.$$

From Equation (10), in order to predict $u_{dm}(t + T(t))$, we have to predict $x_s(t + T_R(t))$, and $\dot{x}_s(t + T_R(t))$ by running the model into the future with $T_R(t)$ units:

$$\dot{\bar{x}}_s(t + T_R(t)) = (1 + \dot{T}_R(t))A_s\bar{x}_s(t + T_R(t))$$
$$+(1 + \dot{T}_R(t))B_s\bar{u}_d(t + T_R(t)) \quad (17)$$
$$\bar{y}(t + T_R(t)) = C_s\bar{x}_s(t + T_R(t))$$

However, it is difficult to estimate $\dot{T}_R(t)$. According to the characteristic of Internet, we can model $T_R(t)$ as uncertain parameters and Equation (17) becomes:

$$\dot{\bar{x}}_s(t + T_R(t)) = (A_s + \Delta A_s)\bar{x}_s(t + T_R(t))$$
$$+(B_s + \Delta B_s)\bar{u}_d(t + T_R(t)) \quad (18)$$
$$\bar{y}(t + T_R(t)) = C_s\bar{x}_s(t + T_R(t))$$

where $\Delta t = \dot{T}_R(t)$, ΔA_s and ΔB_s are modeled as uncertain parameters, $\Delta A_s = \Delta t A_s$ and $\Delta B_s = \Delta t B_s$. If the time delay is constant, the variable delay term, $\dot{T}_R(t)$, in Equation (18) disappears.

Value of $\bar{x}_s(t + T_R(t))$ can be also calculated directly by the current master dynamics $x_m(t)$, $\dot{x}_m(t)$, and $f_h(t)$ via the following observer:

$$\dot{z}(t) = A_s z(t) + B_s(f_{11}\bar{x}_m(t) + c_{11}f_h(t))$$
$$+L(\bar{y}(t + T_R(t)) - y_s(t)) \quad (19)$$
$$y_s(t) = C_s z(t)$$

where L is the observer gain, $\bar{x}_m(t) = \begin{bmatrix} x_m(t) \\ \dot{x}_m(t) \end{bmatrix}$,

and $z(t) = \begin{bmatrix} \hat{x}_s(t) \\ \dot{\hat{x}}_s(t) \end{bmatrix}$.

Since the master output, $\bar{y}(t + T_R(t))$, cannot be measured, we can use the available adjustment of $t - T_L(t)$ as follows:

$$\dot{z}(t) = A_s z(t) + B_s(f_{11}\bar{x}_m(t) + c_{11}f_h(t))$$
$$+L(\bar{y}(t - T_L(t)) - y_s(t - T_R(t) - T_L(t))) \quad (20)$$
$$y_s(t) = C_s z(t)$$

Let the observing error be $e(t) = \bar{x}_s(t + T_R(t) - z(t))$, then:

$$\dot{e}(t) = A_s e(t) + \Delta A_s \bar{x}_s(t + T_R(t))$$
$$-LC_s e(t - T_R(t) - T_L(t)) \tag{21}$$
$$+\Delta B_s(f_{11}\bar{x}_m(t) + c_{11}f_h(t))$$

RBF will calculate in real time the error based on the best fitting environment model as selected bay the multiple neural network. Conditions for a stabilized and transparent controller are developed in the next section.

4. CONTROLLER DESIGN

In this part the controller design with guaranteed conditions for the system transparency and stability is established. The time forward state calculation for the slave state at $(t + T_R(t))$ has been performed in the previous section. Using $\hat{u}_{dm}(t + T(t))$ in Equation (10) to take place $u_{dm}(t)$ of in Equation (1), we have:

$$f_h(t) - (f_{21}x_m(t) + f_{22}\dot{x}_m(t) + f_{23}\hat{x}_s(t + T_R(t))$$
$$+f_{24}\dot{\hat{x}}_s(t + T_R(t)) + c_{21}f_h(t) + c_{22}\hat{f}_e(t + T_R(t))) \tag{22}$$
$$= m_m\ddot{x}_m(t) + b_m\dot{x}_m(t) + k_m x_m(t)$$

Re-arranging Equation (22) yields:

$$m_m\ddot{x}_m(t) + (b_m + f_{22})\dot{x}_m(t)$$
$$+(k_m + f_{21})x_m(t) = -(f_{23}\hat{x}_s(t + T_R(t)) \tag{23}$$
$$+f_{24}\dot{\hat{x}}_s(t + T_R(t)) + c_{22}\hat{f}_e(t + T_R(t)))$$

Substituting Equation (11) into Equation (23) and modifying, we have:

$$m_m\ddot{x}_m(t) + (b_m + f_{22})\dot{x}_m(t) + (k_m + f_{21})x_m(t)$$
$$= -((f_{23} + c_{22}k_e)\hat{x}_s(t + T_R(t))$$
$$+(f_{24} + c_{22}b_e)\dot{\hat{x}}_s(t + T_R(t)) \tag{24}$$
$$+c_{22}m_e\ddot{\hat{x}}_s(t + T_R(t))) + (1 - c_{21})f_h(t)$$

Using $z(t)$ in Equation (19) replacing $\hat{x}_s(t)$ in Equation (24), we have:

$$\ddot{x}_m(t) + \frac{(b_m + f_{22})}{m_m}\dot{x}_m(t) + \frac{(k_m + f_{21})}{m_m}x_m(t) =$$
$$\frac{(1 - c_{21})}{m_m}f_h(t) - (A_e z(t) + B_e \dot{z}(t)) \tag{25}$$

where

$$A_e = \left[\frac{(f_{23} + c_{22}k_e)}{m_m} \quad \frac{(f_{24} + c_{22}b_e)}{m_m} \right],$$

and

$$B_e = \left[0 \quad \frac{c_{22}m_e}{m_m} \right].$$

Substituting Equation (20) into (25) and is being altered, we have:

$$\ddot{x}_m(t) + \frac{(b_m + f_{22})}{m_m}\dot{x}_m(t) + \frac{(k_m + f_{21})}{m_m}x_m(t)$$
$$= -(A_e + B_e A_s)z(t) + B_e B_s(f_{11}\bar{x}_m(t) + c_{11}f_h(t)) \tag{26}$$
$$+B_e LC_s e(t - T(t)) + \frac{(1 - c_{21})}{m_m}f_h(t)$$

or:

$$\ddot{x}_m(t) + \left[\frac{(b_m + f_{22})}{m_m} + B_e B_s f_{12} \right]\dot{x}_m(t)$$
$$+ \left[\frac{(k_m + f_{21})}{m_m} + B_e B_s f_{11} \right]x_m(t)$$
$$= \left[\frac{(1 - c_{21})}{m_m} + B_e B_s c_{11} \right]f_h(t) \tag{27}$$
$$-((A_e + B_e A_s)z(t) + B_e LC_s e(t - T(t)))$$

Converting Equation (27) into state-space form, we have:

$$\dot{\bar{x}}_m(t) = A_m \bar{x}_m(t) + B_m f_h(t)$$
$$- \left(A_z \left(\bar{x}_s(t + T_R(t)) - e(t) \right) + B_z e(t - T(t)) \right) \quad (28)$$
$$y_m(t) = C_m \bar{x}_m(t)$$

where

$$A_m =$$
$$\begin{bmatrix} 0 & 1 \\ -\left(\dfrac{k_m + f_{21}}{m_m} + B_e B_s f_{11} \right) & -\left(\dfrac{b_m + f_{22}}{m_m} + B_e B_s f_{12} \right) \end{bmatrix},$$

$$C_m = \begin{bmatrix} 1 & 0 \end{bmatrix},$$

$$B_m = \begin{bmatrix} 0 \\ \dfrac{1 - c_{11}}{m_m} + B_e B_s c_{11} \end{bmatrix},$$

$$A_z = \begin{bmatrix} 0 \\ A_e + B_e A_s \end{bmatrix},$$

and

$$B_z = \begin{bmatrix} 0 \\ B_e L C_s \end{bmatrix}.$$

From Equations (18), (21) and (28), the model of the entire system is established as:

$$\dot{x}(t) = A x(t) + A_t x(t - T(t)) + B f_h(t)$$
$$y(t) = C x(t) \quad (29)$$

where

$$x(t) = \begin{bmatrix} \bar{x}_m(t) \\ \bar{x}_s(t + T_R(t)) \\ e(t) \end{bmatrix},$$

$$A = \begin{bmatrix} A_m & -A_z & A_z \\ (B_s + \Delta B_s) f_{11} & A_s + \Delta A_s & 0 \\ \Delta B_s f_{11} & \Delta A_s & A_s \end{bmatrix},$$

$$C = \begin{bmatrix} C_m & 0 & 0 \\ 0 & C_s & 0 \\ 0 & 0 & 0 \end{bmatrix},$$

$$B = \begin{bmatrix} B_m \\ (B_s + \Delta B_s) c_{11} \\ \Delta B_s c_{11} \end{bmatrix},$$

and

$$A_t = \begin{bmatrix} 0 & 0 & -B_z \\ 0 & 0 & 0 \\ 0 & 0 & -LC_s \end{bmatrix}.$$

The system in Equation (29) is asymptotically stable as per Park J.H. *et al.* (2009) for any time-varying delay satisfying $h_1 \leq T(t) \leq h_2$ and $-\mu \leq \dot{T}(t) \leq \mu$ where $h_2 > h_1$ and $\mu \geq 0$ if there exist matrices $P > 0$, $Z_1 > 0$, $Z_2 > 0$, and $Q_J > 0$ for $J = 1, 2, 3, 4$ in the following Lyapunov function, such that:

$$V(T) = x'(t) P x(t) + \int_0^{T(t)} f_h(s) ds +$$
$$f'(x'(s) Q_2 x(s)) +$$
$$\int_{t-T(t)}^{t-h_1} x'(s) Q_1 x(s) + \int_{t-T(t)}^{t} \int_{t-h_1}^{t} x'(s) Q_3 x(s) +$$
$$\int_{t-h_2}^{t-h_1} x'(s) Q_4 x(s) + \int_{-h_1}^{0} \int_{t+\theta}^{t} \dot{x}'(s) Z_1 \dot{x}(s) ds d\theta +$$
$$\int_{-h_2}^{-h_1} \int_{t+\theta}^{t} \dot{x}'(s) Z_2 \dot{x}(s) ds d\theta \quad (30)$$

181

The system is now subject to conditions for the transparency of the form developed in Lawrence (1993), the system in Equation (29) achieves the ideal transparency if the following relation is maintained:

$$\frac{F_e(s)}{X_s(s)} = \frac{F_h(s)}{X_m(s)} \tag{31}$$

where $F_e(s)$, $X_s(s)$, $F_h(s)$, and $X_m(s)$ are the Laplace transform of $f_e(t)$, $x_s(t)$, $f_h(t)$, and $x_m(t)$ respectively.

Substituting Equations (2), (3), (7), and (9) into Equation (31) establishes the following conditions for transparency:

$$((1 - c_{21})(m_s + (1 - c_{12})m_e) - c_{11}c_{22}m_e)m_e = (m_s + (1 - c_{12})m_e)m_m \tag{32}$$

$$((1 - c_{21})(m_s + (1 - c_{12})m_e) - c_{11}c_{22}m_e)b_e$$
$$+((1 - c_{12})(b_s - f_{14} + (1 - c_{12})b_e$$
$$-c_{11}(f_{24} + c_{22}b_e))m_e$$
$$= (b_s - f_{14} + (1 - c_{12})b_e)m_m$$
$$+(m_s + (1 - c_{12})m_e)(b_m + f_{22}) + c_{22}m_e f_{12} \tag{33}$$

$$((1 - c_{21})(m_s + (1 - c_{12})m_e) - c_{11}c_{22}m_e)k_e$$
$$+((1 - c_{12})(b_s - f_{14} + (1 - c_{12})b_e$$
$$-c_{11}(f_{24} + c_{22}b_e))b_e + (1 - c_{21})$$
$$(k_s - f_{13} + (1 - c_{12})k_e) - c_{11}(f_{23} + c_{22}k_e))m_e$$
$$= (k_s - f_{13} + (1 - c_{12})k_e)m_m$$
$$+(b_s - f_{14} + (1 - c_{12})b_e)(b_m + f_{22})$$
$$+(m_s + (1 - c_{12})m_e)(k_m + f_{21})$$
$$+c_{22}m_e f_{11} + (f_{24} + c_{22}b_e)f_{12} \tag{34}$$

$$((1 - c_{21})(b_s - f_{14} + (1 - c_{12})b_e)$$
$$-c_{11}(f_{24} + c_{22}b_e))k_e + (1 - c_{21})$$
$$(k_s - f_{13} + (1 - c_{12})k_e) - c_{11}(f_{23} + c_{22}k_e))b_e$$
$$= (k_s - f_{13} + (1 - c_{12})k_e)(b_m + f_{22})$$
$$+(b_s - f_{14} + (1 - c_{12})b_e)(b_m + f_{21})$$
$$+(f_{24} + c_{22}b_e)f_{11} + (f_{23} + c_{22}k_e)f_{12} \tag{35}$$

$$((1 - c_{21})(k_s - f_{13} + (1 - c_{12})k_e)$$
$$-c_{11}(f_{23} + c_{22}b_e))k_e = (k_s - f_{13} + (1 - c_{12})k_e)$$
$$(k_m + f_{21}) + (f_{23} + c_{22}k_e)f_{11} \tag{36}$$

Consequently, if Equations (30), (32), (33), (34), (35), and (36) are feasible, the system will achieve the stability and ideal transparency. Simulations for this controller are examined in the next section.

5. SIMULATION RESULTS

Simulation of this proposed model is developed with Matlab/Simulink for time-variant delays and environment uncertainties. The controller is also tested for the ability to reject disturbances while maintaining its closed-loop stability.

In the simulation, the parameters of the master and slave manipulator are selected as $m_m = 1.5kg$, $b_m = 0.45Ns/m$, $k_m = 1N/m$, $m_s = 1.5kg$, $b_s = 0.45Ns/m$, and $k_s = 1N/m$. It is assumed that the environment uncertainties consist of three models M_1, M_2, and M_3 activated randomly.

The parameters of environment model M_1 are selected as: $m_{e1} = 1.1Kg$, $b_{e1} = 0.6Ns/m$, and $k_{e1} = 0.6N/m$. The initial controller parameters are chosen as: $c_{11} = 30$, $f_{11} = 2$, $f_{12} = 1$, $f_{21} = 1456$, and $f_{22} = 987$. To assure the system transparency, other controller parameters are calculated as solutions of Equations (31), (32), (33), (34), and (35) as: $c_{12} = -27.636$, $c_{21} = -29$, $c_{22} = 28.636$, $f_{13} = -1.8182$, $f_{14} = -1.3682$, $f_{23} = -1456.2$, and $f_{24} = -986.63$.

182

Similarly, the parameters of environment model M_2 are selected as: $m_{e2} = 1.0Kg$, $b_{e2} = 0.4Ns / m$, and $k_{e2} = 0.8N / m$. Then, other initial controller parameters are chosen as: $c_{11} = 20$, $f_{11} = 2$, $f_{12} = 1$, $f_{21} = 1500$, and $f_{22} = 1200$. For transparency conditions, the remaining controller parameters are calculated as: $c_{12} = -17.5$, $c_{21} = -19$, $c_{22} = 18.5$, $f_{13} = -2.2$, $f_{14} = -1.15$, $f_{23} = -1499.8$, and $f_{24} = -1199.8$.

The parameters of environment model M_3 are selected as: $m_{e3} = 1.5Kg$, $b_{e3} = 0.7Ns / m$, and $k_{e3} = 0.5N / m$. The initial controller parameters in this case are chosen as: $c_{11} = 30$, $f_{11} = 2$, $f_{12} = 1$, $f_{21} = 1497$, and $f_{22} = 1356$. Conditions for the system transparency lead to other controller parameters as: $c_{12} = -28$, $c_{22} = 29$, $f_{13} = -1.5$, $f_{14} = -1.25$, $f_{23} = -1497.5$, and $f_{24} = -1355.8$.

The time variant delays in the communication channel are randomly selected with $0 < T_R(t) \leq 3s$ and $0 < T_L(t) \leq 3s$. The three environmental uncertainties, M_1, M_2, and M_3 are also randomly activated and the RBF network is used to select the best fitted model. The maximum allowable bound of time-delay, T_{Max_Delay}, for guaranteeing stability of the system therefore is the sum of the maximum of forward time delay, $T_{Max\_R}(t)$, and the maximum of backward time delay, $T_{Max\_L}(t)$,

$$T_{Max\_Delay} \leq T_{Max\_R}(t) + T_{Max\_L}(t) = 6s.$$

The input human force, f_h, with pulse generator activates the movement of the master device, x_m. Three environment models, M_1, M_2, and M_3 are activated at time from timeline from 0-30s, 31-60s, and 61-90s, respectively. RBF identifies the environmental parameters and activate the suitable adaptive controllers. White noise is also injected to the system to test the ability of the system to maintain its stability as shown in Figure 5.

Figure 6 shows the simulation results for the time-variant delays and the environmental uncertainties. The subplots are indicated the forces and movements of the master and the slave. The adaptive predictive controller was accurate, the system is stable and in good performance since the slave profiles tracking well to the master amid the time-variant delays and the environment uncertainties.

Figure 5. Matlab simulink diagram

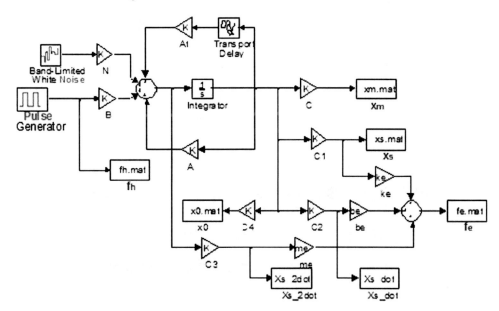

Figure 6. Position of master and slave with time-variant delays and environmental uncertainties

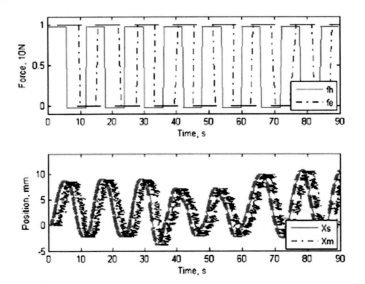

Lastly, the model is tested for the ability to maintain its closed-loop stability with noise disturbances. Simulation results for the time-variant delays and the environmental uncertainties with noises are shown in Figure 7.

Simulation results show that system is stable under the white-noise disturbances. The outputs can track the input properly without steady-state error. The controller is robust to the disturbances.

6. CONCLUSION

This chapter presents a novel theoretical framework to design a teleoperation system dealing with time-variant delays and environmental uncertainties. Using predictive strategies and RBF networks, the stability and transparency of the system are guaranteed amid the noise disturbance. The RBF neural network is trained offline using a set of environmental models and selects the best fitted

Figure 7. Position of master and slave with time-variant delays and noises

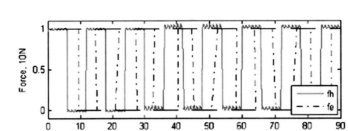

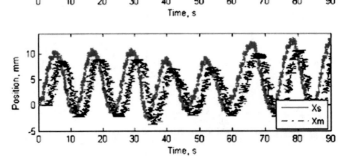

model in the current environment dynamics. Issue of communication uncertain delays in teleoperation is also addressed. A neural network-based multiple model adaptive controller is proposed to design the time forward state observer. Simulations show that the system is stable and has good performance. Although, the system becomes unstable if conditions for the stability and transparency are not feasible. Further analysis is needed for the effectiveness with respect to the achievable performance and reliability of this design.

ACKNOWLEDGMENT

The author would like to thank IGI Global for providing the opportunity to print this book chapter. Special thanks are also due to the reviewers and editors, who have contributed to the preparation of this book chapter. Lastly, the author would like to thank the Tallinn University of Technology (TTU) for supporting the preparation of this book chapter.

REFERENCES

Boukhnifer, M., & Ferreira. (2006). Stability and transparency for scaled teleoperation system. In *Proceedings of the 2006 IEEE/RSJ International Conference on Intelligent Robots and Systems*, (pp. 4217-4222). Bejing, China: IEEE.

Chen, J., Sun, J., Liu, G., & Rees, D. (2010). new delay dependent stability criteria for neural networks with time varying interval delay. *Physics Letters. [Part A]*, *374*, 4397–4405. doi:10.1016/j.physleta.2010.08.070.

Chen, Q. Quan, & Xia. (2007). Neural network-based multiple model adaptive predictive control for teleoperation system. In *Proceeding of the 4th International Symposium on Neural Networks: Advances in Newral Networks*, (pp. 64-69). Nanjing, China: IEEE.

Hashtrudi-Zaad, K., & Salcudean. (2002). Transparency in time-delayed systems and the effect of local force feedback for transparent teleoperation. *IEEE Transactions on Robotics and Automation*, *18*(1), 108–114. doi:10.1109/70.988981.

Kikuchi, K. Takeo, & Kosuge. (1998). Teleoperation system via computer network for dynamic environment. In *Proceedings of the IEEE International Conference on Robotics and Automation*, (pp. 3534-3539). Leuven, Belgium: IEEE.

Lawrence, D. A. (1993). Stability and transparency in bilateral teleoperation. *IEEE Transactions on Robotics and Automation*, *9*(5), 2649–2655. doi:10.1109/70.258054.

Minh, V. T., Nitin, A., & Mansor, W. (2007). *Fault detection and control of process systems* (p. 80321). Article, ID: Mathematical Problems in Engineering. doi:10.1155/2007/80321.

Park, J. H., & Kwon, O. M. (2009). Delay-dependent stability criterion for bidirectional associative memory neural networks with interval time-varying delays. *Modern Physics Letters B*, *23*(1), 35–46. doi:10.1142/S0217984909017807.

Slawinski, E., & Mut, W. (2008). Transparency in time for teleoperation systems. In *Proceedings in IEEE International Conference on Robotics and Automation*. Pasadena, CA: IEEE.

Smith, A.C., & Hashtrudi-Zaad. (2005). Adaptive teleoperation using newral network-based predictive control. In *Proceedings of the 2005 IEEE Conference on Control Applications*, (pp. 1269-1274). Toronto, Canada: IEEE.

Suebromran, A., & Parnichkun. (2005). Disturbance observer-based hybrid control of displacement and force in a medical tele-analyzer. *International Journal of Control, Automation, and Systems*, *3*(1), 70–78.

Zhu, W.H., & Salcudean. (1999). Teleoperation with adaptive motion/force control. In *Proceedings of the IEEE International Conference on Robotics and Automation*, (pp. 231-237). Detroit, MI: IEEE.

Chapter 12
Design and Development of Teleoperation for Forest Machines:
An Overview

Bart Milne
University of Canterbury, New Zealand

Chris Hann
University of Canterbury, New Zealand

XiaoQi Chen
University of Canterbury, New Zealand

Richard Parker
Scion, New Zealand

Paul Milliken
Scion, New Zealand

ABSTRACT

Teleoperation of forestry machinery is a difficult problem. The difficulties arise because forestry machines are primarily used in unstructured and uncontrolled environments. However, improvements in technology are making implementation of teleoperation for forestry machines feasible with off-the-shelf computing and networking hardware. The state-of-the-art in teleoperation of forestry machinery is reviewed as well as teleoperation in similarly unstructured and uncontrolled environments such as mining and underwater. Haptic feedback in a general sense is also reviewed, as while haptic feedback has been implemented on some types of heavy machinery it has not yet been implemented on forestry machinery.

1. INTRODUCTION

Teleoperation of forestry machinery is a difficult problem. Difficulties arise because forestry machines are primarily used in unstructured and uncontrolled environments. Improving technology is easing the difficulties and making teleoperation of forestry machines feasible with off-the-shelf computing and networking hardware.

Effective teleoperation in unstructured and uncontrolled environments requires sensing and modelling combined with a good operator

DOI: 10.4018/978-1-4666-4225-6.ch012

interface. Good feedback and controls creates *telepresence*, where an operator can operate a machine as well as if they were controlling it directly. Some types of feedback of interest include haptic (touch-force), orientation, audio, 3D mapping and binocular vision. Other unstructured and uncontrolled environments include mining (Hainsworth 2001), horticulture (Murakami, Ito et al. 2008; Billingsley, Oetomo et al. 2009), and aircraft control (Army UAS CoE Staff 2010; Pounds, Mahony et al. 2010).

Teleoperation may be thought of as an extension of remote control. Remote controlled systems are cheaper and easier to implement but are limited to situations where the operator can directly observe the machine. Effective remote control becomes difficult or impossible in conditions like underwater (Boyle, McMaster et al. 1995; Balchen 1996; Lin and Kuo 1999; Hirabayashi, Akizono et al. 2006), nuclear reactors (DeJong, Faulring et al. 2006; Basanez, Surrez et al. 2009), underground mining (Duff, Caris et al. 2009), and where the operator has to be isolated from the machine (Hellström, Lärkeryd et al. 2008; Ringdahl 2011; Komatsu 2012).

In summary, some possible benefits of teleoperation of forestry machines based on other field machines are:

- Improved comfort for the operator (Boyle, McMaster et al. 1995; Balchen 1996; Hainsworth 2001; Hellström 2005; DeJong, Faulring et al. 2006; Hirabayashi, Akizono et al. 2006; Duff, Usher et al. 2007; Hellström, Lärkeryd et al. 2008; Murakami, Ito et al. 2008; Brace 2009; Duff, Caris et al. 2009; Army UAS CoE Staff 2010).
- Greater profits by improving operator productivity (Hellström, Lärkeryd et al. 2008; Murakami, Ito et al. 2008; Brace 2009; Ringdahl, Lindroos et al. 2011; Komatsu 2012).

- Implementation of interfaces and control algorithms that enhance operator skills (Lawton, Schoppers et al. 1989; Balchen 1996; Lin and Kuo 1999; Fong and Thorpe 2001; Brander, Eriksson et al. 2004; Hellström 2005; DeJong, Faulring et al. 2006; Hirabayashi, Akizono et al. 2006; Hellström, Lärkeryd et al. 2008; Murakami, Ito et al. 2008; Westerberg, Manchester et al. 2008; Brace 2009; Osafo-Yeboah and Jiang 2009; van der Zee 2009; Gunn and Zhu 2010; Hansson and Servin 2010; Hayn and Schwarzmann 2010; Yoon and Manurung 2010; Komatsu 2012).

As outlined previously teleoperation has many significant potential benefits. The main problems that must be addressed in the development of teleoperation are:

- Determining the sensors required on the target machine.
- Defining the user interface requirements and creating a suitable user interface.
- Ensuring the target machine remains safe at all times.
- Ensuring the operator has sufficient situational awareness.

Section 2 is a summary of the forestry industry in New Zealand. Section 3 is a survey of existing machinery in mining and forestry. Section 4 goes into the aspects of the design of Teleoperation Test and Development Systems. Section 5 covers some test and experimental systems that are useful for developing teleoperation.

2. FORESTRY IN NEW ZEALAND

The economy of New Zealand is highly dependent on primary industries such as agriculture, fishing and forestry. In the year ending June 2011 primary industries earned NZ$31.5 billion in export receipts (71% of the total) (New Zealand Ministry of Primary Industries 2012). Forestry contributed NZ$4.5 billion (10%) in the same period.

Planted forests constitute 1.74 million hectares (7%) of New Zealand's total land area. Radiata pine plantings are the largest, totalling 1.56 million hectares (90% of planted area), with Douglas-fir plantings the second largest with 110,000 hectares (6.3%). The total wood harvest for the year ending 31st March 2010 was 20.7 million m^3. Radiata pine trees are harvested at an average age of 28.4 years and the average clear fell yield is 473 m^3 per hectare. Each individual tree is on average 2.3 m^3 in volume for sawlogs and 1.6 m^3 for structural timber (New Zealand Forestry Owners Association 2011).

A large amount of marginal (steep and remote) farm land was converted to forestry in the 1980s and 1990s. These forests are nearing maturity, and will be harvested for the first time over the coming years. A method of harvesting that maximizes profits must be found (Raymond 2012). Solutions used in other countries have limited applicability to New Zealand. Scandinavia has one of the most heavily mechanised forestry industries in the world but the machinery is designed for smaller trees, flat land and a very high level of automation (Billingsley, Arto et al. 2007). Japan has forestry on steeply sloped marginal land, but the forestry industry in Japan is in decline due to cheaper wood from overseas and is more focused on small scale production and environmental management (MacDonncadha 1997; Nishino 2000; Ota 2001).

In 2009 an estimated 6,000 people were employed in New Zealand's forestry and forestry services industry, including 3,310 people employed in logging. Forestry work has a very high yearly fatality rate of 22.4 per 100,000 workers. Although this results in only about four fatalities per year, the serious harm injury rate is 18 for every 1000 workers per year, or nearly two serious harm injuries per week. The serious harm injury rate is more than double the next danger-

ous industries in New Zealand such as mining and utilities (New Zealand Ministry of Primary Industries 2012). Manual forest harvest work in New Zealand is perceived as low status, poorly paid, difficult and dangerous.

Mechanical harvesting carries costs of machinery, fuel and maintenance but the productivity improvements cause the cost of harvesting to drop by up to 50% (Raymond 2012). Conventional mechanised harvesting equipment is not allowed on slopes steeper than 22° in New Zealand (Occupational Safety and Health Service 1998). Cable extraction of logs is being used with great success on steep terrain but harvesting is still performed manually. Scion Research (Scion), Future Forests Research (FFR) and the University of Canterbury have awarded a PhD research scholarship to develop a teleoperated steep slope (up to 45°) teleoperated forestry harvester. The desired outcome of the project is to create a harvester that improves productivity and removes the worker from an extremely hazardous environment.

Trinder Engineers of Nelson, New Zealand have been developing a cable-tethered steep slope harvester for several years (Evanson and Amishev 2010). As of 2012 Trinder Engineers have developed a "beta" prototype (ClimbMAX) that uses a winched cable tether for stabilization on slopes steeper than 22° (Raymond 2012). At present the ClimbMAX harvester (Figure 1) requires an operator in the cab, but implementing teleoperation will enable the harvester to be operated without an operator in the cab, removing the hazard to the operator of being in heavy equipment on steep slopes.

2.1. Example of Steep Slope Harvesting in New Zealand

Figure 2 shows a typical steep slope harvesting site located at Okuku, about 30 km north-west of Christchurch, New Zealand. In the foreground trees have just been harvested, and in the background some of the trees are being felled. All the

Figure 1. Trinder ClimbMAX winch harvester beta prototype (Evanson and Amishev, 2010; Raymond, 2012)

Figure 2. Steep slope forestry harvesting site in New Zealand

steep slope felling was done manually. Due to the steepness of the terrain the trees were hauled up the slope before being cut to length.

In sites with easier terrain, logs are cut to length before being bunched. For the case of a cable hauler it is more practical to work with full length logs and cut them up at the landing. The logs are held by wire rope strops (chokers) that self-tighten as the logs are pulled up the slope. Figure 3 shows the cable hauler at Okuku Forest extracting a single log, but multiple logs can be extracted at once (Figure 3).

Figure 3. Cable hauler bringing up a log

The next step is to sort the logs by size and quality at a landing, including cutting to desired lengths (Figure 4).

In more gently sloping sites many of the tasks performed by separate machines can be combined. Figure 5 is from a harvesting site in Charteris Bay, on Banks Peninsula near Christchurch in New Zealand. The slope is shallower and the trees are smaller so mechanised harvesting is in use. The trees were felled and bunched at the harvest site and sent to be sorted and classified in a log forwarder. On steep slopes it is not possible to use a forwarder, so the steep slope harvester's functionality may be limited to felling and bunching, with cutting to length being performed after the log has been delivered to the sorting yard.

Figure 4. Log cutting and sorting

Figure 5. Feller, buncher, and log forwarder at Charteris Bay, near Christchurch, New Zealand

2.2. Robotisation of Tree Harvesting in New Zealand

A cable-assisted steep slope harvester has been developed in New Zealand (Evanson and Amishev 2010) but it still requires the presence of an operator in the cab on a steep slope. A PhD project in conjunction with Scion Research and Future Forests Research has been started to improve the productivity and safety of cable-assisted excavators and other steep slope harvesters through teleoperation. Hardware will be developed and retrofitted to the excavator to enable teleoperation. Further improvements in productivity will be sought through the inclusion of features such as:

- Haptic feedback for the remote operator.
- Overlaying IMU information onto the operator's display.
- Semi-automation such as using laser-triangulation sensor to autonomously move the felling head to a position where it can grasp a tree.
- Using laser-triangulation sensors for collision avoidance and localised terrain mapping.
- Binocular vision for improved depth perception.

The increasing proportion of land planted for forestry will increase the demand for forestry workers. The work is hazardous and has a long training time. Robotisation is being explored to isolate the worker from hazards, increase productivity, and attract more new workers. Many areas of New Zealand are physically suitable for forestry but the lack of infrastructure and high labour costs make harvesting uneconomic (Amishev, Evanson et al. 2009; Raymond 2012). The motivations, research issues and challenges are (Chen 2012):

- **Motivations:**
  - **Automation of Forest Operations:** felling, collecting.
  - Improvements in safety.
  - Improvements in productivity.
- **Research Issues:**
  - Machine mobility and stability on steep and soft soil country.
  - Navigation of harvesting machines.
  - Sensor fusion and teleoperation.
- **Challenges:**
  - Highly unstructured, obstructive outdoor environment.
  - Steep and soft soil forestry peculiar to NZ.
  - New Zealand trees are big.

Many of the objectives prior can be pursued in parallel. For example, usage of forestry harvesters on steep slopes is being actively researched (Amishev, Evanson et al. 2009; Parker 2009; Amishev and Evanson 2011).

3. OVERVIEW OF FIELD ROBOTICS FOR MINING AND FORESTRY

Field robots have different characteristics compared to fixed-based industrial robots. These are mobility and use in an unstructured natural environment. Mining robotics have some commonalities with forestry robotics, so mining robots are also reviewed here.

3.1. Forestry

Teleoperation in forestry is under active development, especially in Scandinavia (Hellström 2005; Hellström, Lärkeryd et al. 2008; Hellström, Lärkeryd et al. 2009; Ringdahl, Lindroos et al. 2011). Teleoperated forestry harvesting is challenging due to the unstructured and uncontrolled environment. The machine's control system must operate in the presence of uncertainty about tree size and environmental hazards like wind, weather, sunstrike, and soil conditions.

Simulation based control systems like Skogforsk's virtual harvester platform are useful for developing teleoperation (Brander, Eriksson et al. 2004). Skogforsk developed the simulator to test the effect of different forestry harvester control methods on operator productivity. Although this platform was not a teleoperated system, simulators are useful for delivering teleoperation user interfaces. For example, if different equipment is designed to control the input/output of the simulated environment, then the system is effectively teleoperated, even if the design did not have that specific objective.

The "Besten and Kuriren" system is an example of a remote controlled tree harvester (Bergkvist, Nordén et al. 2006). The Besten is designed for use with multiple log forwarders (Figure 6). The

Figure 6. "Besten" remote controlled harvester. Forwarder on right (Bergkvist, Nordén, et al., 2006).

log forwarders' operator controls the Besten, effectively turning each log forwarder into a harvester. Each log forwarder controls the Besten in turn when it returns to the harvest site. Under certain conditions of tree size and stand density the Besten can improve productivity by 15% but the productivity improvement is dependent on having enough log forwarders to keep the Besten in operation at all times (Bergkvist, Nordén et al. 2006). This solution is not suitable for steep slopes as it requires direct forwarder access to the harvester.

Log forwarders transport cut logs from the harvesting site to the sorting area. Teleoperation for log forwarders is easier to develop due to the more structured and controlled operating environment. A single worker can supervise multiple machines in a similar manner to autonomous load-haul-dump trucks in mining. Automating the loading and unloading of logs is more difficult but the operator can always switch to manual mode if required. The log forwarder's crane arm can then be teleoperated using enhanced control methods such as mixed reality and "point and click" (Westerberg, Manchester et al. 2008). Westerberg's paper only dealt with controlling the arm, it did not deal with the entire machine.

Differential GPS (DGPS) is specified as the accuracy is better than having GPS receivers on the forwarder. In particular the forest cover and terrain can attenuate the GPS signals, causing signal dropouts and producing errors of tens of metres (Emde, Krahwinkler et al. 2012). The DGPS base station is placed in a fixed location optimized for satellite reception. Extra satellite positioning networks such as GLONASS or QZSS can be used to further enhance accuracy and reliability of positioning.

DGPS has been used to remotely control the position and path of a log forwarder (Ringdahl, Lindroos et al. 2011). If DGPS coverage is not available the gyro, steering angle data and machine speed data provides estimates of position and heading. Wheel odometry was considered for position estimation but in a forest environment it is too vulnerable to accumulated inaccuracy from wheel slippage. Other methods of distance and speed measurement such as radar and laser scanning can be used to overcome the accuracy problems of wheel odometry (Ringdahl, Lindroos et al. 2011).

A recent innovation in forestry harvesting is to harvest trees that have been flooded during dam construction. The interest in underwater harvesting comes from eliminating hazards to shipping and the untapped supply of highly valuable species such as old growth red cedar and tropical hardwoods. Triton estimates there is US$50 billion worth of wood that can be harvested this way (Triton Logging 2012). Triton Logging has developed two types of underwater harvesters: an excavator based harvester that is operated from a barge ("Sharc", See Figure 7) and a harvester that is operated entirely underwater ("Sawfish"). The Sharc's harvesting head is on a specially designed telescopic arm and can work in depths of up to 37 metres. The operator is located in the excavator cab but uses cameras and sonar sensors to locate the harvesting head.

The Sawfish (Figure 8) is designed to work entirely under water. The Sawfish uses air bags to float the log to the surface after being cut. The airbag system enables the underwater harvester to work continually.

Figure 7. Triton "sharc" underwater harvester (Triton Logging, 2012)

Figure 8. Triton "sawfish" underwater logging vehicle (Triton Logging, 2012)

3.2. Mining

Teleoperated machines have been used for many years in mining. Rio Tinto has several different types of teleoperated machines in use, including autonomous load-haul-dump trucks and mining plant equipment (Duff, Usher et al. 2007; Duff, Caris et al. 2009; Gunn and Zhu 2010; Komatsu 2012). Mining provides many opportunities for using teleoperation due to the variety of operating conditions from highly hazardous to highly structured. Load-haul-dump (LHD) trucks remove material from the mining site and take it to be processed. LHD trucks are very suitable for supervisory control as they work on preplanned routes in a highly structured environment. Komatsu have implemented their "FrontRunner" system for running LHD trucks autonomously at West Angelas Mine (Figure 9). The control system architecture is very similar to the autonomous forwarder control system as described in (Ringdahl, Lindroos et al. 2011).

Some work does not require the operator's presence in the immediate environment. An ore crusher at West Angelas mine in Western Australia is used to break up lumps of iron ore that are too large for further processing (Figure 10). The ore crusher was modified for teleoperation as part of a feasibility study (Duff, Caris et al. 2009; Zhu, Gedeon et al. 2011). The ore crusher was successfully operated from 1000 km away in Perth. Although the system worked it was thought that haptic feedback provided a better experience by allowing the operator to feel the behaviour of the tip of the boom (Gunn and Zhu 2010).

Figure 9. Autonomous load-haul-dump system (Komatsu, 2012)

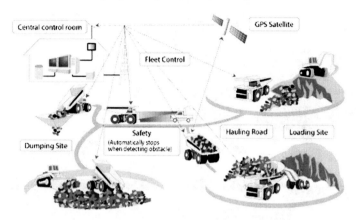

Figure 10. Teleoperated ore crusher (Duff, Caris, et al., 2009; Zhu, Gedeon, et al., 2011)

Teleoperation is desirable in mining for isolation of the user from operational hazards. For example, high wall coal mining has the advantage of removing less overburden, but the operational hazards are significant: a high and possibly unstable cliff and drilling into unknown ground (See Figure 11).

The coal seam itself also has the risk of collapse and seepage of explosive gases. Teleoperation isolates the operator from the collapse and explosion hazards. In some cases the mining drill does not have any cameras so a model of the coal seal and position information from the drilling head is used to generate the drill head graphic. Other information shown includes production, depth, and explosion hazard (Figure 12).

Figure 11. High wall coal mining machine (Hainsworth, 2001)

Figure 12. High wall coal miner status screen (Hainsworth, 2001)

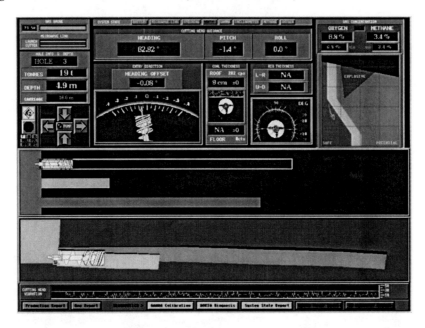

The Numbat is a mine vehicle developed both for mining rescue and as a research and development platform (Figure 13). In the event of a mining accident, the Numbat can be sent in before rescue teams to check on conditions such as visibility and explosive gas concentrations. The control system and methods are quite simple, as the Numbat was designed to move slowly enough that the operator could work out the machine dynamics in real time. Mine rescue is quite a difficult and time consuming activity but reducing the speed of operation may not be suitable in other applications due to loss of productivity.

The Numbat has a dedicated status screen (Figure 14), showing data like temperature and explosive gas concentrations. The temperature sensors appear to read below absolute zero when they are disconnected. If this is the case, it is better to have an explicit indication of sensor status to prevent confusing the operator.

4. TELEOPERATION AND CONTROL SYSTEMS

The operating environment of a robot is a critical factor in determining how to implement its sensing, locomotion and user interface. Environmental limits may be imposed by industrial processes such as reflow soldering, or the robot may fixed or operated solely within a highly constrained environment. Highly structured and controlled environments are typically found in factories. Robots operating under strictly controlled conditions

Figure 13. "Numbat" teleoperated mine vehicle (Hainsworth, 2001)

Figure 14. Numbat status screen (Hainsworth, 2001)

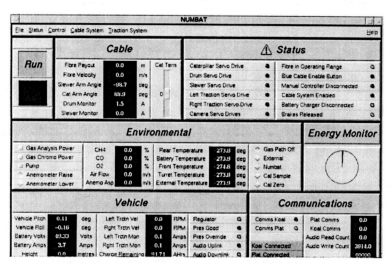

can stop and sound alarms if any vital condition moves out of an acceptable range. It is assumed in the design of the robot that the non-compliant environment will be corrected manually.

Uncontrolled environments have more significant disturbances and the control system must be robust enough to handle the disturbances, particularly disturbances that impact on personal safety (Bard 1986; Nagamachi 1986; Parsons 1986; Hainsworth 2001; Dede and Tosunoglu 2006; Murakami, Ito et al. 2008). A simple example is a heater in a building. The temperature controller must compensate for the effects of the opening and closing of doors. The system is non-linear because the heater can positively force the temperature but not negatively force it. If a linear controller is used the performance of the system will be compromised. In particular, careless use of the "I" integral component in a PID controller could result in severe problems with integral windup. Limiting the magnitude of the "I" component may help but the control system has been made non-linear as well and harder to analyse.

A summary of the different types of operating environments is given:

- **Controlled Environments:** Operating conditions are tightly controlled and known in advance. Dealing with disturbances may not be required at all except to stop and set off an alarm. These sorts of controlled environments are common in factories.

- **Semi-Controlled Environments:** Some operating conditions are known and controlled but others are not e.g. autonomous load-haul-dump trucks on pre-planned routes (Komatsu 2012).

- **Uncontrolled Environments:** Little control over operating conditions. General parameters are known but cannot be predicted. The control system must be able to handle significant disturbances. Examples of uncontrolled environments include underwater (Boyle, McMaster et al. 1995) and battlefield conditions (Army UAS CoE Staff 2010).

4.1. Communications

There is no requirement to use the Internet for teleoperation communication, but the Internet is an attractive medium to use due to the widespread availability of Internet-based communications hardware (Hu, Yu et al. 2001; Álvares and Ferreira 2006; Duff, Usher et al. 2007; Rodríguez and Guerrero 2010). Alternative communications protocols may require more work to implement error checking and data transmission protocols and methods, and may be difficult to extend beyond the work site. The Internet is very widely available, but it has unpredictable time delay that can cause major difficulties when implementing control systems (See Figure 16). Internet-based control systems require delay compensation that can accommodate varying time delays and ensure stability and controllability (Yokokohji, Imaida et al. 2000; Hokayem and Spong 2006; Slawiński, Postigo et al. 2007; Chopra, Berestesky et al. 2008).

Wireless communications are the most useful for teleoperation of mobile machines. Different wireless protocols have distinct advantages and disadvantages, with a general three-way trade-off between speed, range and error tolerance (Figure 15). A communications protocol that already takes care of flow control, signal timing and error correction is most desirable. The main WiFi standards are 802.11b and 802.11g. 802.11b has longer range but lower data rate, the opposite is true with 802.11g (See Figure 15). In (Thompson, Harmison et al. 2006), two 802.11b adapters were used, as well as an 802.11g adapter and a MeshNetworks WiFi wireless modem. The range performance of the MeshNetworks adaptor was the least erratic and approximately the same as the 802.11b (See

Figure 16). Both the MeshNetworks and the 802.11b adaptor had a range of 50 metres. The performance of the 802.11g adaptor dropped off very quickly as it reached its maximum 25m range (Thompson, Harmison et al. 2006). The 802.11b adaptor had the least delay within its range (60ms), but the delay began to get erratic outside a 15m range. The MeshNetworks adaptor had the most constant delay behaviour (160ms), and its delay only started becoming erratic at the edge of its signal range. The 802.11g adaptor has the shortest delay of all (40ms) but as the signal goes out of range the delay increases to as much as 1000ms until the signal drops out altogether (Thompson, Harmison et al. 2006). The behaviours suggest that the MeshNetworks adapter has a stronger error detection and correction algorithm but at the cost of lower peak performance.

A shortcoming of TCP/IP is that the transmission delay is not predictable, which is made much worse when it is sent over wireless. Other communications methods such as ISDN have more predictable communications delays than TCP/IP (Figure 17). In one experiment in data transfer over an ISDN the setup time was 60ms with a transmission delay of 20ms per 1000 bytes (Figure 17). While specialised networks like ISDN can overcome problems of unpredictable delay,

Figure 15. Transmission speed behaviour of different wireless Ethernet devices. 802.11-type device transmit powers are 20 dBm, MeshNetworks 23 dBm (Thompson, Harmison, et al., 2006).

Figure 16. Average delays of various WiFi equipment (Thompson, Harmison, et al., 2006)

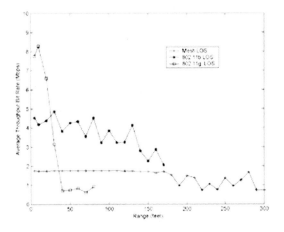

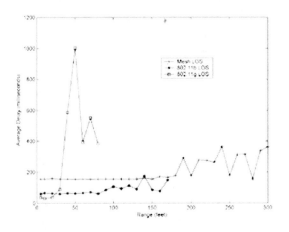

Figure 17. Delay behaviour of ISDN compared to TCP/IP (Chong, Kotoku, et al., 2003)

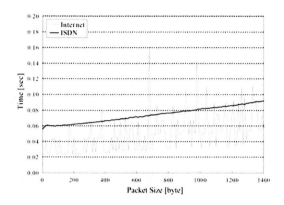

4.2. Operator Interfaces and Control Methods

Telepresence is when the operator is able to control a teleoperated machine as effectively as if they were operating it directly (Sheridan 1995; Hokayem and Spong 2006). Telepresence requires a combination of good sensing, good modelling and a good user interface. User interfaces are often manufactured from existing pieces of standard hardware such as television screens and handheld controllers for use in the field (Hainsworth 2001). The control console is part of a specially made seat, and the feedback is visual via monitors. The control system appears to be a radio control console that has been mounted in front of the seat.

An advanced interface with highly customized and integrated hardware is not necessarily a requirement in teleoperation. For example, using carefully selected pre-existing hardware and software may reduce development costs substantially.

Argonne National Laboratory has been developing teleoperated equipment for work in nuclear reactors since the 1950s (Ferguson 1978; Basanez, Surrez et al. 2009). A more recent system is a teleoperated pipe saw intended for use when dismantling or performing maintenance on nuclear reactors (DeJong, Faulring et al. 2006). During usability testing it was found that the operators had great trouble keeping the saw blade planar, which resulted in jamming and breaking blades. Constraining the motion of the saw to a plane helped prevent blade breakage. For general movements the operator is able to use "free form" mode, where the saw movement is unconstrained. When the operator is cutting the pipe they can select the mode to constrain the movement of the saw in one plane only. The horizontal controls are recognizable as the r-axis and θ-axis in cylindrical coordinates.

A "virtual cylinder" is used to restrict radial movement of the saw. In some respects the radial constraint is unnecessary, as the blade guard and the pipe will naturally restrict the radial movement

interoperability and availability of hardware may be difficult. The Internet is also desirable since it often has a much larger available bandwidth than an ISDN. An alternative approach is to increase the autonomy so that the control system is less dependent on receiving instructions in real time (Álvares and Ferreira 2006).

For audio and video available bandwidth is a vital consideration. A typical audio bandwidth requirement might be 128,000 bits per second for stereo sound. The video bandwidth requirements are dependent on codec but a simple format like M-JPEG (motion JPEG) is preferred for its simplicity and ability to recover from dropouts. For ease of control a minimum frame rate of 10 frames per second is desirable (Chen and Thropp 2007). For stereo vision this translates to 20 frames of video per second. A high quality 640×480 pixel JPEG image is about 110 kilobytes, and a medium quality image is about 50 kB. For stereo vision and sound at 10 frames per second the required bandwidths are approximately 18 megabits per second for high quality and 8.3 megabits per second for medium quality. The bandwidth may be more than what WiFi equipment can deliver. It may be possible to use more advanced video formats such as MPEG formats but the extra computational overhead and processing time adds to any lag present due to communications delays.

of the saw. Keeping the saw blade in the same plane vastly reduces the chance of blade breakage. The user interface for the pipe say includes a screen that shows the position of the virtual surface in real time (DeJong, Faulring et al. 2006).

Control using virtual constraints and haptic feedback is applicable to forestry harvesting. The harvesting head can be constrained to movement in an exact horizontal plane, regardless of the actual angles the harvester may be on.

5. TELEOPERATION CONCEPTS, CONTROL SYSTEMS, AND EXPERIMENTAL PLATFORMS

There is no inherent requirement to use a test system for developing teleoperation. However, using a test system is desirable for the following reasons:

- The machine is too large to fit into available laboratory space (Duff, Usher et al. 2007; Duff, Caris et al. 2009).
- The machine may be in active production use and not available for developmental use.
- The machine may operate in an environment that is not reproducible in the laboratory, such as underwater (Boyle, McMaster et al. 1995; Hirabayashi, Akizono et al. 2006).
- The machine may be located far away from a suitable location for development (Duff, Caris et al. 2009).
- A failure of the development system may cause considerable damage and down time. The machine may be a "one-off" that cannot be replaced (Duff, Caris et al. 2009).
- It is not necessary to have all of the machine as only a part of the machine is being teleoperated (Hansson and Servin 2010; Yoon and Manurung 2010).

5.1. Teleoperation Development Systems by Purpose

Practically any aspect of the behaviour or implementation of the development system is at the discretion of the designer. However, the more closely the test system replicates the behaviour of the field machine the easier it will be to apply results and design features.

Example teleoperation development scenarios:

- **Proof of Concept:** Teleoperation is being developed but without retrofitting a base machine. The control algorithm for a large machine is to be studied or refined. A part of the machine is isolated and used in a test laboratory (Chong, Kotoku et al. 2003; Hansson and Servin 2010)
- **Prototype Design:** A "base platform" exists but requires modification or refinement for the specific application. The detailed mechanical design will be performed in concert with the teleoperation, or the teleoperation system is being used to influence the mechanical design. The mechanical design can be altered to accommodate teleoperation.
- **Design for Retrofit:** An existing fully functional machine is to be retrofitted for teleoperation. All the mechanical design work has been done and the teleoperation system needs to be designed to accommodate the machines mechanical behaviours (Hirabayashi, Akizono et al. 2006).

The purpose of the teleoperation will define what information needs to be gathered for processing and presenting to the operator. Sensors for a lab scale model may be unsuitable for use on the field machine but are still useful for development if they use the same communications protocols and have similar speed and accuracy to field machine sensors. Models are useful if the operating conditions such as dust, underwater or underground make visual feedback difficult. The model could be built from pre-prepared data with sensor data incorporated into it.

5.2. Teleoperation for Uncontrolled Environments

Uncontrolled environments make control system design challenging due to the increased likelihood of significant disturbances. Disturbances can cause actuators to be driven into non-linear regions or cause a desired position to become physically impossible to reach. Good control systems will keep the target machine safe and controllable despite the disturbances and non-linearity.

Overcoming the challenges of developing teleoperation for field machines provides an opportunity for significant research and development. A key to gaining insight into this field is to first understand teleoperation at a smaller scale. However, most importantly the testing protocols and methods developed at the smaller scale need to be completely transferrable to the larger scale. This approach is thus focussed on developing a flexible method, rather than a "one size fits all" approach that required a large amount of testing and cost. For example, a small number of controlled step responses could be applied on the development system to tune a given model, and tested on a wide variety of responses. If this small number is adequate to characterise the dynamics it will give confidence that a similar approach will work on the full scale where only limited experiments can be performed.

In addition, the hardware and software architecture used for the small scale should be a similar system to the full scale. Hence, any control algorithms developed in this small scale will be immediately applicable to large machines and thus significantly reduce development time and cost. Finally, a small scale machine, for example an excavator, is typically much harder to control than a large excavator. Hence, if control algorithms are developed to successfully operate a small excavator, the problem of controlling the larger excavator will become much simpler. There are a number of specific options for the specific type of small scale system including:

- A simulation tool using models validated from small scale experiments.
- A completely different kind of machine for example using a Kuka robot for validating a control method without any particular target machine in mind.
- A smaller machine that works similarly to the full size machine and contains similar hardware and software interfaces including instrumentation setups.
- A component of the full size target machine for example using the arm of an excavator but not including the tracks or chassis in the development system.

5.2.1. Teleoperation Test and Development Systems in Practice

Teleoperation for robots in controlled environments is relatively straight forward because the operating environment is known and predictable in advance. Environmental controls reduce the chance of damage to the machine. For example, circuit board soldering by factory machines is performed under tightly controlled conditions where replacement of failed components is easy.

In forestry harvesting uncontrollable disturbances include sun strike, weather and unknown ground conditions. The exact magnitude and timing of these disturbances cannot be accurately predicted in advance. For example, sensors and actuators may be driven to the end of their range, introducing an unexpected non-linearity. The control algorithm must be able to automatically handle this non-linearity and other disturbances to ensure the operator maintains good control.

Some devices have hardware that is usable within a laboratory environment. An example is a log grabber test platform for developing semi-autonomous control. The log grabber has a telescopic arm and has been used in the laboratory with real logs (Hansson and Servin 2010). Only the log grabber arm has been recreated in the laboratory. The target machine already exists so it makes sense to use the same hardware architecture (Figure 18, Figure 19).

The driving hardware for the development system may be different from the target machine. Different hardware will reduce the transferability of system responses but if the development system hardware uses the same communications protocols the communications system can stay the same, reducing the time and complexity of transferring the system to the target machine. It is essential the results from the development system are applied to the target machine properly. Knowledge of the mechanical behaviour of the target machine is required to do this properly.

Sometimes the field machine is small enough to be used in the laboratory. The field machine can have work done in the laboratory and outside in more structured and controlled conditions. A small excavator was fitted out for teleoperation with arm position was small enough for use within a laboratory (Kim, Kim et al. 2009). The hydraulic control valves have been converted to electrical operation from the original hand levers. Once a control method is more fully developed the machine can be tested in a controlled version of its usual operational environment (Figure 20, Figure 21). The arm-controlled excavator was tested outside with piles of loose dirt (Kim, Kim et al. 2009). Operating in a more structured and controlled environment reduces the risk and consequence of damage but also reduces the realism of the environment it operates in and can potentially affect the quality of the results.

A fixed excavator arm was created in the laboratory to determine how to make a two-lever excavator control more intuitive (Yoon and Manurung 2010). The investigation only required the arm and bucket (Figure 22). However, in a full track control hydraulic system, there are interac-

Figure 18. Concept teleoperation system (Hansson & Servin, 2010)

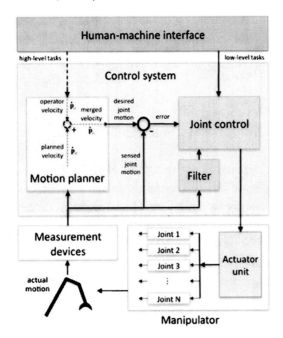

Figure 19. Log grabber test laboratory (Hansson & Servin, 2010)

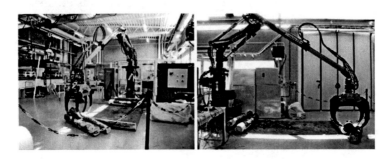

Figure 20. Excavator fitted out for teleoperation (Kim, Kim, et al., 2009)

Figure 21. Excavator being field tested (Kim, Kim, et al., 2009)

tions through changes in pressure that would not be seen in the simple system of the type. Hence, to extend the results of this study to the full system may require significant modifications, further demonstrating the significant challenge in developing teleoperation for large machines.

The size, cost and availability of field robots has important consequences for developing teleoperation. It is a development system for Argonne National Laboratory's DAWP system (DeJong, Faulring et al. 2006). The test system implements

Figure 22. Excavator test system (Yoon & Manurung, 2010)

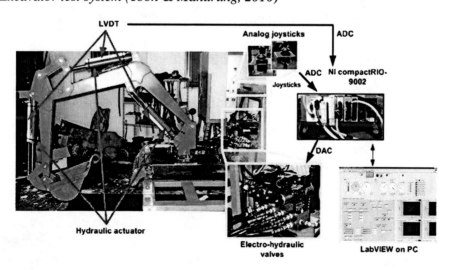

a limited subset of functionality but the part being tested is the saw itself, teleoperation, haptic feedback and the control interface. The sensor behaviour and configuration on the development system determines how easily the saw can be used with the DAWP. Obviously it is desirable to have to do as little redesign as possible. Modularising the teleoperation system design is helpful for ensuring the control system is relatively simple to repair or upgrade (Hansson and Servin 2010). Separating the teleoperation system components can require more design effort but makes it easier to transfer teleoperation system components and concepts between different devices.

5.3. Haptic Feedback

Haptic feedback is useful for providing supplementary information to the operator. Force reflection is one possibility but haptic feedback has been used to synthesise virtual force fields (DeJong, Faulring et al. 2006; Ghanbari, Abdi et al. 2010). There are commercially available systems such as the Phantom OMNI, and haptic control systems have used this to control excavators (Osafo-Yeboah and Jiang 2009) and it is also possible to design and build custom made haptic devices (van der Zee 2009). Force-velocity feedback can have problems with stability due to delays in the feedback loop contributing energy to the system. Passivity techniques like wave variable transforms may help alleviate stability problems (Hokayem and Spong 2006; Aziminejad, Tavakoli et al. 2008; Chopra, Berestesky et al. 2008; Alise, Roberts et al. 2009). Most excavator arms operate in a manner easily adaptable to cylindrical coordinates, and it is easy to map the movement of devices such as a Phantom OMNI haptic device to the arm on an excavator (Figure 23).

A Phantom OMNI haptic device has been used for use controlling a backhoe (Hayn and Schwarzmann 2010). The haptic device was used for trajectory guidance when performing manoeuvres shifting soil using the excavator bucket. The

Phantom OMNI was set up with the arm in the same orientation as the excavator arm (Figure 24). The disadvantage of the set up is that the base of the Phantom gets in the way of the operator's arm. An alternative set up is to have the Phantom OMNI arm facing back towards the operator. Appropriate feedback can eliminate any confusion about the movement of the Phantom Omni arm relative to the excavator arm.

Figure 23. Example mapping of a phantom OMNI haptic device to excavator boom motions (Hayn & Schwarzmann, 2010)

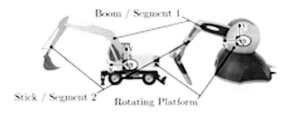

Figure 24. Excavator teleoperation with haptic feedback (Hayn & Schwarzmann, 2010)

CONCLUSION

The project was started in February 2012 so it is still in an early stage. Further development of hardware controls and teleoperation will be possible on the delivery of the hydraulic test system. The control system will be deployed on a small excavator once the communications and control methods are proven on the hydraulic test system. If the trials on the small excavator are successful the control will be deployed on a larger excavator for field testing on a forestry harvesting site.

The project duration is from February 2012 to February 2015. The duration includes system development, field testing, documentation, and thesis write-up. Teleoperation has large potential benefits for the New Zealand forestry industry in safety and productivity. Injuries and deaths from forestry are an ongoing problem and robotisation is a solution to these problems without reducing the land usable for forestry. Teleoperated forestry harvesting has been developed in other countries, but the forests are either of high enough value to justify the extra costs or the harvesting conditions are much easier.

The continual increase in processing ability and improvements in sensor and communications technology is broadening the scope of applications for teleoperation. Using a less expensive and resource-intensive minimal modelling and rapid development approach will allow development of sophisticated teleoperation for smaller, lower cost machines.

The mining industry has adopted teleoperation as a means of improving productivity and safety. Mining provides a pathway for forestry, as they are both extractive industries that use heavy machinery. Both industries have hazards that are difficult or impossible to reduce or eliminate, meaning that isolation is the only solution that both reduces the risk to human life and enables the resource to be extracted.

There are many opportunities for future work in this area. One of the foremost challenges is developing a teleoperation system that will function properly in an unstructured and uncontrolled environment while also keeping development and product costs low enough for the machine to be useful in industry. This research project will address these issues starting with the hydraulic test system. This system will develop the major communications and control methods before implementation on a small excavator. Once the concept is proven on the small excavator, the methods will be extended to a larger excavator for field testing on a forestry harvesting site.

ACKNOWLEDGMENT

The authors acknowledge Future Forests Research and Scion for their funding and assistance with materials and proof reading.

REFERENCES

Alise, M., Roberts, et al. (2009). On extending the wave variable method to multiple-DOF teleoperation systems. *IEEE/ASME Transactions on Mechatronics, 14*(1), 55–63. doi:10.1109/TMECH.2008.2006181.

Álvares, A. J., & Ferreira. (2006). Webturning: Teleoperation of a CNC turning center through the internet. *Journal of Materials Processing Technology, 179*, 251–259. doi:10.1016/j.jmatprotec.2006.03.096.

Amishev, D., Evanson, et al. (2009). Felling and bunching on steep terrain - A review of the literature. *FFR Technical Note, 1*(7).

Amishev, D., & Evanson. (2011). Walking machines in forest operations. *FFR Technical Note, 3*(9).

Army, U. A. S. CoE Staff. (2010). U. S. army roadmap for unmanned aircraft systems 2010-2035. Fort Rucker, AL: U. S. Army UAS Center of Excellence (ATZQ-CDI-C).

Aziminejad, A., Tavakoli, et al. (2008). Transparent time-delayed bilateral teleoperation using wave variables. *IEEE Transactions on Control Systems Technology, 16*(3), 548–555. doi:10.1109/TCST.2007.908222.

Balchen, J. G. (1996). Model based teleoperation of untethered underwater vehicles with manipulators. *Modeling. Identification and Control, 17*, 37–45. doi:10.4173/mic.1996.1.4.

Bard, J. F. (1986). An assessment of industrial robots: Capabilities, economics, and impacts. *Journal of Operations Management, 6*, 99–124. doi:10.1016/0272-6963(86)90020-3.

Basanez, L. (2009). *Surrez, et al.* Teleoperation.

Bergkvist, I., Nordén, et al. (2006). Innovative unmanned harvester system. *Skogforsk Results*.

Billingsley, J., Arto, et al. (2007). *Robotics in agriculture and forestry*.

Billingsley, J., Oetomo, et al. (2009). Agricultural robotics. *IEEE Robotics & Automation Magazine, 19*.

Boyle, B. G., McMaster, et al. (1995). Concept evaluation trials of teleoperation system for control of an underwater robotic arm by graphical simulation techniques. *Transactions of the Institute of Measurement and Control, 17*(5), 242–250. doi:10.1177/014233129501700504.

Brace, M. (2009). Automating the mine. *Earthmatters,* (19).

Brander, M., Eriksson, et al. (2004). Automation of knuckleboom work can increase productivity. *Skogforsk Results*.

Chen, J. Y. C., & Thropp. (2007). Review of low frame rate effects on human performance. *IEEE Transactions on Systems, Man, and Cybernetics. Part A, Systems and Humans, 37*(6), 1063–1076. doi:10.1109/TSMCA.2007.904779.

Chen, X. (2012). *Mobile robotics in agriculture and horticulture*.

Chong, N. Y., Kotoku, et al. (2003). A collaborative multi-site teleoperation over an ISDN. *Mechatronics,* (13): 957–979. doi:10.1016/S0957-4158(03)00010-2.

Chopra, N., Berestesky, et al. (2008). Bilateral teleoperation over unreliable communication networks. *IEEE Transactions on Control Systems Technology, 16*, 304–313. doi:10.1109/TCST.2007.903397.

Dede, M., & Tosunoglu. (2006). Fault-tolerant teleoperation systems design. *Industrial Robot: An International Journal, 33*, 365–372. doi:10.1108/01439910610685034.

DeJong, B. P., Faulring, et al. (2006). Lessons learned from a novel teleoperation testbed. *Industrial Robot: An International Journal, 33*(3), 187–193. doi:10.1108/01439910610659097.

Duff, E., Usher, et al. (2007). *Web-based telerobotics revisited*.

Duff, E., Caris, et al. (2009). *The development of a telerobotic rock breaker*.

Emde, M., Krahwinkler, et al. (2012). Sensor fusion in forestry. *GPS World, 21*, 48.

Evanson, T., & Amishev. (2010). A steep slope excavator feller buncher. *FFR Technical Note, 3*, 1-8.

Ferguson, K. R. (1978). *Past and future challenges in developing remote systems technology*.

Fong, T., & Thorpe. (2001). Vehicle teleoperation interfaces. *Autonomous Robots*, 9–18. doi:10.1023/A:1011295826834.

Ghanbari, A. Abdi, et al. (2010). Haptic guidance for microrobotic intracellular injection. In *Proceedings of the 2010 3rd IEEE RAS & EMBS*. IEEE.

Gunn, C., & Zhu. (2010). *Haptic tele-operation of industrial equipment*.

Hainsworth, D. W. (2001). Teleoperation user interfaces for mining robotics. *Autonomous Robots*, *11*, 19–28. doi:10.1023/A:1011299910904.

Hansson, A., & Servin. (2010). Semi-autonomous shared control of large-scale manipulator arms. *Control Engineering Practice*, *18*(9), 1069–1076. doi:10.1016/j.conengprac.2010.05.015.

Hayn, H., & Schwarzmann. (2010). A haptically enhanced operational concept for a hydraulic excavator. *Robert Bosch GmbH*, 199-220.

Hellström, T. (2005). *Intelligent vehicles in forestry*.

Hellström, T., Lärkeryd, et al. (2008). *Autonomous forest machines - Past, present and future*.

Hellström, T., Lärkeryd, et al. (2009). Autonomous forest vehicles: Historic, envisioned, and state-of-the-art. *International Journal of Forest Engineering*, *20*(1).

Hirabayashi, T., Akizono, et al. (2006). Teleoperation of construction machines with haptic information for underwater applications. *Automation in Construction*, (15): 563–570. doi:10.1016/j.autcon.2005.07.008.

Hokayem, P. F., & Spong. (2006). Bilateral teleoperation: An historical survey. *Automatica*, 2035–2057. doi:10.1016/j.automatica.2006.06.027.

Hu, H., Yu, et al. (2001). Internet-based robotic systems for teleoperation. *Assembly Automation*, *21*, 143–151. doi:10.1108/01445150110388513.

Kim, D., Kim, et al. (2009). Excavator teleoperation system using a human arm. *Automation in Construction*, *18*(2), 173–182. doi:10.1016/j.autcon.2008.07.002.

Komatsu. (2012). *KOMATSU: Autonomous haulage system - Komatsu's pioneering technology deployed at Rio Tinto Mine in Australia*.

Lawton, D. T., Schoppers, et al. (1989). *Interactive model based vision system for telerobotic vehicles*.

Lin, Q., & Kuo. (1999). Assisting the teleoperation of an unmanned underwater vehicle using a synthetic subsea scenario. *Presence (Cambridge, Mass.)*, *8*(5), 520–530. doi:10.1162/105474699566431.

MacDonncadha, M. (1997). *Japanese forestry and forest harvesting techniques: With emphasis on the potential use of Japanese harvesting techniques on steep & sensitive sites in Ireland*.

Murakami, N., Ito, et al. (2008). Development of a teleoperation system for agricultural vehicles. *Computers and Electronics in Agriculture*, *63*, 81–88. doi:10.1016/j.compag.2008.01.015.

Nagamachi, M. (1986). Human factors of industrial robots and robot safety management in Japan. *Applied Ergonomics*, *17*, 9–18. doi:10.1016/0003-6870(86)90187-0 PMID:15676565.

New Zealand Forestry Owners Association. (2011). *New Zealand Plantation forest industry - Facts & figures 2010/2011*.

New Zealand Ministry of Primary Industries. (2012). *New Zealand ministry of primary industries - International trade statistics*. Retrieved from http://www.mpi.govt.nz/news-resources/statistics-forecasting/international-trade.aspx

Nishino, T. (2000). Forestry principles in Japan. Encyclopedia of Life Support Systems.

Occupational Safety and Health Service. (1998). *Safety And health in forest operations - Approved code of practice*. Occupational Safety and Health Service.

Osafo-Yeboah, B., & Jiang. (2009). Usability evaluation of a haptically controlled backhoe excavator simulation. In *Proceedings of IIE Annual Conference*, (pp. 961-966). IIE.

Ota, I. (2001). The economic situation of small scale forestry in Japan. *EFI Proceedings, 36*, 29-41.

Parker, R. (2009). Robotics for Steep country tree felling. *FFR Technical Note, 2*(1).

Parsons, M. H. (1986). Human factors in industrial robot safety. *Journal of Occupational Accidents, 8*, 25–47. doi:10.1016/0376-6349(86)90028-3.

Pounds, P., Mahony, et al. (2010). Modelling and control of a large quadrotor robot. *Control Engineering Practice, 18*(7), 691–699. doi:10.1016/j.conengprac.2010.02.008.

Raymond, K. (2012). Innovation to increase profitability of steep terrain harvesting in New Zealand. *New Zealand Journal of Forestry, 57*(2), 19–23.

Ringdahl, O. (2011). *Automation in forestry - Development of unmanned forwarders*. Umeå University.

Ringdahl, O., Lindroos, et al. (2011). Path tracking in forest terrain by an autonomous forwarder. *Scandinavian Journal of Forest Research, 26*(4), 350–359. doi:10.1080/02827581.2011.566889.

Rodríguez, C. A., & Guerrero. (2010). Wireless robot teleoperation via internet using IPv6 over a bluetooth personal area network. *Rev. Fac. Ing. Univ. Antioquia*, 172-184.

Sheridan, T. B. (1995). Teleoperation, telerobotics and telepresence: A progress report. *Control Engineering Practice, 3*, 205–214. doi:10.1016/0967-0661(94)00078-U.

Slawiñski, E., Postigo, et al. (2007). Bilateral teleoperation through the internet. *Robotics and Autonomous Systems, 55*, 205–215. doi:10.1016/j.robot.2006.09.002.

Thompson, E. A., Harmison, et al. (2006). Robot teleoperation featuring commercially available wireless network cards. *Journal of Network and Computer Applications, 29*(1), 11–24. doi:10.1016/j.jnca.2004.11.001.

Triton Logging. (2012). *Triton logging website*.

van der Zee, L. F. (2009). *Design of a haptic controller for excavators*. Department of Electrical Engineering, University of Stellenbosch.

Westerberg, S., Manchester, et al. (2008). *Virtual environment teleoperation of a hydraulic forestry crane*.

Yokokohji, Y., Imaida, et al. (2000). *Bilateral teleoperation: Towards fine manipulation with large time delay*.

Yoon, J., & Manurung. (2010). Development of an intuitive user interface for a hydraulic backhoe. *Automation in Construction, 19*(6), 779–790. doi:10.1016/j.autcon.2010.04.002.

Zhu, D., Gedeon, et al. (2011). Moving to the centre: A gaze-driven remote camera control for teleoperation. *Interacting with Computers*, (23): 85–95. doi:10.1016/j.intcom.2010.10.003.

KEY TERMS AND DEFINITIONS

Development System: A scaled down system or system with reduced functionality that allows some aspects of the behaviour to be simulated in a more structured and controlled environment.

Disturbance: An input to a control system that is not directly related to operator commands.

Haptic Feedback: Feedback that uses the sense of touch.

Teleoperation: An enhancement of remote control where the operator can operate a device without observing it directly.

Telepresence: The feeling that an operator is actually in the same location as a teleoperated machine by the use of different types of visual, auditory and haptic feedback and information from simulations.

Uncontrolled Environment: An environment that can change in a way that the robot operator cannot control (e.g. weather).

Unstructured Environment: An environment that is not designed explicitly for automated robot operation.

Chapter 13
Time Delay and Uncertainty Compensation in Bilateral Telerobotic Systems:
State-of-Art with Case Studies

Spyros G. Tzafestas
National Technical University of Athens, Greece

Andreas-Ioannis Mantelos
National Technical University of Athens, Greece

ABSTRACT

This chapter presents the state-of-art of the bilateral teleoperation field. It starts with a discusion of the early class of techniques, which are based on passivity and scattering theory. The main issue in bilateral telerobotic systems is the communication delay between the operator and the remote site (environment), which (if not treated) can lead the system to instability. The chapter continues by presenting the evolution of modern control techniques for stabilization and compensation of the time delay consequences. These techniques include predictive control, adaptive control, sliding-mode robust control, neural learning control, fuzzy control, and neurofuzzy control. Four case studies are reviewed that show what kind of results can be obtained.

1. INTRODUCTION

Telerobotics is the subfield of robotics which is concerned with the design operation, and control of teleoperators (Batsomboon & Tosunoglou, 1996; Buss & Schmidt, 1999; Niemeyer, Preusche, & Hirzinger, 2008; Sheridan, 1992). The word "*tele*" comes from the Greek word "*τηλε = tele*" which means "*distant.*" All complex words which have as prefix the word "tele" (e.g., tele-learning, tele-medicine, tele-communication, etc.) have the meaning that the respective operation or task

DOI: 10.4018/978-1-4666-4225-6.ch013

is performed by two distant systems or humans which are physically separated. This separation may be very small (e.g., when an operator and a robot are located in the same room) or very large with a long distance separating them. For convenience of the reader, we first define the basic terms that are used in the general field of telerobotics, namely: teleoperators, telerobot, teleoperation, telemanipulation, local site, remote site, direct or manual control, supervisory control, shared control, unilateral control, bilateral control, telepresence-transparency.

- **Teleoperator:** A system or device which enables a human operator to manipulate and sense objects at a distance (i.e., beyond his/her reach).
- **Telerobot:** A robot that works using human instructions from a distance and performs live actions at a distant environment via sensors and effectors.
- **Teleoperation:** Stresses the task level operations via human intelligence and human-machine interfaces.
- **Telemanipulation:** A remote robot (slave), working in a difficult or dangerous environment, tracks the motion of a master manipulator. To this end, two distinct processes are needed, namely: interaction between the operator and the master robot, and the interaction between the slave robot and its environment.
- **Local Site or Master Site:** The site where the human operator and the master robot are located.
- **Remote Site or Slave Site**: The site where the slave robot and its environment are located. The complete system composed by the local and remote robots is known as *master-slave* system.
- **Direct (Manual) Control**: The operator is controlling the motion of the robot directly without the help of any automatic device.

- **Supervisory Control:** The operator is controlling the robot with high-level commands via high-level feedback and substantial intelligence and/or autonomy.
- **Shared Control:** A control type that belongs between the two extremes of direct control and supervisory control, i.e., the robot is not fully directly controlled but some degree of intelligent autonomous help is available to assist the operator.
- **Unilateral Control:** The type of control where the human operator cannot feel any accurate force from the remote site, i.e., the operator can feel neither the hardness nor the texture of the remote environment (object).
- **Bilateral Control:** The control where there exists force feedback, as e.g., in surgical telerobotic systems, where the surgeon can remotely sense the quality of suture and locate veins and bones beneath the tissue, and so is able to exert a proper force control during cutting and puncturing.
- **Telepresence:** The ability of the operator not only to manipulate the remote environment, but also to perceive the environment as if encountered directly. In other words, the operator feels himself/herself to be physically present at the remote site. Sheridan called telepresence a *compelling illusion* and a *subjective sensation* (Sheridan, 1992).
- **Transparency:** The property that specifies how close the operator's perceived mechanical impedance (force/velocity ratio) comes to recreating the true environment impedance. The medium of the operator-environment interaction is the master-slave system. Obviously, if we have full (perfect) transparency the medium can be considered as absent. It is remarked that the ultimate goal in designing a bilateral control system is to get full transparency.

The design of telerobotic systems needs the synergy of a variety of scientific fields within the general mechatronics discipline described in a holistic way in Habib (2007, 2008). Control technology, communication technology, information technology, sensor technology, and human-machine interaction and coexistence technology, among others, are all needed for the design and construction of accurate, dexterous, and intelligent telerobotic systems (Abbot & Okamura, 2003; Hannaford, 1989; Isermann, 1997; Schweitzer, 1996; Tzafestas, 2009; van Brussel, 1989).

The purpose of the present chapter is to provide the historical evolution of the telemanipulation/ telerobotics field, discussing a representative set of control techniques developed over the years for achieving bilateral system stability and transparency. Equipped with the definitions of the fundamental concepts of telerobotics given previously, the chapter proceeds as follows. Section 2 discusses the general architecture of bilateral systems. Section 3 provides a list of the alternative control techniques focusing on the passivity and wave variable control, the predictive/adaptive control, and the sliding-mode robust control. Section 4 deals with neural, fuzzy and neuro-fuzzy bilateral control. Finally, section 5 provides four case studies which are concerned with the application of the prior techniques. The results obtained in these case studies are encouraging and verify the theoretical expectations concerning stability and force-reflection (transparency) performance.

2. GENERAL ARCHITECTURE OF BILATERAL TELEROBOTIC SYSTEMS

A general non-detailed operational diagram of bilateral telerobotic systems has the form shown in Figure 1. The meaning of the various components follows from the discussion of section 1. The transmition of the master signals to the slave site and vice versa is performed through the communication channel which typically is a computer communication network (such as the Internet). This transmission involves unavoidable time delays, which, as it has been proved theoretically and verified experimentaly, may lead to instablility and degradation of transparency. This fact has motivated strong research over the years for minimizing the effects of the time delay, especially for the case where the communication medium is the Internet. It is noted that due to the availability, low cost and widespread of the Internet, emphasis is curently given to Internet-based teleoperation.

In general, the Quality-of-Service (QoS) offered by a computer communication network, such as the Internet, is characterised by the following four parameters:

- Time Delay
- Jitter
- Bandwidth
- Packet Loss

Figure 1. General diagram of a typical bilateral telerobotic system

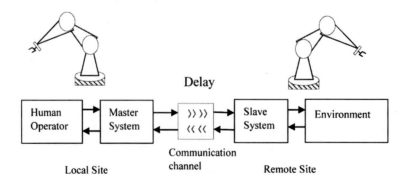

The most critical parameter is the time delay (called *end-to-end delay*) which involves a component consisting of the propagation delay and transmission (or medium access) delay. The trasnmission delay depends on transmission speed and packet size. The propagation delay represents a constant component of the total delay tha expresses the signal propagation time, which depends on the speed of the electrical or optical signal. Other components of the end-to-end (e-2-e) delay are: packet setting delay, processing delay, and queuing delay which varies with traffic load on the router and is actually the dominant component of e-2-e delay. The impact of the other parameters (jitter, bandwidth, packet loss) can be faced via many alternative methods which usually involve a trade off with the time delay. A more detailed diagram of a bilateral telepresence system is shown in Figure 2 (Buss & Schmidt, 1999).

The operator gives multi-modal command inputs to the human-system interface, i.e., visual, acoustic, tactile and haptic commands, which are transmitted to the executing teleoperator on the remote site, across the communication channels (barriers). If the operator gives symbolic commands to the teleoperator pushing buttons and watching the respective action in the remote environment it is said that the operator-remote site coupling is weak. The coupling is considered strong in the case of kinesthetic modality of a bilateral teleoperation system. Here, the motion (force) of the operator is measured, communi-

cated and imposed on the teleoperator, and vice versa, the resulting forces (motion) of the teleoperator are sensed and fed back to the human operator.

The early teleoperation systems were of the unilateral type in order to avoid the difficulties of creating local autonomy. This means that the operator was specifying the robot's motion by him/herself through a joystic (usually spring centered) or some other master mechanism serving as input device. Here, we may have unilateral acceleration or rate control, but the rate control was mostly used because it allows more accurate position control in reaching and holding a target location (Conti & Khatib, 2005; Massimoni, Sheridan, & Roseborough, 1989). In this chapter we will study the bilateral telerobotic control systems which involve force feedback from the slave to the master.

3. BILATERAL TELEROBOTIC CONTROL PROBLEMS AND METHODS

3.1. General Issues

Typically, all robotic control systems involve modelling uncertainties, and input uncertainities (disturbances). The first uncertainties arise primarly because of the existence of non-modeled higher frequency dynamics, whereas the second uncertainties are usually due to nonlinear fric-

Figure 2. Pictorial illustration of a multimodal telepresence-telemanipulation system

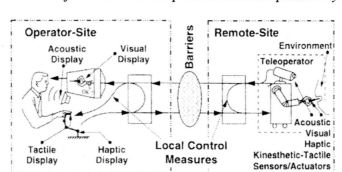

tion effects and payload variations. In telerobotics there is the additional issue of end-to-end time delay which in practice is varying and uncertain. Therefore, to assure stability and transparency all these uncertainties must be compensated. To this end, throughout the years many alternative and complementary methods have been developed. These include the following (Artigas, Preusche, & Hirzinger, 2006; Lozano, Chopra, & Spong, 2002):

- Passivity-Based Method (Anderson & Spong, 1989; Arcara & Melchiori, 2002)
- Scattering Theory-Based Method (wave variables) (Stramiglioli, van der Schaft, Maschke, & Melchiori, 2002)
- Predictive Control Methods (Ganjefar, Momeni, Sharifi, & Behesthi Hamidi, 2003; Smith & Hashtrudi-Zaad, 2006)
- Adaptive Control Methods (Lee & Chung, 1998; Polushin, Tayebi, & Marquez, 2005)
- Robust Control Methods (Cho & Park, 2003; Natori, Tsuji, & Ohnisi, 2004; Tzafestas & Prokopiou, 1997)
- Neural Network-Based Control Methods (Chen, Quan, & Xia, 2007; Huang, 2002; Pongaen, 2008; Tzafestas, Prokopiou, & Tzafestas, 2001)
- Fuzzy Logic Control Methods (Hu, Ren, Thompson, & Sheridan, 1996; Wang, 2011)
- Neuro-Fuzzy Control Methods (Lin, Tsai, & Liu, 2001; Pongaen, 2008)

3.2. Passivity and Wave Variables

Passivation of telepresence systems has been proposed in (Anderson & Spong, 1989) and (Niemeyer & Slotine, 1991). Passivity is a property dealing with the energies of a system. A system with initial energy is called passive if:

$$E(t) = \int_0^t P(\tau)d\tau + E_i \geq 0 \qquad (1)$$

where $E(t)$ is the total energy of the system at time t, and $P(\tau)$ denotes the net power at input and output ports. Passivity is a nesessary and sufficient condition for a two port, coupled to an arbitrary network, to be stable, and implies that the energy in the two-port system is dissipated instead of generated. A telerobotic 2-port system has the form shown in Figure 3.

Here, v denotes velocity $v = \dot{x}$ and $F(t)$ the corresponding force. The input power $P_1(t)$ at time t is given by $P_1(t) = F_1(t)v_1(t)$ and the output power by $P_2(t) = F_2(t)v_2(t)$. Therefore, assuming that the initial energy E_i is zero, the passivity condition becomes:

$$E_{total}(t) = \sum_{j=1}^{m} E_j(t) \geq 0 \qquad (2)$$

If the slave interacts with passive environments (which do not involve any active device such as a motor, etc.), the system stability is assured when the human is not contained in it. In actual telerobotic systems the human operator on the master site closes the loop taking into account the stability requirements using one of the techniques listed in section 3.1. Since the master and slave robots are passive (as it happens to all mechanical systems with spring-like damping), in case that no delay occurs in the master-slave communication, a position-position control scheme is passive and thus stable. However, passivity-based control, is conservative although it can face some level of uncertainty. Another drawback of passivity-based controllers is that it cannot hide the dynamics of

Figure 3. 2-port representation of a telerobotic system S

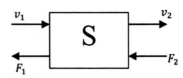

the slave robot. On the contrary the position-force scheme hides the slave inertia and friction from the operator. Therefore if the operator introduces energy into the master without feeling any resistance, the system itself creates and injects the kinetic energy to the slave, a fact that violates passivity and leads to potential instability.

Passivity is not assured by its own if the communication suffers from time delay. This is because the energy entering into the communication channel and exiting the output side do not add up. Rather energy may be generated inside the channel, which contributes to instability. To overcome this, two techniques have been proposed, namely: (1) use of scattering matrices (instead of admittance matrices) which can be tolerate delays (Stramiglioli, van der Schaft, Maschke, & Melchiori, 2002), and (2) observing and imposing passivity to assure stability (Iqbal, Roth, & Abu-Zaitoon, 2005; Ryu, Preusche, Hannaford, & Hirzinger, 2005). On the basis that wave phenomena circumvent the delay caused instability the wave variable system representation was introduced in (Anderson & Spong, 1989) and later presented in more intuitive passivity-based physical representation in (Niemeyer & Slotine, 1997). Actually, the wave variables $\mathbf{u}$ and $\mathbf{v}$ provide a modification or extension to the theory of passivity and replace the standard power variable $\dot{\mathbf{x}}$ (velocity) and $\mathbf{F}$ (force). The relation of $(\mathbf{u}, \mathbf{v})$ with $(\dot{\mathbf{x}}, \mathbf{F})$ is obtained by considering the power flowing through the system and separating the power moving forward and returning, i.e.:

$$P = \dot{\mathbf{x}}^T \mathbf{F} = \frac{1}{2}\mathbf{u}^T\mathbf{u} - \frac{1}{2}\mathbf{v}^T\mathbf{v} \qquad (3)$$

which leads to the transformations:

$$\mathbf{u} = \frac{b\dot{\mathbf{x}} + \mathbf{F}}{\sqrt{2b}}, \mathbf{v} = \frac{b\dot{\mathbf{x}} - \mathbf{F}}{\sqrt{2b}} \qquad (4)$$

Indeed using (4) we obtain:

$$\mathbf{u}^T\mathbf{u} = \left(b^2\dot{\mathbf{x}}^T\dot{\mathbf{x}} + 2b\dot{\mathbf{x}}\mathbf{F} + \mathbf{F}^T\mathbf{F}\right)\big/2b$$

$$\mathbf{v}^T\mathbf{v} = \left(b^2\dot{\mathbf{x}}^T\dot{\mathbf{x}} - 2b\dot{\mathbf{x}}\mathbf{F} + \mathbf{F}^T\mathbf{F}\right)\big/2b$$

from which Equation (3) follows. It can be easily verified that the transformations (4) can be pictorially represented as shown in Figure 4, where $\dot{\mathbf{x}}$ and $\mathbf{v}$ are used as inputs.

The wave variable method was applied to force feedback (bilateral) teleoperation systems with success, but it suffers from the fact that wave variables cannot be physically measured and are not as intuitive as the velocity and force variables. This has led the researchers in the field to develop more successful adaptive, predictive and robust methods as it will be explained bellow. Note that b is the wave impedance which relates velocity to force and offers a tuning knob to the operator. Two enhanced wave-variable control techniques that incorporate position-position and position-force structure were presented in (Munir & Book, 2002) and (Tanner & Niemeyer, 2006).

3.3. Predictive and Adaptive Control

The design of predictive and adaptive controllers for stable and transparent bilateral teleoperation has received a great attention by telerobotics researchers. There two classes of methods:

Figure 4. Pictorial illustration of the wave transformation, which relates force, velocity, and right/ left moving waves u and v

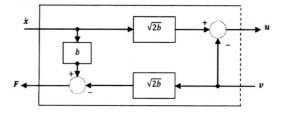

- Wave Variables-Based Methods
- Dynamic Master-Slave Model Methods

In (Berestesky, Chopra, & Spong, 2004) a solution to the problem of compensating the passivity loss due to time-varying delay and packet loss is provided. This solution is based on the digital reconstruction wave variable and introduces the use of a buffering and interpolation scheme, which conserves the passivity and improves stability and transparency.

In (Lee & Spong, 2006) a proportional Plus Derivative (PD) control scheme is proposed. This scheme involves damping compensation that assures the passivity of the closed-loop teleoperator in the presence of constant time delay and parametric uncertainty.

In (Munir & Book, 2002) a control scheme is proposed that predicts the incoming wave variables from the slave minimizing the negative effects of the transmission delays. This scheme combines a Smith predictor with a Kalman filter for estimating the slave's behavior and compensating at the master site for delays and losses in the communication channel.

In (Hashemzadeh, Hassanzadeh, Tavakoli, & Alizadeh, 2000) the effects of time-varying delays on the teleoperator's passivity are faced via modeling the time-varying delay by a constant delay and an additive output disturbance. To this end, proper disturbance estimator blocks are added in both the feedforward and feedback paths to estimate and compensate these disturbances.

In (Zhu & Salcudean, 2000) a control scheme is proposed which employs an adaptive motion/force controller. In this scheme the system is considered as a free floating mass with a linear damper determined by the control and scaling parameters (Lee & Chung, 1998)

In (Hung, Narikiyo, & Tuan, 2003) a virtual manipulator is employed for designing a nonlinear controller that can assure asymptotic (position/velocity) tracking and acceptable force tracking even if acceleration measurements, accurate dynamic parameters, and human/environment accurate models are not available.

In (Yoshida & Namerikawa, 2009) the problem of environment estimation –based predictive control for delayed teleoperation is considered. The goal is actually to achieve delayed position tracking and non-delayed force feedback. In this scheme the master controller implements force control with environmental force prediction while the slave controller implements position and force control. This scheme follows a 3-channel architecture (Kubo, Iiyama, Natori, Ohnishi, & Furukawa, 2007).

In (Polushin, Tayebi, & Marquez, 2005) two adaptive control schemes are proposed that make the closed-loop teleoperator system Input-to-State Stable (ISS) with respect to external forces for any constant communication delay. To this end, a new version of the ISS small-gain theorem is developed and used.

As an introduction to adaptive telerobotic controller for stability and transparency, let us consider, without loss of generality, one degree of freedom master and slave robots described by linear dynamic models, namely:

$$\textbf{Master}: M_m \ddot{x}_m + B_m \dot{x}_m = u_m + f_m$$

$$\textbf{Slave}: M_s \ddot{x}_s + B_s \dot{x}_s = u_s - f_s$$

where the indexes m and s refer to the master and slave robot, M denotes mass (inertia) and B represents viscosity (friction) coefficients (See Figure 6). The variables f_m and f_s represent the force applied by the operator to the master, and the force applied by the slave to the environment, respectively. Similarly, the variables u_m and u_s are the controls (driving forces) of the master and slave.

The problem is to design an adaptive controller which assures that the position x_s of the slave robot's tip tracks a desired master's positions x_m and at the same time adaptively estimates the unknown dynamic parameters of the slave robot-environment system. To this end, use will be made of the Lyapunov stabilization method (Slotine & Li, 1991), (Spong & Vidyasagar, 1989).

The combined slave robot-environment system is described by the following dynamic model:

$$M_s \ddot{x}_s + B_o \dot{x}_s + K_e x_s = u_s \tag{5}$$

where M_s is the unknown slave's mass (assumed $M_s > 0$), B_o is the unknown total slave-environment friction coefficient $B_o = B_s + B_e$, and K_e is the unknown stiffness coefficient of the environment. Let x_{sd} be the desired slave position, and define the error variables $s(t)$ and $e(t)$ as:

$$s = \dot{x}_s - \dot{x}_{sd}, e = x_s - x_m \tag{6}$$

where $\dot{x}_{sd} = \dot{x}_m - \Omega e, \Omega > 0$.

Now, we write the dynamic equation in the regression form:

$$u_s = \mathbf{H}^T(\ddot{x}_s, \dot{x}_s, x_s)\xi \tag{7}$$

where $\mathbf{H}(\ddot{x}_s, \dot{x}_s, x_s) = [\ddot{x}_s, \dot{x}_s, x_s]^T$, and $\xi = [M_s, B_o, K_e]^T$ is the unknown parameters vector. Denoting by $\hat{\xi}$ the estimated value of ξ, provided by the adaptive estimator, the estimation error $\tilde{\xi}$ is given by:

$$\tilde{\xi} = \hat{\xi} - \xi \tag{8}$$

To assure the asymptotic convergence of s and $\tilde{\xi}$ to a vicinity of zero we select the following Lyapunov function:

$$V = \frac{1}{2}(M_s s^2 + \tilde{\xi}^T \mathbf{\Gamma}^{-1} \tilde{\xi})$$

where the diagonal matrix $\mathbf{\Gamma} = diag[\gamma_1, \gamma_2, \gamma_3]$ is positive definite. The adaptive controller-estimator must be selected such that $\dot{V} \leq 0$. To this end, we differentiate V with respect to time and get:

$$\dot{V} = sM_s\dot{s} + \tilde{\xi}^T\mathbf{\Gamma}^{-1}\dot{\tilde{\xi}} = sM_s(\ddot{x}_s - \ddot{x}_{sd}) + \tilde{\xi}^T\mathbf{\Gamma}^{-1}\dot{\tilde{\xi}}$$

$$(since\ \dot{\tilde{\xi}} = 0)$$

$$= s(u_s - B_o\dot{x}_s - k_e x_s - M_s\ddot{x}_{sd}) + \tilde{\xi}^T\mathbf{\Gamma}^{-1}\dot{\tilde{\xi}}$$

It is easy to show that choosing the parameter adaptation law and the controller as:

$$\dot{\hat{\xi}} = -\Gamma[\mathbf{H}(\ddot{x}_{sd}, \dot{x}_{sd}, x_s)s + \mathbf{H}_f e_f] \tag{9}$$

$$u_s = \mathbf{H}^T(\ddot{x}_{sd}, \dot{x}_{sd}, x_s)\hat{\xi} - K_v s \tag{10}$$

where $K_v > 0$ is a positive velocity gain, and $\mathbf{H}_f, e_f$ and u_{sf} are filtered quantities given by:

$$\dot{\mathbf{H}}_f = -\alpha\mathbf{H}_f + \alpha\mathbf{H}(\ddot{x}_s, \dot{x}_s, x_s)$$

$$e_f = \mathbf{H}_f^T\hat{\xi} + u_{sf} \tag{11}$$

$$\dot{u}_{sf} = -\alpha u_{sf} + \alpha u_s$$

The time derivative of the Lyapunov function becomes:

$$\dot{V} = -[(B_o + K_v)s^2 + \tilde{\xi}^T\mathbf{H}_f\mathbf{H}_f^T\tilde{\xi}] \leq 0 \tag{12}$$

for $B_o + K_V > 0$ and any filtering matrix $\mathbf{H}_f$. The inequality (12) assures, by Lyapunov theorem and Barbalat matrix, the asymptotic convergence of s and $\tilde{\xi}$ to respective small regions around zero.

In the previous derivation the dynamic parameters of the master were assumed to be precisely known. If this is not the case, a robust master controller (e.g., a sliding-mode controller) must be used, as it will be shown in the following. It is noted that in master-slave systems, stability must be assured for the whole system (master and slave site). In the literature many methods were proposed, e.g., (Lee & Chung, 1998). In (Moh, 2011) the problem of facing dynamic disturbances in master-slave systems is solved using a disturbance observer, and the resulting "disturbance observer-based controller" assures global asymptotic force tracking and global exponential position and disturbance tracking.

3.4. Sliding-Mode Robust Control

Sliding-mode robust controller design was studied in (Garcia-Valdovinos, Parra-Vega, & Arteaga, 2007; Leblebici, Calli, Unel, Sabanovic, Bogosyan, & Gokasan, 2011); Park & Cho, 1999; Tzafestas & Prokopiou, 1997). In (Tzafestas & Prokopiou, 1997) the teleoperation scheme of (Lee & Lee, 1993) and (Lee, 1994) is enhanced by a sliding-mode controller in order to be able to compensate for modeling uncertainties arising after the slave picks up an object of unknown mass and shape, or caused by hardware malfunctioning, accidental deformation or deliberate model simplification. As in all cases of sliding-mode controllers, the bounds of uncertainties must be available. In (Park & Cho, 1999) a sliding-mode controller for the slave, and impedance controller for the master is proposed, where the non-linear gain can be set independently of the time delay variation. This controller compensates the performance deterioration due to the existence of the varying and unknown communication delay. In (Leblebici, Calli, Unel, Sabanovic, Bogosyan, & Gokasan, 2011) a sliding-mode observer is designed in conjunction with a disturbance observer to predict the states of the slave system. These states are then used for the controller con-

struction. In the following we present an outline of the sliding-mode methodology (Slotine & Li, 1991; Spong & Vidyasagar, 1989; Tzafestas & Prokopiou, 1997).

Consider an n-degree-of-freedom robotic manipulator whose dynamics are estimated as:

$$\hat{\mathbf{D}}(\mathbf{q})\ddot{\mathbf{q}} + \hat{\mathbf{h}}(\mathbf{q}, \dot{\mathbf{q}}) = \ddot{\mathbf{A}}, \ \mathbf{q} \in \mathbb{R}^n \qquad (13)$$

where $\mathbf{D}(\mathbf{q})$ is the $n \times n$ inertial matrix (which is a positive definite invertible matrix), $\hat{\mathbf{h}}(\mathbf{q}, \dot{\mathbf{q}})$ involves the centrifugal, Coriolis and gravitational terms, and $\ddot{\mathbf{A}} \in \mathbb{R}^n$ is the input generalized force vector. We select a separate sliding-surface for each link, i.e.:

$$s_i = \dot{\tilde{q}}_i + 2\Omega \tilde{q}_i + \Omega^2 s_t \int_t \tilde{q}_i(\tau)d\tau = 0 \qquad (14)$$

where $i = 1, 2, ..., n$, $\tilde{q}_i = q_i - q_{i,d}$ is the joint tracking error, and Ω is the equivalent control bandwidth. The control law is chosen such that outside the sliding surface $s_i = 0$, the following sliding condition holds:

$$\frac{1}{2}\dot{s}_i^2 \leq -\eta |s_i| \qquad (15)$$

where η is a positive constant. A suitable control that satisfies (15) is:

$$\ddot{\mathbf{A}} = \hat{\mathbf{D}}u + \hat{\mathbf{h}}$$

$$u_i = G_i(q_i)[\hat{u}_i - \overline{k}_i(q_i, \dot{q}_i)sat(s_i / \phi_i)] \qquad (16)$$

$$\hat{u}_i = \ddot{\tilde{q}}_{di} - 2\Omega \dot{\tilde{q}}_i - \Omega^2 \tilde{q}_i$$

where $G_i, \overline{k}_i$ and ϕ_i are parameters depending on the uncertainty and the joint position and velocity. The prior controllers use the "sat" function instead of the "sgn" function in order to avoid extremely high control activity (chattering). The teleoperator control system with the sliding mode controller has the form shown in Figure 5.

In Figure 5, x_m is the master position, $x_{ds} = k_{sc} x_m$ is the scaled position sent to the slave, F_h is the force applied by the operator to the master robot via his/her muscle dynamics. An estimator provides $\hat{F}_h$ which together with the desired trajectory is sent to the sliding controller.

The force F_{es} produced by the interaction of the slave with the environment is equal to $F_{es} = Z_e(x_e - x_s)$, where x_e, x_s are the positions of the environment and slave, respectively, and Z_e is the generalized mechanical impedance of the environment. For the details of this scheme the reader is referred to (Lee & Lee, 1993) and

(Tzafestas & Prokopiou, 1997). Another approach to sliding-mode control is presented in (Lu, 1995) where the sliding surface is defined using the difference of the actual and desired impedance, instead of the tracking error $\tilde{q}$.

The time-varying delay robust sliding mode controller of (Park & Cho, 1999) can be summarized as follows. The overall master-slave system is depicted in Figure 6. For simplicity, the dynamic equations are again considered one-dimensional:

Master : $M_m \dot{v}_m + B_m v_m = u_m + f_m$ (17)

Slave : $M_s \dot{v}_s + B_s v_s = u_s - f_s$ (18)

where $v_m = \dot{x}_m$ and $v_s = \dot{x}_s$ are the velocities of the master and slave respectively.

The control scheme provides impedance control for the master and sliding-mode control for the slave. Use is made of the following scaling relations between the master and the slave in position and force.

$$x_s = K_p x_{md}, f_m = K_f f_{ed}$$ (19)

where x_s, x_{md} are the desired position command from the slave and master respectively, f_{ed} is the external force command from the slave to the master and K_p, K_f are human-specified scaling factors. The impact force imposed to the human operator is equal to $f_m - K_f f_e$. Therefore the dynamic equation that specifies the desired human-master impedance characteristics is:

$$M\ddot{x}_m + B\dot{x}_m + K x_m = f_m - K_f f_e$$ (20)

where M, B and K are the parameters (inertia, friction coefficient, stiffness) of the desired impedance.

Figure 5. Teleoperator with sliding-mode controllers

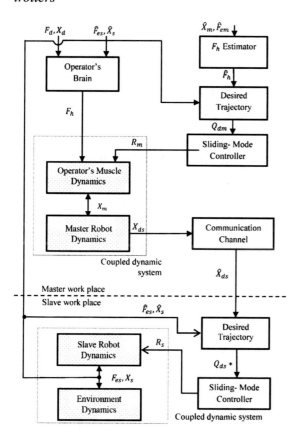

Figure 6. Pictorial illustration of the master-slave telerobotic system

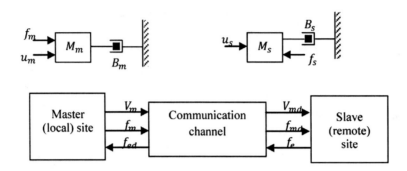

Assuming that f_m can be measured, and eliminating $\ddot{x}_m = \dot{v}_m$ from (17) and (20) we get:

$$u_m = [B_m - (M_m / M)B]v_m + (M_m / M - 1)f_m$$
$$- (M_m / M)(K_f f_e + K x_m) \tag{21}$$

The sliding surface $s(t)$ is defined as:

$$s(t) = \dot{\tilde{x}}(t) + \Omega \tilde{x}(t) \tag{22}$$

which gives:

$$\dot{s}(t) = (\ddot{x}_s - K_p \ddot{x}_m) + \Omega(\dot{x}_s - K_p \dot{x}_m)$$

Setting $\dot{s}(t) = 0$, solving for $\ddot{x}_s$, and replacing in (18) we get:

$$u_s = B_s v_s + f_e - (K_p M_s / M)$$
$$(Bv_m - f_m + K_f f_e + K x_m) - M_s \Omega \dot{\tilde{x}} \tag{23}$$

In the absence of modeling (parameter) uncertainties and delay, the control law (23) assures that the master reflects properly the contact force of the slave, and that the slave will track exactly the position of the master. If there is time delay T_m from the master to the slave and T_s from the slave to the master, then we have:

$$v_{md}(t) = v_m(t - T_m(t))$$
$$f_{md}(t) = f_m(t - T_m(t)) \tag{24}$$
$$f_{ed}(t) = f_e(t - T_s(t))$$

and so the controllers (21) and (23) take the form:

$$u_m(t) = [B_m - (M_m / M)B]v_m(t)$$
$$+(M_m / M - 1)f_m(t) \tag{25}$$
$$- (M_m / M)(K_f f_{ed}(t) + K x_m(t))$$

$$u_s(t) = B_s v_s(t) + f_e(t)$$
$$-(k_p M_s / M)\{Bv_{md}(t) - f_{md}(t)$$
$$+ K_f f_e(t) + K x_{md}(t)\} \tag{26}$$
$$- M_s \Omega \dot{\tilde{x}}_d(t) - KG sat(s_d / \phi)$$

where $s_d(t)$ is the new sliding surface:

$$s_d(t) = \dot{\tilde{x}}_d + \Omega \tilde{x}_d, \tilde{x}_d(t) = x_s(t) - K_p x_{md}(t) \tag{27}$$

Using (18) and (26) we get:

$$\dot{s}_d(t) + (K_p K_f / M)[f_e(t)$$
$$- f_e(t - T_m - T_s)] + (G / M_s)sat(s_d / \phi) = 0 \tag{28}$$

Therefore to satisfy the sliding condition (16) for $s_d(t)$, i.e., to have:

$$\frac{1}{2}\dot{s}_d^2 = s_d \dot{s}_d \leq \eta |s_d|$$

The gain G must be selected as:

$$G \geq M_s \eta + K_p K_f (M_s / M) \\ |f_e(t) - f_e(t - T_m - T_s)| \tag{29}$$

Obviously, when the round trip delay $T_o = T_m + T_s$ increases, to satisfy this condition the controller gain G must be increased, which may lead to stronger chattering. To overcome this problem the term $K_f f_e(t)$ in (26) is replaced by $K_f f_e(t - T_o)$. Clearly, if no delay occurs the resulting new controller is identical to (26). But if there is a round trip delay $T_o = T_m + T_s$ with the modified controller the sliding surface Equation (28) reduces, to:

$$\dot{s}_d(t) + (G / M_s) sat(s_d / \phi) = 0$$

which gives a gain:

$$G \geq M_s \eta \tag{30}$$

not depending on the delay.

The block diagram of the overall closed loop teleoperation system is shown in Figure 7.

4. NEURAL, FUZZY, AND NEURO-FUZZY CONTROL

4.1. Neural Learning Teleoperator Control

Neural Networks (NN) were applied in many different engineering or nonengineering fields, such as signal and image processing, control, robotics, forecasting, transportation, and economic systems etc. Their use is based on their fundamental capability to approximate efficiently a wide range of non-linear functions via a simple learning that uses input-output data. This is the so-called *universal approximation* capability of NNs. Especially, in large complex systems such as wheeled mobile robots, legged robots, cooperating manipulators and telemanipulators, where the models under treatment become very complicated, they can offer simple and valuable solutions. Examples of the application of NNs in the robotics fields can be found in (Guez & Selinsky, 1989; Kawato, Uno, Isobe, & Suzuki, 1988; Lewis, Liu, & Yesildirek, 1995; Tzafestas, 1995). In particular, NNs were applied to teleoperators in (Ang & Riviere, 2001; Chen, Quan, & Xia, 2007; Pongaen, 2008; Smith & Mohashtrudi-Zaad, 2005; Tzafestas, Prokopiou, & Tzafestas, 2001).

In (Tzafestas, Prokopiou, & Tzafestas, 2001) a method is provided in which the whole controller is divided in three subnetworks, each one identify-

Figure 7. Structure of the overall sliding-mode delayed telemanipulator system

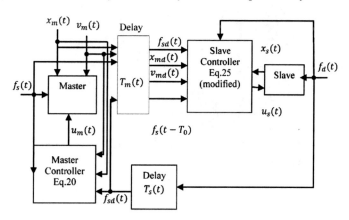

ing a part of the robot dynamics. To simplify the online training a special Heuristic Error Distribution method (the *HERD* method) is designed, which reduces the confusion on the network's role when large modeling errors arise as, e.g., when the robot picks up an object or after an accident. Especially, for a general robotic system (**J** is the Jacobian matrix):

$$\mathbf{D}(\mathbf{q})\ddot{\mathbf{q}} + \mathbf{C}(\mathbf{q},\dot{\mathbf{q}})\dot{\mathbf{q}} + \mathbf{g}(\mathbf{q}) = \ddot{\mathbf{A}} - \mathbf{J}^T(\mathbf{q})\mathbf{f}_e \quad (31)$$

Each NN (multilayer perceptron) is used to learn and identify the matrices $\mathbf{D}(\mathbf{q}), \mathbf{C}(\mathbf{q},\dot{\mathbf{q}})$ and $\mathbf{g}(\mathbf{q})$ separately, employing the universal NN approximation representation (See Figure 8).

$$\mathbf{y} = \mathbf{W}^T\tilde{\mathbf{A}}(\mathbf{V}^T\mathbf{x}) \quad (32)$$

where V, W are weight matrices for the hidden and output layers, respectively, $\tilde{\mathbf{A}}(\cdot)$ is a matrix with elements the NN sigmoid function, and x,y are the input and output vectors, respectively. The final algorithm employs both measurable data $(\mathbf{q},\dot{\mathbf{q}},\ddot{\mathbf{q}})$ as well as information based on the structure of the method. Denoting by **A** the overall torque, and e the overall error we have:

$$\ddot{\mathbf{A}} = \ddot{\mathbf{A}}_D + \ddot{\mathbf{A}}_C + \ddot{\mathbf{A}}_g, \mathbf{e} = \mathbf{e}_D + \mathbf{e}_C + \mathbf{e}_g \quad (33)$$

where $\ddot{\mathbf{A}}_D, \ddot{\mathbf{A}}_C, \ddot{\mathbf{A}}_g$ and $\mathbf{e}_D, \mathbf{e}_C, \mathbf{e}_g$ are the torques and errors corresponding to **D**,**C** and **g** respectively. Therefore, using (32) we get:

$$\ddot{\mathbf{A}}_D = \mathbf{W}_D^T\tilde{\mathbf{A}}(\mathbf{V}_D^T\mathbf{x}_D)$$

$$\ddot{\mathbf{A}}_C = \mathbf{W}_C^T\tilde{\mathbf{A}}(\mathbf{V}_C^T\mathbf{x}_C) \quad (34)$$

$$\ddot{\mathbf{A}}_g = \mathbf{W}_g^T\tilde{\mathbf{A}}(\mathbf{V}_g^T\mathbf{x}_g)$$

and the overall NN approximation scheme has the form of Figure 9.

The structure of the neurocontroller employed is shown in Figure 10 which belongs to the Model Reference Adaptive Control (MRAC) class.

The overall closed-loop teleoperator system has exactly the structure of Figure 5, where the blocks "sliding controller" are replaced by neuro controllers of the previous form (Figure 10). The slave neurocontroller produces the control torque for the coupled system "slave robot-environment". Similarly the master neurocontroller produces the torque for the "Operator muscle-master robot" system. The block "desired trajectory" in Figure

Figure 8. Universal NN approximator based on multi-layer perceptron (MLP) with two hidden layers

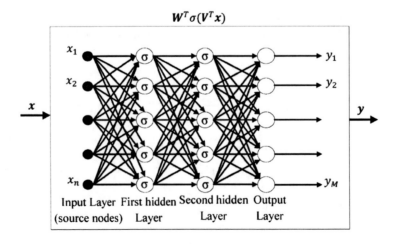

Figure 9. Partitioned NN robot identification scheme

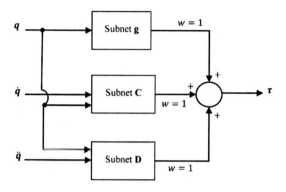

5 generates the command to the respective controller according to the master command and the desired generalized impedance. If instead of the Multilayer Perceptron (MLP) neural network we use Radial-Basis Function (RBF) NN, the universal approximator (32) is replaced by:

$$y = \phi^T(x)W \qquad (35)$$

where W is a weight matrix and $\phi(x)$ is a nonlinear matrix "radial basis" function. The derivations are the same.

The prior neural controller teleoperator scheme has also the ability to face a certain extend of time delay, although it is smaller than the delay that can be compensated by the sliding mode control scheme (Figure 5).

A neurocontroller scheme designed especially for compensating a constant time delay, in addition to disturbances and model uncertainties, was presented in (Huang, 2002). A brief summary of this scheme which is based on the Smith predictor follows. Consider the typical linear closed-loop time-delay system of Figure 11.

The symbol $\bar{f}(s)$ denotes the Laplace transform of the time signal $f(t)$, i.e.,:

$$\bar{f}(s) = \int_0^\infty e^{-st} f(t)dt$$

and $G_c(s), G_p(s), G_d(s) = e^{-sT}$ are the transfer functions of the controller, the plant, and the delay block, respectively. The closed transfer function (for $\bar{d}(s) = 0$), is:

$$\frac{\bar{y}(s)}{\bar{u}(s)} = \frac{G_c(s)G_p(s)e^{-sT}}{1 + G_c(s)G_p(s)e^{-sT}}$$

Figure 10. Model reference adaptive neuro controller

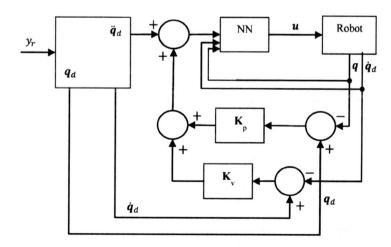

Figure 11. Typical feedback time-delay systems with a disturbance d(s)

Figure 11. Typical feedback time-delay systems with a disturbance d(s)

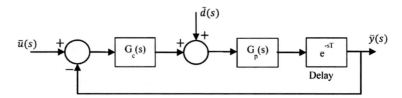

It is seen that the delay transfer function appears in the denominator (characteristic) polynomial, which is undesired. To eliminate the term exp(-sT) from the denominator the modified *Smith predictor* scheme of Figure 12 can be used (Hua, 2002).

With $\bar{d}(s) = 0$, the closed-loop transfer function of this system is found to be:

$$\frac{\bar{y}(s)}{\bar{u}(s)}$$

$$= \frac{G_c(s)G_p(s)e^{-sT}}{1 + G_c(s)\hat{G}(s) - G_c(s)\hat{G}(s)e^{-sT} + G_c(s)G_p(s)e^{-sT}}$$

where $\hat{G}(s)$ is a prediction (approximation) of $G_p(s)$. If $\hat{G}(s) = G_p(s)$ exactly, then we get:

$$\frac{\bar{y}(s)}{\bar{u}(s)} = \frac{G_c(s)G_p(s)e^{-sT}}{1 + G_c(s)G(s)}$$

which shows that the delay has been eliminated from the transfer function denominator as desired. This idea is used in combination with a NN approximator of the slave robot and a linear Smith-type predictor at the human (master) site. To compensate the nonlinear part of the slave we use the controller:

$$\ddot{\mathbf{A}}_s(t) = \mathbf{D}_s(\mathbf{q}_s(t))[\mathbf{q}_m(t - T) - \hat{\mathbf{W}}_i^!(\mathbf{q}_s(t))] \quad (36)$$

To robustify the control system against modeling and input uncertainties and increase the efficiency of the Smith predictor, a robustifying term r(t) is used as shown by dashed lines in Figure 13 designed using the sliding mode technique. In (Huang, 2002) a full proof of asymptotic stability of the complete master-slave controlled system in a vicinity of zero is provided. The NN block can be an MLP, an RBF, or a Recurrent Neural Network (RNN).

Figure 12. The modified Smith predictor control scheme

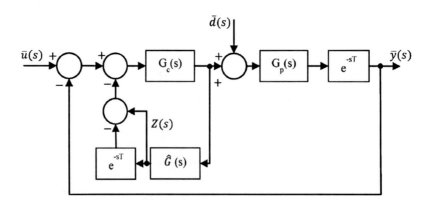

Figure 13. Overall structure of the NN telerobotic system with time-delay

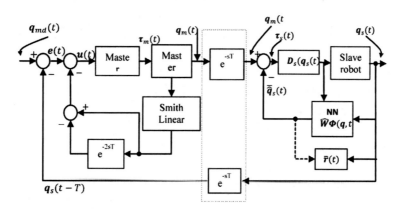

In (Ang & Riviere, 2001) a cascaded correlation NN trained via extended Kalman filtering (Watanabe & Tzafestas, 1990) and (Watanabe, Fukuda, & Tzafestas, 1992) was used to provide a surgeon's desired motion suppressing any erroneous components. In (Smith & Mohashtrudi-Zaad, 2005) a technique using Smith predictor and neural network is proposed for online estimating the dynamics of the slave and environment which allows replication of the environment contact force at the master using a similar NN. The architecture of this method is shown in Figure 14.

The technique is similar to that of (Huang, 2002) with the difference that a NN estimator/predictor is also used at the master and the robustifying term is absent.

4.2. Fuzzy and Neurofuzzy Teleoperator Control

4.2.1 Fuzzy Modeling and Control

Fuzzy modeling uses fuzzy logic which tolerates a kind of vagueness and ambiguity in modeling a system using linguistic (word) variables. Fuzzy controllers use human process knowledge and human control strategies stored in the form of fuzzy **IF-THEN** rules (Teranno, Asai, & Sugeno, 1992). A fuzzy controller involves the following units:

- Fuzzification Unit (FU)
- Fuzzy Rule Base (FRB)
- Fuzzy Inference (composition) Unit (FIU)

Figure 14. The force-position predictive teleoperator control architecture

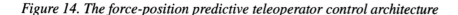

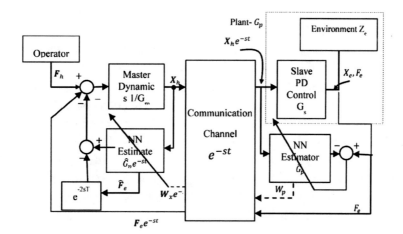

- Defuzzification Unit (DU)

The fuzzification unit converts standard numerical (crisp) values to fuzzy sets (i.e., fuzzy or linguistic values), the rule base consists of a set of a fuzzy rules, the inference unit combines all fuzzy variables to a single fuzzy set using the max or union fuzzy composition, and finally, the defuzzification converts the resulting fuzzy set (fuzzy value) to a crisp value u(t) which is the controller's output imposed to the system under control. This is done by calculating the "*Center Of Gravity*" (COG) or equivalently the "*Center Of Area*" (COA), i.e.,

$$u(t) = \sum_i x_i \mu(x_i) \bigg/ \sum_i \mu(x_i)$$

where x_i is the ith element of the output variable's universe and $\mu(x_i)$ is the associated membership function. Typically, the "min" or "product" operation is used for inference, and the "max" for composition leading to the so-called "*min-max*" or "*product-max*" method respectively (Wang, 1994).

The fuzzy modeling of a complex nonlinear systems is based on the fuzzy universal approximation property:

$$y(\mathbf{x} \mid \boldsymbol{\phi}) = \mathbf{w}^T(\mathbf{x})\boldsymbol{\phi} \qquad (37)$$

where:

$$\mathbf{w}(\mathbf{x}) = \left[\frac{\lambda_1}{\sum\limits_{k=1}^{M} \lambda_k}, \frac{\lambda_2}{\sum\limits_{k=1}^{M} \lambda_k}, ..., \frac{\lambda_M}{\sum\limits_{k=1}^{M} \lambda_k} \right]^T \qquad (38)$$

and $\boldsymbol{\phi}$ is a parameter vector with components the crisp values provided at the outputs of the fuzzy system with COG defuzzification. The fuzzy rules in the fuzzy rule base have the following form:

RULE **k: IF** x_1 is A_1^k **AND** ... **AND** x_n is A_n^k
    THEN $y_k = B_k$ $\left(k = 1,2,...M\right)$

where A_i^k is a fuzzy set in the input universe of discourse (superset) $X_i \in \mathbb{R}$, B_k is a fuzzy set in the output universe of discourse $Y_k \in \mathbb{R}$, $k = 1,2,...,M$ and $x_i(i = 1,2,...,n), y_k(k = 1,2,...M)$ are the input and output variables.

It was shown in (Wang, 1994) that using singleton fuzzifier and Gaussian membership functions the formula (37) can approximate any real continuous function to any desired accuracy. Thus, similarly to the NNs, (37) is called a *fuzzy universal approximator*. Actually, (37) coincide with (35) when the membership functions are Gaussian:

$$\phi_j(x_j) = \rho_i^j \, \overline{\exp}\left[-\frac{1}{2}(x_i - \overline{x}_i^j)^2 \bigg/ (\sigma_i^j)^2 \right] \qquad (39)$$

where $\rho_i^j, \overline{x}_i^j$ and σ_i^j are real-valued parameters with, $0 \leq \rho_i^j \leq 1$, and

$$\mathbf{x} = [x_1, x_2, ..., x_n]^T, \mathbf{w}$$
$$= [w_1, w_2, ..., w_n]^T, \boldsymbol{\phi}(\mathbf{x}_i)$$
$$= [\phi_1(x_i), \phi_2(x_i), ..., \phi_M(x_i)]^T$$

Actually, all fuzzy teleoperator controllers are based on (37) and, in principle, they are similar to corresponding neural controllers. As an example it is mentioned that the neurocontrol systems discussed in Section 4.1 can be implemented in fuzzy form by replacing the NN modelling (approximation) blocks by corresponding fuzzy modelling blocks. In this spirit the fuzzy teleoperator systems of (Hu, Ren, Thompson, & Sheridan, 1996; Wang, 2011; Yi & Chung, 1998) were designed producing very encouraging stability and transparency performance.

4.2.2. Neuro Fuzzy Control

Pure fuzzy systems do not offer a certain method for selecting the form of the membership function, or a learning method of the firing strength of the rules. The previous drawbacks can be eliminated if we use NN learning for selecting membership functions (i.e., the fuzzy sets of the rules) automatically. A neuro-fuzzy system of this type is known as *Adaptive Neurofuzzy Inference System* (ANFIS). In particular, if the form of the membership functions is given, then the NN has to learn from the input-output data, and provide their parameters. For example, if the Gaussian membership functions (39) are selected, then the NN has to provide the parameters $\rho_i^j, \bar{x}_i^j$ and σ_i^j that are optimal with respect to the input-output data of the application at hand. In the literature of tele-manipulation the capability of NNs to learn membership functions has been utilized in various ways in connection to the master of the slave or both. In (Cha, Cho, & Kim, 1996) a force-reflection gain selecting algorithm in a position-force type bilateral teleoperation system is proposed using NNs and fuzzy-logic. The NN is used to learn the characteristics of the master arm and the environment, and the fuzzy-logic estimator is used to estimate the force-reflection gain. In this way the overall performance is improved since a too small gain leads to poor task performance, and a too large gain results in system instability. In (Manifar, 2010) the slave controller is designed using a combined classical controller and neuro-fuzzy controller. The training of the ANFIS controller is performed with the aid of a genetic algorithm which leads to an online neurofuzzy gain scheduling controller. In (Pongaen, Bicker, Hu, & Burn, 2004) a first-order Sugeno neurofuzzy model is used to obtain the appropriate parameters of a position/force telerobotic controller under various conditions. The estimation algorithm combines the root mean-square estimator rule with the back-propagation rule. This hybrid estimator is designed to adjust the membership functions and the liner polynomial equations of fuzzy inference.

In the following, we provide a brief review of the neurofuzzy bilateral telerobotic controller presented in (Cha & Cho, 2002). The bilateral telerobotic system has the typical structure shown in Figure 15.

When contact force from the environment-slave interaction occurs, the compliance controller accepts this force as input and gives at its output the corrective motion x_c. The modified reference (desired) trajectory x_{d2} is given by $x_{d2} = x_{d1} - x_c$ where x_{d1} is the master arm trajectory. At the same time, the contact force F_e is also reflected back to the human operator through the force

Figure 15. Structure of position-force bilateral teleoperator system

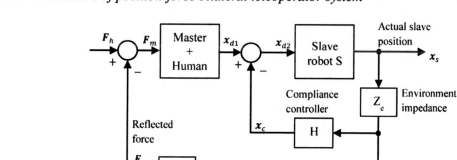

reflection gain K_f. By the small gain theorem, "Bounded Input-Bounded Output" (BIBO) stability is assured if the loop gain is less than unity, i.e.:

$$G_{dc} = H_o Z_{e,o} S_o \leq 1$$

where $H_o, Z_{e,o}$ and S_o are the DC gain of the respective blocks. This gives the BIBO stability condition:

$$H_o \leq 1/S_o Z_{e,o} \qquad (40)$$

This condition suggests that the compliance controller must be selected taking into account the characteristics of the slave and the environment. Clearly, better stability is obtained with smaller compliance. The compliance controller is implemented by the neuro-fuzzy system shown in Figure 16.

The NN adjusts the membership functions of the fuzzy rule base in accordance to information received about the contact force through a failure detector (this kind of learning control is known as *reinforced control*) (Sutton & Barto, 1998). The purpose of this detector is to see whether the desired task is performed successfully or not. The reflected force F_r is computed as $F_r = K_f F_s$ and

converted to input torques via the transpose of the Jacobian matrix. The force reflection gain must always satisfy the stability condition (4)

5. CASE STUDIES

Here, four representative case studies will be reviewed which show the effectiveness of the corresponding bilateral teleoperator control schemes. These case studies were performed using *adaptive control, sliding-mode control, neurocontrol,* and *neuro-fuzzy control*. Case studies 2 and 3 present pure simulation experimental results. Case studies 1 and 4 contain real physical experimental results. Case study 4 concerns a master/slave telerobotic system where a 3-DOF force-vertical articulated reflector master arm is cooperating with a 6-DOF industrial robot (SAMSUNG FARA A1-U). The telerobotic system controller was implemented by a PC486DX2-66 which involves a kinematics routine, a workspace mapping routine, a slave arm inverse kinematics routine, a compliance controller and a force reflecting controller (Cha & Cho, 2002).

5.1. Case Study 1: Adaptive Control

This experimental case study concerns the application of the adaptive control scheme described in

Figure 16. The neuro-fuzzy compliance controller

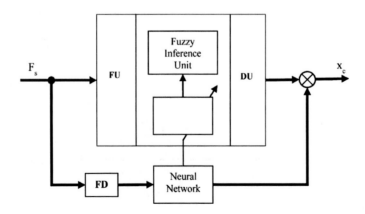

Section 3.3 (Equations (9)-(11)). Two simulation and hardware experiments were performed as follows (Lee & Chung, 1998):

Simulation 1:

Master Impedance :

$$Z_m = 2s + 0.2(Ns / m)$$

Slave Impedance :

$$Z_s = 2s + 0.2(Ns / m)$$

Operator Impedance :

$$Z_{op} = s + 50 + \frac{2000}{s}(Ns / m)$$

where s is the generalized complex variable $s = \alpha + j\omega$ of the Laplace transform representation. The slave robot is touching a hard object (environment) with $B_e = 50(Ns / m)$ and $K_e = 1000(N / m)$ without losing the contact. The human operator applies a force:

$$f_{op} = 200$$
$$\left\{ 1 - \cos\left(\frac{\pi}{2}t\right) + 50\left[\sin\left(\frac{\pi}{4}t\right) + \sin\left(\frac{\pi}{8}t\right)\right]\right\}(N)$$

The system was assumed initially at rest. The parameters of the controller were selected as: $\Omega = 100, a = 100, K_v = 100,$ " $= diag[2, 30, 2000]$. If the exact dynamics of the master is canceled, i.e., $C_m = -Z_m$ the resulting tracking performance and transparency are satisfactory (Figure 17a) (Lee & Chung, 1998). If the master dynamics is overcompensated, e.g., $C_m = -3s + 0.5$ the adaptation results show an oscillatory performance (Figure 17b). If the dynamic parameters of the master are not known we set $C_m = 0$ and the intervening impedance $Z_i = Z_m$ is used. This makes the performance more robust. In all cases the estimated parameters

Figure 17. Force tracking performance

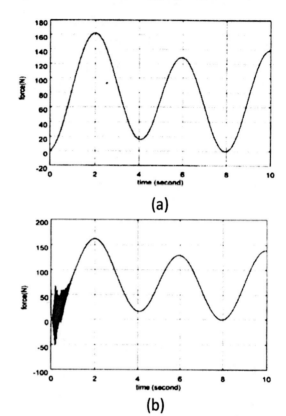

(a)

(b)

Figure 18. Convergence of the adaptive estimator for the mass. Similarar convergence results are obtained for the viscous coefficient and the stiffness parameters

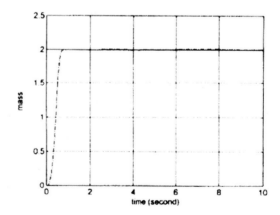

converge to the true values in 1 sec, with zero initial values (Figure 18) (Lee & Chung, 1998).

Simulation 2:

All parameters are the same as in simulation 1 except that $C_m = 0$. The task of the slave is to move in free space, starting from $x_s = 0$, toward an object located at $x_s = 0.05(m)$, and interact with it afterwards. The position and force performance is shown in (Figure 19).

Figure 19. Position and force tracking performace

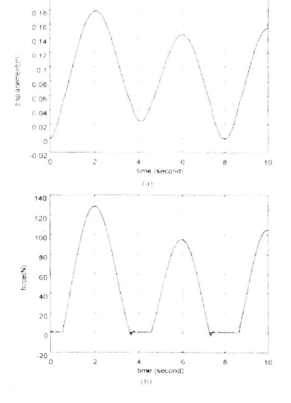

(a) $C_m = -Z_m$, (b) $C_m = -3s+0.5$

Experiment:

An experiment was also performed in (Lee & Chung, 1998) which has the block diagram shown in Figure 20. The robots used are identical 1 DOF gripper type.

The slave started at $x_s = 0$ and moved toward a relatively hard plastic object located at $x_s = 6.3(cm)$. After the contact, the slave robot moved away from the object, and then changed its direction of motion toward the object. The controller's parameters used are:

$$\Omega = 10000, a = 62.8, \mathrm{K}_v$$
$$= 0.001, \text{"} = \begin{bmatrix} 0.0001, 0.0001, 0.1 \end{bmatrix}$$

The results obtained are shown in Figure 21 (Lee & Chung, 1998).

Figure 21. Force response of the slave with the controller

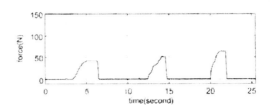

Figure 20. Structure of the hardware system

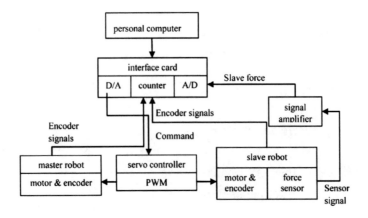

5.2. Case Study 2: Sliding-Mode Control

Some results obtained using the sliding-mode control scheme of Figure 5 will be presented. Two identical 2-DOF revolute manipulators were used as master and slave. Table 1 shows their parameters and the modeling errors considered (Tzafestas & Prokopiou, 1997)

Tasks A and B were performed in free space, and tasks C and D are contact tasks. The slave and master were initially stationary at free space at the same position. Then the operator tries to move the slave in position x_d and feels the reflected force f_d. This will be felt if he penetrates inside the object as much as the last column of

Table 2 shows. The object in contact tasks is modeled as an elastic immobile "wall" of infinite dimensions. The response for task B, obtained by the non-robust controller of (Lee & Lee, 1993), is shown in Figure 22. Ideal is regarded the response obtained directly with the real generalized impedance (Tzafestas, Prokopiou, & Tzafestas, 2001).

The corresponding results for task B, obtained using the sliding-mode control, are shown in Figure 23 (Tzafestas, Prokopiou, & Tzafestas, 2001).

The improvement in position tracking with the robust controller is directly seen, but the input torque at the master is subject to strong activity (chattering), which is more intense in contact tasks.

Table 1. System parameters and, modeling errors

Operator
$\mathbf{M}_h = \begin{bmatrix} 0.15 & 0 \\ 0 & 0.15 \end{bmatrix}, \mathbf{B}_{hu} = \begin{bmatrix} 0.5 & 0 \\ 0 & 0.5 \end{bmatrix}, \mathbf{K}_{hu} = \begin{bmatrix} 7.0 & 0 \\ 0 & 7.0 \end{bmatrix}, \mathbf{B}_{hf} = \begin{bmatrix} 0.5 & 0 \\ 0.5 & 0 \end{bmatrix}, \mathbf{K}_{hf} = \begin{bmatrix} 1.0 & 0 \\ 0 & 1.0 \end{bmatrix}$
Master and Slave Robots
Link 1: m=10, l=1, l$_c$=0.5, I=0.8, link 2: m=5, l=1, l$_c$=0.5, I=0.45 Units (m, kg, kg/m²). Symbols: m, mass; l, length; l$_c$,center of mass position from previous joint; I, moment of inertia.
Generalized Impedance
$\mathbf{M}_{dm} = \begin{bmatrix} 0.05 & 0 \\ 0 & 0.05 \end{bmatrix}, \mathbf{B}_{dm} = \begin{bmatrix} 4.0 & 0 \\ 0 & 4.0 \end{bmatrix}$ $\mathbf{M}_{ds} = \begin{bmatrix} 0.1 & 0 \\ 0 & 0.1 \end{bmatrix}, \mathbf{B}_{ds} = \begin{bmatrix} 8.0 & 0 \\ 0 & 8.0 \end{bmatrix}, \mathbf{K}_{ds}$ $= \begin{bmatrix} 10.0 & 0 \\ 0 & 10.0 \end{bmatrix}, \mathbf{B}_{fs} = \begin{bmatrix} 0.75 & 0 \\ 0 & 0.75 \end{bmatrix}, \mathbf{K}_{fs} = \begin{bmatrix} 0.0 & 0 \\ 0 & 0.0 \end{bmatrix}$
Communications Channel and G$_{rp}$
$\mathbf{K}_{sc} = \begin{bmatrix} 1.0 & 0 \\ 0 & 1.0 \end{bmatrix}, \mathbf{K}_{cs} = \begin{bmatrix} 0.2 & 0 \\ 0 & 0.2 \end{bmatrix}, \mathbf{G}_{rp} = \begin{bmatrix} 1.0 & 0 \\ 0 & 1.0 \end{bmatrix}$
Modeling Errors
Unless otherwise stated, 20% and 50% for the slave and master, respectively, on the mass and moment of inertia of the last link.

Table 2. Free-space and contact simulation tasks

Task	$\mathbf{x}_o$	$\mathbf{x}_d$	$\mathbf{f}_{es,d}$	$\overline{}_c$	Penetration
A	(0.8, 0.8)	(0.8105, 0.8105)	-		
B	(0.8, 0.8)	(0.85, 0.85)	-		
C	(0.8, 0.8)	(0.8105, 0.8105)	(1, 1)	$\begin{bmatrix} 20 & 20 \\ 20 & 20 \end{bmatrix}$	0.0354
D	(0.8, 0.8)	(0.8105, 0.8105)	(1, 1)	$\begin{bmatrix} 500 & 500 \\ 500 & 500 \end{bmatrix}$	0.0354

Subscripts o and d represents starting and goal values, respectively. All distances are in meters, forces in Newtons. The object's border is given by the equation: $y = 1.617 - x$. A point on the border is $X_e = (0.8085, 0.8085)$ and a unity vector vertical to it $(0.7071, 0.7071)$. For task D: $\mathbf{M}_{ds} = \begin{bmatrix} 1.0 & 0 \\ 0 & 1.0 \end{bmatrix}$

Figure 22. Results for task B with non-robust control (a) slave trajectory, (b) master trajectory, (c) input-torque for the slave's first joint, (d) input torque for the master's first joint (ideal, with full computation delay period 1msec)

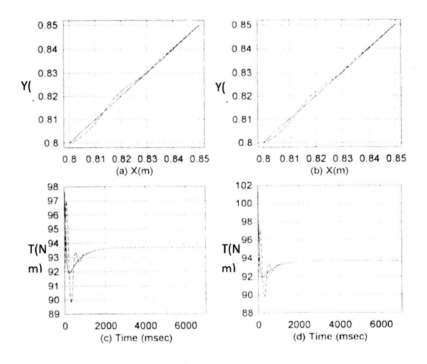

Figure 23. Task B results with the sliding-mode control for the model with uncertainties (a) slave trajectory, (b) distance from target, (c) input torque at the slave's first joint, (d) input-torque for the slave's first joint (detail)

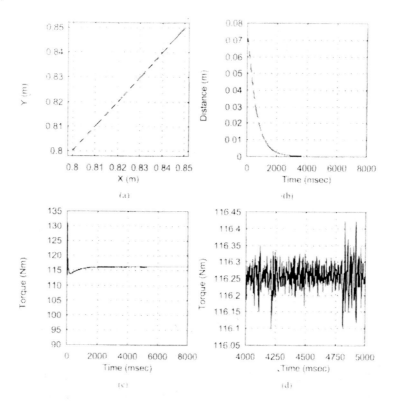

5.3. Case Study 3: Neural Control

The first NN-learning control method of Section 4.1 with partitioned identification has been applied at the same tasks (A, B, C, D) discussed in Section 5.2. The results are very encouraging and are of about the same quality as the sliding-mode results. These results for task B are shown in Figure 24 (Tzafestas, Prokopiou, & Tzafestas, 2001).

(a) Position x(-master, ---slave)

(b) Force f (-master, --- slave)

With the exception of the few milliseconds, during which the subnets learn the new dynamics, the control torque has almost no chattering. The peaks of the torque mark the beginning of the phases of training and movement.

5.4. Case Study 4: Neurofuzzy Control

The present case study concerns the application of the neurofuzzy control scheme represented by Figure 15 and Figure 16. The fuzzy system units (fuzzifier, rule base, inference unit, and defuzzifier) were implemented in the standard way. The input variables are contact forces and absolute values of change of contact forces. The output variable is the corrective motion. This means that the controller is a fuzzy PD controller. The rule base contains the rules.

$$RULE\ \mathbf{R}_1^k : \ \mathbf{IF}\ f_1\ \text{is}\ F^k\ \mathbf{AND}\ f_2$$
$$\text{is}\ G^k\ \mathbf{AND}\ f_3\ \text{is}\ G^k$$
$$\mathbf{AND}\ \Delta f_1\ \text{is}\ \Delta \mathbf{F}^k\ \mathbf{AND}\ \hat{y}_1\ \text{is}\ \hat{Y}_1^k$$
$$\mathbf{THEN}\ y_1\ \text{is}\ Y_1^k$$

Figure 24. Results for the B with MLP neural universal approximator for free movement (a) slave trajectory, (b) distance from the target, (c) input torque at the slave's first link, (d) input torque at the slave's first link (detail)

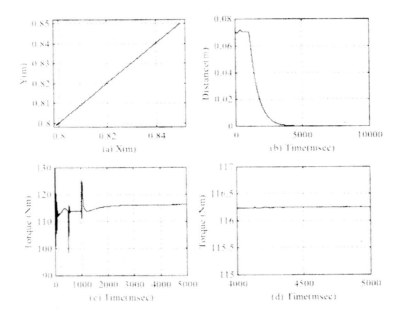

RULE $\mathbf{R}_2^k$: **IF** f_1 is G^k **AND** f_2

is F^k **AND** f_3 is G^k

AND Δf_1 is ΔF^k **AND** $\hat{y}_2$ is $\hat{Y}_2^k$

THEN y_2 is Y_2^k

Table 2 shows the tasks at which the system was tested.

RULE $\mathbf{R}_3^k$: **IF** f_1 is G^k **AND** f_2

is F^k **AND** f_3 is G^k

AND Δf_3 is ΔF^k **AND** $\hat{y}_3$ is $\hat{Y}_3^k$

THEN y_3 is Y_3^k

where k denotes the kth rule, $f_i (i = 1, 2, 3)$ are the three forces at the three joints of the manipulator, Δf_i are the corresponding changes, $\hat{y}_i$ are one-step ahead corrective forces, and y_i are the three current forces corrective forces for $i = x, y$ and z. Each set of rules $R_i^k (i = 1, 2, 3)$ involves $5 \times 2 \times 2 \times 2 \times 3 = 120$ rules. The control goal

is to make the slave arm as compliant as possible in order to maximize the force reflection gain. Therefore, the failure detector (*critic unit*) is the following:

$\mathbf{R}_{c1}$: **IF** $f(t) > f_{max}$ **THEN** fail

$\mathbf{R}_{c2}$: **ELSE** $\left| f(t) - f(t-1) \right| > \Delta f_{max}$

THEN fail

$\mathbf{R}_{c3}$: **ELSE** success

where $f(t) = \sqrt{f_1^2 + f_2^2 + f_3^2}$, $f_{max} = 50N$, $\Delta f_{max} = 3N$, *failure value* $= -0.055$, and *success value* $= 0.055$. The fuzzy set for $F^k, G^k, \Delta F^k$ and y have membership function of the triangular and trapezoidal form as shown in Figure 25 (Cha & Cho, 2002).

The controller output (change of maximum contact force) in the z-direction, for two wall-type objects (plastic, steel) is as shown in Figure 26 (Cha & Cho, 2002).

Figure 25. Membership functions of $F^k, G^k, \Delta F^k$ and y^k

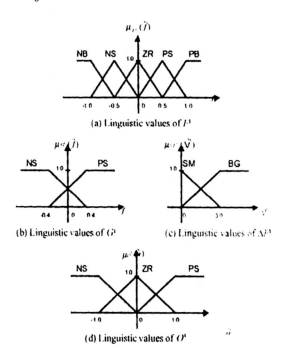

Figure 26. Change of maximum contact forces in the z-direction

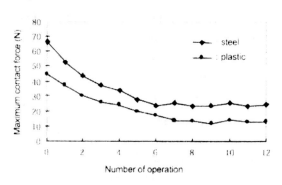

The results for the plastic-wall contact experiment are shown in Figure 27.

In this experiment, the objective was to keep a constant contact force of 20 N in the z-direction. The contact took place at about 1.7 sec. The controller gain was selected to be 0.3 which is the maximum allowable gain for stability. The results achieved (Figure 27a) are superior to those ob-

tained by the method of (Kim, 1989) with the maximum allowed gain 0.2. The latter showed a moderate oscillatory behavior stabilized at $\sim 20N$ after 5 sec (Figure 27b).

6. CONCLUSION

Telerobotic systems represent one of the most important fields of robotics. Teleoperation systems have a long history and during the last three decades have received a strong attention by researchers and practitioners for both their challenging problems and their important applications. They have been used to allow human operators to perform in hazardous environments or unreachable places, such as handling radioactive materials, nuclear power reactor maintainance, volcano exploration, planet exploration, sea-bed exploration, and tele-surgery. Teleoperation systems suffer, in addition to standard modelling and input uncertainties of robotic manipulators, by time delays especially when the medium of master-slave communication is the

Figure 27. Contact forces against time, (a) NFCM-based bilateral control, (b) Kim's advanced bilateral control

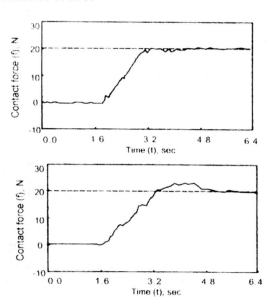

Internet. This chapter has presented a comprehensive overview of the existing techniques for facing both modelling/input uncertainties and small or large time delays, constant or varying. Especially, for the delays occurring over the Internet a large set of techniques has been developed for modelling, studying and estimating the delays. These techniques include, among others the following:

- Queuing Network Modeling
- Time Series Modeling
- Multimodel Approach
- Kalman and Particle Filtering (Lehmann, 2006; Reed, Feintuch, & Bershold, 1981)
- Neural Network Modeling (Belhai & Tagina, 2009; Parlos, 2002)

They can be used, and some of them have been used, in combination with standard controllers for compensating the time delay effectss in Internet bilateral teleoperator systems.

REFERENCES

Abbot, J. J., & Okamura, A. M. (2003). Virtual fixture architecture for telemanipulation. In *Proceedings of the IEEE International Conference on Robotics and Automation*, (pp. 2798-2805). IEEE.

Anderson, R. J., & Spong, M. W. (1989). Bilateral control of teleoperators with time delay. *IEEE Transactions on Automatic Control*, *34*, 494–501. doi:10.1109/9.24201.

Ang, W. T., & Riviere, G. N. (2001). Neural network methods for error canceling in human-machine manipulation. In *Proceedings of the 23rd Annual International Conference of Engineering in Medicine and Biology*, (pp. 3462-3465). IEEE.

Arcara, P., & Melchiori, C. (2002). Control schemes for teleoperation with time delay: A comparative study. *Robotics and Autonomous Systems*, *28*, 49–64. doi:10.1016/S0921-8890(01)00164-6.

Artigas, J., Preusche, G., & Hirzinger, G. (2006). Time domain passivity-based telepresence with time delay. In *Proceedings of IEEE/RSJ International Conference Intelligent Robotics Systems (IROS)*, (pp. 4205-4210). IEEE.

Batsomboon, P., & Tosunoglou, S. (1996). A review of teleoperation and telesensation system. In *Proceedings of the 1996 Florida Conference on Recent Advances in Robotics*. Florida Atlantic University.

Belhai, S., & Tagina, M. (2009). Modeling and prediction of the Internet delay using recurrent neural networks. *Journal of Networks*, *4*, 528–535.

Berestesky, P., Chopra, N., & Spong, M. W. (2004). Discrete-time passivity in bilateral teleoperation over the internet. In *Proceedings IEEE International Conference in Robotics and Automation*, (pp. 4557-4564). IEEE.

Buss, M., & Schmidt, G. (1999). Control problems in multi-modal telepresence systems. In *Proceedings of the European Control Conference (ECC'99)*, (pp. 65-101). ECC.

Cha, D.-H., & Cho, H. S. (2002). A neurofuzzy algorithm-based advanced bilateral controller for telerobot systems. *ICASE: The Institute of Control Automation and Systems Engineers*, *4*, 100–107.

Cha, D. H., Cho, H. S., & Kim, S. (1996). Design of a force reflection controller for telerobot systems using neural network and fuzzy logic. *Journal of Intelligent & Robotic Systems*, *16*, 1–24. doi:10.1007/BF00309653.

Chen, Q., Quan, J., & Xia, J. (2007). Neural network based multiple model adaptive predictive control for teleoperation. In *Proceedings 4th International Symposium on Neural Networks: Advances in Neural Networks*, (pp. 64-69). Berlin: Springer.

Cho, Y. C., & Park, J. H. (2003). Stable bilateral teleoperation under a time delay using a robust impedance control. *Journal of Mechatronics, 15*, 1–10.

Conti, F., & Khatib, O. (2005). Spanning large work spaces using small haptic devices. In *Proceedings 1ˢᵗ Joint Eurohaptics Conference*, (pp. 183-188). Pisa, Italy: Eurohaptics.

Flemmer, H. (2004). *Aspects of using passivity in bilateral telemanipulation* (Tech. Report No. TRITA-MMK 2004:16). Stockholm, Sweden: Royal Institute of Technology.

Ganjefar, S., Momeni, H., Sharifi, F. J., & Behesthi Hamidi, M. T. (2003). Behaviour of Smith predictor in teleoperation systems with modeling and delay time error. In *Proceedings IEEE Conference on Control Applications*, (pp. 1176-1180). IEEE.

Garcia-Valdovinos, L. G., Parra-Vega, V., & Arteaga, M. A. (2007). Observer-based sliding-mode impedance control of bilateral teleoperation under constant unknown time-delay. *International Journal of Robotics and Autonomous Systems, 55*, 609–617. doi:10.1016/j.robot.2007.05.011.

Guez, A., & Selinsky, J. (1989). Neuro-controller design via supervised and unsupervised learning. *Journal of Intelligent & Robotic Systems, 2*, 307–335. doi:10.1007/BF00238695.

Habib, M. K. (2007). Mechatronics: A unifying interdisciplinary and intelligent engineering paradigm. *IEEE Industrial Electronics Magazine, 1*(2), 12–24. doi:10.1109/MIE.2007.901480.

Habib, M. K. (2008). Interdisciplinary mechatronics: Problem solving, creative thinking and concurrent design synergy. *International Journal of Mechatronics and Manufacturing Systems, 1*(1), 264–269. doi:10.1504/IJMMS.2008.018272.

Hannaford, B. (1989). A design framework for teleoperators with kinesthetic feedback. *IEEE Transactions on Robotics and Automation, 5*(4), 426–434. doi:10.1109/70.88057.

Hashemzadeh, F., Hassanzadeh, I., Tavakoli, M., & Alizadeh, G. (2000). A new method for bilateral teleoperation passivity under varying stability. In *Mathematical Problems in Engineering*. Hindawi Publ. Corp..

Hu, J., Ren, J., Thompson, J., & Sheridan, T. (1996). Fuzzy sliding control of a force reflecting teleoperator system. In *Proceedings 5ᵗʰ IEEE International Conference on Fuzzy Sytems*, (pp. 2162-2167). New Orleans, LA: IEEE.

Huang, J. (2002). Neurocontrol of telerobotic systems with time delays. In Lewis, F. L., Campos, J., & Selmic, R. (Eds.), *Neuro-Fuzzy Control of Industrial Systems with Actuator Nonlinearities*. Philadelphia: SIAM Publishing. doi:10.1137/1.9780898717563.ch8.

Hung, N., Narikiyo, T., & Tuan, H. (2003). Nonlinear adaptive control of master slave system in teleoperation. *Control Engineering Practice, 11*, 1–10. doi:10.1016/S0967-0661(02)00068-0.

Iqbal, A., Roth, H., & Abu-Zaitoon, M. (2005). Stabilization of delayed teleoperation using predictive time-domain passivity control. In *Proceedings IASTED International Conference on Robotics and Applications*, (pp. 20-25). Cambridge, MA: IASTED.

Isermann, R. (1997). Mechatronics systems: A challenge for control engineering. In *Proceedings American Control Conference*, (pp. 2617-2632). ACC.

Kawato, M., Uno, Y., Isobe, M., & Suzuki, R. (1988). Hierarchical neural network model for voluntary movement with application to robotics. *IEEE Control Systems Magazine*, 8–16. doi:10.1109/37.1867.

Kim, W. S. (1989). Developments of new force reflecting control schemes and an application to a teleoperator training simulator. In *Proceedings IEEE International Conference on Robotics and Automation*, (pp. 1764-1767). IEEE.

Kubo, R., Iiyama, N., Natori, K., Ohnishi, K., & Furukawa, H. (2007). Performance analysis of a three-channel control architecture for bilateral teleoperation with time delay. *IEEJ Transactions, 1A*(127), 1224–1230. doi:10.1541/ieejias.127.1224.

Leblebici, T., Calli, B., Unel, M., Sabanovic, A., Bogosyan, S., & Gokasan, M. (2011). Delay compensation in bilateral control using a sliding mode observer. *Journal of Electrical Engineering and Computer Science, 19*, 851–859.

Lee, D., & Spong, M. W. (2006). Passive bilateral teleoperation with constant time delay. *IEEE Transactions on Robotics, 22*, 269–281. doi:10.1109/TRO.2005.862037.

Lee, H. K., & Chung, M. J. (1998). Adaptive controller of a master-slave system for transparent teleoperation. *Journal of Robotic Systems, 15*, 465–475. doi:10.1002/(SICI)1097-4563(199808)15:8<465::AID-ROB3>3.0.CO;2-J.

Lee, S., & Lee, H. S. (1993). Modeling, design and evaluation of advanced teleoperator control system with short time delay. *IEEE Transactions on Robotics and Automation, 9*, 607–623. doi:10.1109/70.258053.

Lehmann, E. A. (2006). Particle filtering approach to adaptive time-delay estimation. In *Proceedings 2006 International Conference IEEE Acoustic, Speech and Signal Processing*, (pp. 1129-1132). Toulouse, France: IEEE.

Lewis, F. L., Liu, K., & Yesildirek, A. (1995). Neural net robot controller with guaranteed tracking performance. *IEEE Transactions on Neural Networks, 6*, 703–715. doi:10.1109/72.377975 PMID:18263355.

Lin, W., Tsai, C., & Liu, J. (2001). Robust neuro-fuzzy control of multivariable system by tuning consequent membership function. *Fuzzy Sets and Systems, 124*.

Lozano, R., Chopra, N., & Spong, M. W. (2002). Passivation of force reflecting bilateral teleoperators with time varying delay. In *Proceedings Mechatronics '02*. Enschede, The Netherlands: Mechatronics.

Manifar, S. (2010). Application of GA based neuro-fuzzy automatic generation for teleoperation systems. In *Proceedings 6th International Conference on Digital Context, Multimedia Technology and its Applications (IDC-2010)*, (pp. 296-301). IDC.

Massimono, M. J., Sheridan, J. B., & Roseborough, J. B. (1989). One handed tracking in six degrees of freedom. In *Proceedings IEEE International Conference Systems, Man, and Cybernetics*, (vol. 2, pp. 498-503). IEEE.

Munir, S., & Book, W. J. (2002). Internet-based teleoperation using wave variables with prediction. *IEEE/ASME Transactions on Mechatronics, 7*, 119–127. doi:10.1109/TMECH.2002.1011249.

Natori, K., Tsuji, T., & Ohnisi, K. (2004). Robust bilateral control with internet communication. In *Proceedings of the 30th IEEE Annual Conference on Industrial Electronics*, (pp. 2321-2326). Busan, Korea: IEEE.

Nie, J., & Linkens, D. (1995). *Fuzzy-neural control: Principles, algorithms and applications*. Englewood Cliffs, NJ: Prentice Hall.

Niemeyer, G., Preusche, C., & Hirzinger, G. (2008). Telerobotics. In Siciliano & Habib (Eds.), Springer Handbook of Robotics, (pp. 741-757). Berlin: Springer.

Niemeyer, G., & Slotine, J.-E. (1997). Using wave variables for system analysis and robot control. In *Proceedings IEEE International Conference on Robotics and Automation*. Albuquerque, NM: IEEE.

Niemeyer, G., & Slotine, J. J. (1991). Stable adaptive teleoperation. *IEEE Journal of Oceanic Engineering, 16*, 152–162. doi:10.1109/48.64895.

Niemeyer, G., & Slotine, J.-J. E. (2004). Telemanipulation with time delays. *The International Journal of Robotics Research, 23*, 873–890. doi:10.1177/0278364904045563.

Park, J. H., & Cho, H. C. (1999). Sliding-mode controller for bilateral teleoperation with varying time delay. In *Proceedings 1999 IEEE/ASME International Conference on Advanced Intelligent Mechatronics*, (pp. 311-316). Atlanta, GA: IEEE.

Parlos, A. G. (2002). Identification of the internet-end-to-end delay dynamics using multi step neuro predictors. In *Proceedings 2002 International Joint Conference on Neural Networks*, (pp. 2460-2465). Honolulu, HI: IEEE.

Polushin, I. G., Tayebi, A., & Marquez, J. (2005). Adaptive schemes for stable teleoperation with communication delay based on IOS small gain theorem. In *Proceedings 2005 American Control Conference*, (pp. 4143-4148). Portland, OR: ACC.

Pongaen, W. (2008). Using neuro-fuzzy control to enhance maneuverability of master-slave system in position feedback frameworks. In *Proceedings IEEE International Conference on Robotics and Biomimetics (ROBIO)*. IEEE.

Pongaen, W., Bicker, R., Hu, Z., & Burn, K. (2004). Approach to telerobotic control using neuro-fuzzy techniques. In *Proceedings 11ᵗʰ World Congress in Mechanism and Machine Science*, (pp. 1761-1766). IEEE.

Reed, F., Feintuch, P., & Bershold, N. (1981). Time delay estimation using the LMS adaptive filter-static behavior. *IEEE Transactions on ASSP, 29*, 561–571. doi:10.1109/TASSP.1981.1163614.

Ryu, J.-H., Preusche, B., Hannaford, B., & Hirzinger, G. (2005). Time-domain passivity control with reference energy following. *IEEE Transactions on Control Systems Technology, 13*, 737–742. doi:10.1109/TCST.2005.847336.

Schweitzer, G. (1996). Mechatronics for the design of human-oriented machines. *IEEE/ASME Transactions on Mechatronics, 1*(2), 120–126. doi:10.1109/3516.506148.

Sheridan, T. B. (1992). *Telerobotics, automation and human supervisory control*. Cambridge, MA: MIT Press.

Slotine, J. J., & Li, W. (1991). *Applied nonlinear control*. Englewood-Cliffs, NJ: Prentice Hall.

Smith, A., & Mohashtrudi-Zaad, K. (2005). Neural network-based teleoperation using Smith predictors. In *Proceedings IEEE International Conference on Mechatronics and Automation (ICMA 05)*. IEEE.

Smith, A. C., & Hashtrudi-Zaad, K. (2006). Smith predictor type control architectures for time delayed teleoperation. *The International Journal of Robotics Research, 25*, 797–818. doi:10.1177/0278364906068393.

Spong, M. W., & Vidyasagar, M. (1989). *Robot dynamics and control*. New York: Wiley.

Stramiglioli, S., van der Schaft, A., Maschke, B., & Melchiori, C. (2002). Geometric scattering in robotic telemanipulation. *IEEE Transactions on Robotics and Automation, 18*, 588–596. doi:10.1109/TRA.2002.802200.

Sutton, R., & Barto, A. (1998). *Reinforcement learning: An introduction.* Cambridge, MA: MIT Press.

Tanner, N. A., & Niemeyer, G. (2006). High-frequency acceleration feedback in wave variables telerobotics. *IEEE/ASME Transactions on Mechatronics, 11*, 119–127. doi:10.1109/TMECH.2006.871086.

Teranno, T., Asai, K., & Sugeno, M. (1992). *Fuzzy systems theory and its applications.* Boston, MA: Academic Press.

Tzafestas, S. G. (1995). Neural networks in robot control. In Tzafestas, S. G., & Verbruggen, H. B. (Eds.), *Artificial Intelligence in Industrial Decision Making, Control, and Automation* (pp. 327–358). Dordrecht, The Netherlands: Springer. doi:10.1007/978-94-011-0305-3_11.

Tzafestas, S. G. (2009). *Human and nature minding automation.* Berlin: Springer.

Tzafestas, S. G., & Prokopiou, P. A. (1997). Compensation of teleoperator modeling uncertainties with sliding-mode controller. *Robotics and Computer-integrated Manufacturing, 13*, 9–20. doi:10.1016/S0736-5845(96)00030-0.

Tzafestas, S. G., Prokopiou, P. A., & Tzafestas, C. S. (2001). A new partitioned robot neurocontroller: General analysis and application to teleoperator modeling uncertainties. *Machine Intelligence and Robot Control, 3*, 7–26.

van Brussel, H. (1989). The mechatronics approach to motion control. In *Proceedings International Conference on Motion Control: The Mechatronics Approach.* Antwerp, Belgium: IEEE.

Wang, L. X. (1994). *Adaptive fuzzy systems and control: Design and stability.* Englewood Cliffs, NJ: Prentice Hall.

Wang, R. (2011). Bilateral control of teleoperation system with fuzzy singularly perturbed model. In *Proceedings International Conference on Intelligent Control and Information Processing (ICCIP 11)*, (pp. 941-945). ICCIP.

Watanabe, K., Fukuda, T., & Tzafestas, S. G. (1992). An adaptive control for CARMA systems using linear neural networks. *International Journal of Control, 56*, 483–497. doi:10.1080/00207179208934324.

Watanabe, K., & Tzafestas, S. G. (1990). Learning algorithms for neural networks with the Kalman filters. *Journal of Intelligent & Robotic Systems, 3*, 305–319. doi:10.1007/BF00439421.

Yi, S. Y., & Chung, M. J. (1998). Robustness of fuzzy logic for an uncertain dynamic system. *IEEE Transactions on Fuzzy Systems, 6.*

Yoshida, K., & Namerikawa, T. (2009). Stability and tracking properties in predictive control with adaptation for bilateral teleoperation. In *Proceedings 2009, American Control Conference*, (pp. 1323-1328). St. Louis, MO: ACC.

Zhu, W. H., & Salcudean, S. E. (2000). Stability guaranteed teleoperation: An adaptive motion/force control approach. *IEEE Transactions on Automatic Control, 45*, 1951–1959. doi:10.1109/9.887620.

Section 4

Chapter 14
Modelling and Simulation Approaches for Gas Turbine System Optimization

Hamid Asgari
University of Canterbury, New Zealand

XiaoQi Chen
University of Canterbury, New Zealand

Raazesh Sainudiin
University of Canterbury, New Zealand

ABSTRACT

This chapter deals with research activities that have been carried out so far in the field of modelling and simulation of gas turbines for system optimization purposes. It covers major white-box and black-box gas turbine models and their applications to control systems.

1. INTRODUCTION

The rapidly growing knowledge-based industry has been always looking for creative and bright ideas. Mechatronics, as a *multidisciplinary* field of engineering, is one of those innovative phenomena that has contributed many advantages to our industrial society. It represents a unifying paradigm that integrates, permeates, and comprehends fundamental and modern engineering (Habib, 2006). Mechatronics combines a variety of engineering disciplines including mechanics, electronics, computer science, systems design,

and control to fulfill the challenges of modern technology and the demand for innovation (Habib, 2008). It has been a powerful solution to many sophisticated problems in complex industrial systems such as Gas Turbines (GT).

Gas turbine is considered as an internal combustion engine which uses the gaseous energy of air to convert chemical energy of fuel to mechanical energy. Although the story of gas turbines has taken a root in history, it was not until 1930s that the first practical GT was developed by Frank Whittle and his colleagues in Britain for a jet aircraft engine (Kulikov & Thompson, 2004). Gas turbines were

DOI: 10.4018/978-1-4666-4225-6.ch014

developed rapidly after World War II and became the primary choice for many applications. That was especially because of enhancement in different areas of science such as aerodynamics, cooling systems, and high-temperature materials which significantly improved the engine efficiency. Then, it is not surprising if gas turbines have been increasing in popularity year by year.

Today, gas turbines are one of the major parts of modern industry. They have been playing a key role in aeronautical industry, power generation, and main mechanical drivers for large pumps and compressors. They have the ability to provide a reliable and continuous operation. The operation of nearly all available mechanical and electrical equipment and machinery in industrial plants such as petrochemical plants, oil field platforms, gas stations and refineries, depends on the power produced by gas turbines.

Figure 1 shows the main components of a single-shaft gas turbine engine; including compressor, combustion chamber (combustor), and turbine. The set of these components is called engine core or Gas Generator (GG). Compressor and turbine are connected by the central shaft and rotate together.

As the figure shows, air enters the compressor at section 1 and is compressed through passing the compressor. The hot and compressed air enters the combustion chamber (combustor) at section 2. In combustor, fuel is mixed with air and ignited. The hot gases which are the product of combustion are forced into the turbine at section

3 and rotate it. Turbine drives the compressor and the GG mechanical output, which can be an electricity generator in a power plant station, a large pump or a large compressor.

2. GAS TURBINE CYCLE

Gas turbines work based on Brayton cycle. Figure 2 indicates a typical standard Brayton cycle in temperature-entropy frame (Tavakoli et al., 2009). As it can be seen from the figure, the actual processes in the compressor (1-2) and turbine (3-4) are irreversible and non-isentropic. Points 2s and 4s show the ideal situation, when these processes are assumed isentropic. Neglecting pressure loss in the air filters and the combustion chamber, processes 2-3 and 4-1 can be considered isobar (Tavakoli et al., 2009).

Considering Figure 1 and Figure 2, basic thermodynamic equations for the main parts of a single-shaft gas turbine can be written as follows (Al-Hamdan & Ebadi, 2006). For the compressor:

$$T_{02} = T_{01} + \frac{T_{01}}{\eta_c} \left[\left(\frac{P_{02}}{P_{01}} \right)^{\frac{(\gamma_{air} - 1)}{\gamma_{air}}} - 1 \right] \qquad (1)$$

Figure 2. Typical Brayton cycle in temperature-entropy frame (Tavakoli, et al., 2009)

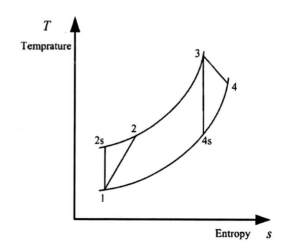

Figure 1. A simple schematic of a typical single-shaft gas turbine

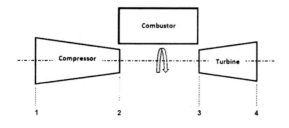

$$\gamma_{air} = C_{P_{air}} / C_{v_{air}} \qquad (2)$$

$$\dot{W}_c = \dot{m}_{air} C_{P_{air}} (T_{02} - T_{01}) \qquad (3)$$

in which:

- P_{01} : Compressor inlet stagnation pressure (Pa)
- T_{01} : Compressor inlet temperature (K)
- P_{02} : Compressor outlet stagnation pressure (Pa)
- T_{02} : Compressor outlet temperature (K)
- $C_{P_{air}}$: Specific heat of air in constant pressure (J/kgK)
- $C_{v_{air}}$: Specific heat of air in constant volume (J/kgK)
- γ_{air} : Ratio of specific heats of air
- η_c : Compressor efficiency
- $\dot{W}_c$: Compressor power (W)
- $\dot{m}_{air}$: Mass flow rate of air (kg/s)

Indices 1 and 2 refer to input and output of the compressor. Similar equations can be written for the turbine:

$$T_{04} = T_{03} - T_{03}\eta_t [1 - \left(\frac{P_{04}}{P_{03}}\right)^{\frac{(\gamma_{gas}-1)}{\gamma_{gas}}}] \qquad (4)$$

$$\gamma_{gas} = C_{P_{gas}} / C_{v_{gas}} \qquad (5)$$

$$\dot{W}_t = \dot{m}_{gas} C_{P_{gas}} (T_{03} - T_{04}) \qquad (6)$$

in which:

- P_{03} : Turbine inlet stagnation pressure (Pa)

- T_{03} : Turbine inlet temperature (K)
- P_{04} : Turbine outlet stagnation pressure (Pa)
- T_{04} : Turbine outlet temperature (K)
- $C_{P_{gas}}$: Specific heat of gas in constant pressure (J/kgK)
- $C_{v_{gas}}$: Specific heat of gas in constant volume (J/kgK)
- γ_{gas} : Ratio of specific heats of gas
- η_t : Turbine efficiency
- $\dot{W}_t$: Turbine power (W)
- $\dot{m}_{gas}$: Mass flow rate of gas (kg/s)

Indices 3 and 4 refer to input and output of the turbine. The following thermodynamic equations can show the relationship between important parameters of the combustion chamber:

$$\frac{1}{F} = \frac{\eta_{cc}(\text{LCV})}{C_{P_{gas}}(T_{03} - T_{02})} - 1 \qquad (7)$$

$$P_{03} = P_{02}(1 - \xi_{cc}) \qquad (8)$$

in which:

- η_{cc} : Combustion chamber efficiency
- F : Fuel to air ratio
- $C_{P_{gas}}$: Specific heat of gas in constant pressure
- LCV : Fuel lower calorific value
- ξ_{cc} : Pressure loss in combustion chamber

3. GAS TURBINE MODELLING AND SIMULATION

Manufacturing of the most efficient, reliable and durable gas turbines is a continuous challenge for scientists and therefore, there is a strong potential for further research in this field to get in-depth understanding of the nonlinear behavior of these systems. This potential has been a powerful motivation for researchers and engineers in different disciplines to continue to design, build, develop, and operate new generations of gas turbines and their related control systems based on the latest developments and achievements.

Making models of gas turbines and their related control system has been a useful technical and cost-saving strategy for performance optimization of the equipment before final design process and manufacturing. A variety of analytical and experimental models of GTs has been built so far. However, the need for optimized models for different objectives and applications has been a strong motivation for researchers to continue to work in this area. Models may also be used online on sites for optimization, condition monitoring, sensor validation, fault detection, trouble shooting, etc. Before starting to make a GT model, some basic factors should be considered. GT type, configuration, modelling methods, model construction approaches, modelling objectives as well as control system type and configuration are among the most important criteria at the beginning of the modelling process.

3.1. Gas Turbine Type

As the first step of modelling, it is necessary to get enough information about the type of gas turbine which is to be modelled. Although there are different types of GT based on their applications in industry, they have the same main common parts including combustion chamber, compressor, and turbine.

Gas turbines are divided into two main categories including aero gas turbines (jet engines) and stationary gas turbines. In aero industry, gas turbine is used as propulsion system to make thrust and to move an airplane through the air. Thrust is usually generated based on the Newton's third law of action and reaction. There are varieties of aero gas turbines including turbojet, turbofan, and turboprop. In stationary gas turbines, GG may be tied to electro generators, large pumps or compressors to make turbo-generators, turbo-pumps, or turbo-compressors respectively. If the main shaft of the GG is connected to an electro generator, it can be used to produce electrical power. *Industrial Power Plant Gas Turbines (IPGTs)* are playing a key role in producing power, especially for the plants which are far away on oil fields and offshore sites where there is no possibility for connecting to the general electricity network.

3.2. Gas Turbine Configuration

Configuration of a gas turbine is an important criterion in GT modelling. Although all gas turbines nearly have the same basic structure and thermodynamic cycle, there are considerable distinctions when they are investigated in details. For instance, to enhance gas turbine cycle, system efficiency or output power, through different methods such as reheating, intercooling or heat exchange, particular GT configurations are utilized. Gas turbines can be categorized based on the type of their shafts. They may be single-shaft or split-shaft. In a single-shaft gas turbine, the same turbine rotor which drives the compressor is connected to the power output shaft through a speed reduction. In a split-shaft gas turbine, the gas generator turbine and the *Power Turbine (PT)* are mechanically disconnected. Gas generator turbine, also called *Compressor Turbine (CT)* or *High Pressure (HP)* turbine, is the component which provides required power for driving the compressor and accessories. However, power turbine, also called *Low Pressure (LP)* turbine, does the usable work. Figure 3 shows a typical twin-shaft gas turbine engine (MAN Diesel & Turbo Co.).

Figure 3. A typical twin-shaft gas turbine engine (MAN Diesel & Turbo Co.)

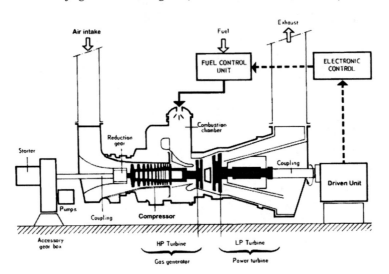

3.3. Gas Turbine Modelling Methods

There are different approaches to model a dynamic system such as a gas turbine. Mathematical modelling is considered as a general methodology for system modelling. It uses mathematical language to describe and predict behavior of a system. Important advances and development of scientific fields may be tied to the quality of mathematical models and their agreement with the results of experimental measurements. Mathematical models can be classified as "linear and nonlinear," "deterministic and stochastic (probabilistic)", "static and dynamic," or "discrete and continuous" models.

3.3.1. Linear and Nonlinear Models

A model is called linear if all objective functions and constraints of the system are represented by linear equations. Otherwise, it is considered as a nonlinear model. Although industrial equipment usually shows nonlinear behavior, in many cases the model is simplified to be analyzed linearly. There are different methods to linearize a nonlinear system. However, in setting up a model which can accurately predict behavior of complex and sensitive systems such as gas turbines, considering nonlinear dynamics is unavoidable.

3.3.2. Deterministic and Stochastic (Probabilistic) Models

In a deterministic model, all variable states are determined uniquely by the parameters in the model and by the sets of previous states of these variables. Therefore, a deterministic model expresses itself without uncertainty due to an exact relationship between measurable and derived variables. Conversely, in a stochastic model, quantities are described using stochastic variables or stochastic processes (Ljung & Glad, 1994). Therefore, in a stochastic model, variable states are described using random probability distributions.

3.3.3. Static and Dynamic Models

The variables which usually characterize a system change with time. If there are direct, instantaneous links among these variables, the system is called static. If the variables of a system change without direct outside influence so that their values depend on earlier applied signals, then the system is called dynamic (Ljung & Glad, 1994).

3.3.4. Discrete and Continuous Models

A mathematical model is called continuous-time when it describes the relationship between continuous time signals. Continuous-time models are shown with a function f (t) that changes over continuous time intervals. A model is called discrete-time when it directly expresses the relationships between the values of the signals at discrete instants of time (Ljung & Glad, 1994). Relationship between signal values is usually expressed by using differential equations. In practical applications, signals are most often obtained in sampled form in discrete time measurements.

3.4. Gas Turbine Model Construction Approaches

There are different ways to construct a mathematical model based on the prior information about the system. These approaches can be classified into three main categories including white-box, black-box, and gray-box models.

3.4.1. White-Box Models

A white-box model is used when there is enough knowledge about the physics of the system. In this case, mathematical equations regarding dynamics of the system are utilized to make a model. This kind of model deals with dynamic equations of the system which are usually coupled and nonlinear (Jelali & Kroll, 2004). To simplify these equations in order to make a satisfactory model, making some assumptions based on ideal conditions and using different methods for linearization of the system is unavoidable. There are different software such as SIMULINK/MATLAB and MATHEMATICA which are really helpful in this case.

3.4.2. Black-Box Models

A black-box model is used when no or little information is available about the physics of the system (Jelali & Kroll, 2004). In this case, the aim is to disclose the relations between variables of the system using the obtained operational input and output data from performance of the system. *Artificial Neural Network (ANN)* is one of the most significant methods in black-box modelling. ANN is a fast-growing method which has been used in different industries during recent years. The main idea for creating ANN which is a subset of artificial intelligence is to provide a simple model of human brain in order to solve complex scientific and industrial problems in a variety of areas.

3.4.3. Gray-Box Models

The phrase gray-box may be also used when an empirical model is improved by utilizing a certain available level of insight about the system (Norgaard et al., 2000). In this approach, experiments can be combined with mathematical model building to improve model accuracy (Jelali & Kroll, 2004).

3.5. Gas Turbine Control System and Configuration

One of the most important factors in modelling and control of gas turbines is the type and configuration of control system of GT. Control system is a vital part of any industrial equipment. Type and configuration of a control system is in a close relationship with the complexity of the system dynamics and the defined tasks during the whole performance period. Lacking a proper control system can lead to serious problems such as compressor surge: overheat, overspeed, etc. (Giampaolo, 2009). The final effect of these problems may be system shutdown and severe damages to the main components of GT.

There are three main functions for the control system of all gas turbine including "startup and shutdown sequencing control," "steady-state or operational control," and "protection control for protection from overheat, overspeed, overload, vibration, flameout and loss of lubrication." In a power network with several gas turbines, all individual control systems are closely connected with a central *Distributed Control System (DCS)* (Boyce, 2002). Control system of gas turbines may be open-loop or closed-loop. In an open-loop control system, the manipulated variable is positioned manually or by using a pre-determined program. However, to control a device in a closed-loop control system, one or more variables of measured data process parameters are used to move the manipulated variable. To keep the closed-loop control system effective and stable, the controller should be properly related to the process parameters (Boyce, 2002). Figure 4 and Figure 5 show open-loop and closed-loop block diagrams for a typical process respectively.

3.6. Gas Turbine Modelling Objectives

There are different goals for making a model of gas turbines such as condition monitoring, fault detection and diagnosis, sensor validation, system identification as well as design and optimization of control system. Thus, a clear statement of the modelling objectives is necessary to make a successful GT model.

Figure 4. Block diagram of an open-loop system (Daenotes, 2012)

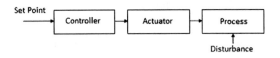

Figure 5. Block diagram of a closed-loop system (Daenotes, 2012)

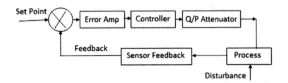

3.6.1. Condition Monitoring

One of the goals of making a GT model may be condition monitoring. Condition monitoring is considered as a major part of predictive maintenance. It assesses the operational health of GTs and indicates potential failure warning(s) in advance which help operators to take the proper action predicted in preventative maintenance schedule (Clifton, 2006). Condition monitoring is a very helpful tool in maintenance planning and can be used to avoid unexpected failures. Lost production, overtime, and expediting costs can be effectively prevented by predicting failures before any serious damage occurs in the system. To minimize the maintenance costs for very important and expensive machines such as gas turbines, it is necessary to monitor the operational conditions of vital and sensitive parts of the equipment and to obtain their related data continuously for further analysis. Good condition monitoring reduces the number of wrong decisions, minimizes the demand for spare parts and reduces maintenance costs. A good maintenance system should be capable of monitoring all vital parameters of a GT such as vibration, temperature, pressure, rotational speed, load, oil level and quality, etc. Besides, it should be able to predict the future state of the system and to prevent unwanted shutdowns as well as fatal breakdowns.

3.6.2. Fault Detection and Diagnosis

A GT model may be created in order to predict and detect faults in the system. Fault diagnosis acts as an important and effective tool when operators want to shift from preventive maintenance to predictive maintenance in order to reduce the maintenance cost (Lee, et al., 2010). It concerns with monitoring a system to identify when a fault has occurred as well as to determine the type and location of the fault.

3.6.3. Sensor Validation

GT models can be used for sensor validation purposes. Sensors are essential parts of any industrial equipment. Without reliable and accurate sensors, monitoring and control system of the equipment cannot work properly. If any of the sensors fails to send signal, a GT may not operate correctly and may even face shutdown. Sensor validation is about detection, isolation, and reconstruction of a faulty sensor. Some sensors may fail to report correct data due to different reasons or may become unavailable during maintenance operation. Sensor validation can improve reliability and availability of the system, and reduce maintenance costs. It enhances reliability for the equipment and safety for the personnel. Sensor validation is also an effective tool to prevent unwarranted maintenance or shutdown. It has a considerable effect in increasing equipment's lifetime and assuring reliable performance. It can strengthen automation of the system by providing valid data for diagnostic and monitoring systems.

3.6.4. System Identification

One the main objective of gas turbine modelling is system identification. System identification infers a mathematical description; a model of a dynamic system from a series of measurements of the system (Norgaard et al., 2000). Despite all significant research carried out in this field during the last decades, there is still a need for GT models with higher degree of accuracy and reliability for system identification purposes. This is because of the nonlinear and complex nature of GT dynamics.

3.6.5. Design and Optimization of Control System

Mathematical models may be created to design or optimize GT control system. It is obvious that any control system should be able to measure the output of the system using sensing devices, and to take required corrective action if the value of measured data deviates from its desired corresponding value (Burns, 2001). Control as a branch of engineering deals with the behavior of dynamical systems. The output performance of the equipment which is under control is measured by sensors. These measurements can be used to give feedback to the input actuators to make corrections toward desired performance. In spite of the significant research in this field, there are still increasing demands for accurate dynamic models and controllers, in order to investigate the system response to disturbances and to improve existing control systems.

4. WHITE-BOX MODELS OF GAS TURBINES

White-box models of gas turbines can be categorized into power plant gas turbine models and aero gas turbine models. In a power plant gas turbine, the mechanical power generated by gas turbine will be used by a generator to produce electrical power. However, in an aero gas turbine, the outgoing gaseous fluid can be utilized to generate thrust.

4.1. White-Box Models of Power Plant Gas Turbines

Rowen is a known name in the field of mathematical modelling and simulation of GTs. He presented a simplified mathematical model of a heavy-duty single-shaft power plant gas turbine (Rowen, 1983). He discussed different issues regarding modelling including parallel and isolated operation, gas and liquid fuel systems as well as isochronous and droop governors. The model could be very useful in studies related to power system dynamics. Rowen's model has been a base for many researchers to build up varieties of gas turbine models using different approaches. Rowen, in another effort, presented a simplified model of single-shaft heavy-duty gas turbines in mechanical drive service under different ambient conditions and *Inlet Guide Vanes (IGVs)* positions (Rowen, 1992). He considered the characteristics of control and fuel systems in the new model. Figure 6 shows the block diagram of Rowen's model including fuel and control systems (Rowen, 1983; Tavakoli et al., 2009).

Najjar investigated performance of GTs in single-shaft and twin-shaft operation modes using a model free power gas turbine driving an electric dynamometer (Najjar, 1994). GT operational data and their related curves for important parameters such as thermal efficiency, specific fuel consumption and net output power were considered in order to estimate GT performance. The results showed that in single-shaft mode of the free gas turbine engine, power was increased significantly at part load. However, in a twin-shaft mode, torque characteristics at part load were improved remarkably. A model for a twin-shaft combustion turbine was estimated by Hannett et al. (1995). They derived the required data for simulation through the tests for the assessment of governor response to disturbances.

A nonlinear state space model of a single-shaft gas turbine for loop-shaping control purposes was developed by Ailer et al. (2001). The main idea was to improve dynamic response of the engine by implementation of a developed nonlinear controller. The model was developed and simulated in SIMULINK/MATLAB software, based on engineering principles, GT dynamics, and constitutive algebraic equations. Model verification was performed by open-loop simulations against qualitative operation experience and engineering intuition. The researchers considered several assumptions during modelling process in order to simplify the complicated nonlinear model and to obtain a low-order dynamic model. Although the assumptions made the model appropriate for control purposes, some important aspects of the GT dynamics were neglected during simplification process. Figure 7 shows dynamic step response of the simulated model for number of revolutions (Ailer et al., 2001). As it can be seen, the gas turbine is stable at this operational point.

A dynamic model for a twin-shaft gas turbine was developed by Ricketts based on a generic methodology (Ricketts, 1997). The model could also be used for designing an appropriate adaptive controller. An investigation for using of exhaust gases of an open-cycle twin-shaft gas turbine was performed by Mostafavi et al. (1998). They carried out a thermodynamic analysis and concluded that at low temperature ratios, pre-cooling could in-

Figure 6. Rowen's model for a heavy-duty gas turbines (Rowen, 1983; Tavakoli, et al., 2009)

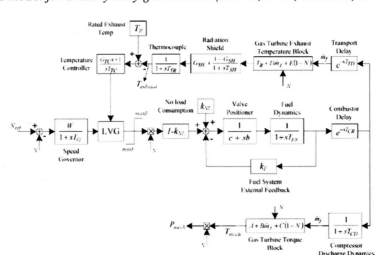

Figure 7. Dynamic step response of the system (1/sec.) (Ailer, et al., 2001)

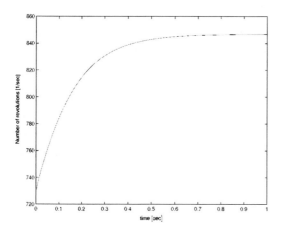

crease the efficiency and specific network of the cycle. Besides, the pre-cooled cycle could operate at a higher compressor pressure ratio and temperature ratio without increasing the maximum cycle temperature. Nagpal et al. presented their field experiences in testing and model validation of turbine dynamic models and their associated governors for Industrial power plant gas turbines (Nagpal et al., 2000). Based on the field measurements, they showed that GAST model which is a widely used model to represent the dynamics of GT governor systems, has two main deficiencies.

Firstly, the model could not predict GT operation accurately at high levels of loads. Secondly, the accurate adjustment of the model parameters, according to the oscillations around the final setting frequency, may not be attained.

Al-Hamdan and Ebaid discussed modelling and simulation of a single-shaft gas turbine engine for power generation based on the dynamic structure and performance of its individual components (Al-Hamdan & Ebadi, 2006). They used basic thermodynamic equations of a single-shaft gas turbine to model the system. They developed a computer program for the engine simulation which could be used as a useful tool to investigate GT performance at off-design conditions and to design an appropriate efficient control system for specific applications. Figure 8 shows variations of temperatures in different sections of the modelled GT versus net power output. T_{02}, T_{03} and T_{04} are output temperatures of compressor, combustor and turbine respectively.

Kaikko et al. presented a steady-state nonlinear model of a twin-shaft industrial gas turbine and its application to online condition monitoring (Kaikko et al., 2002). They evaluated the GT performance parameters in references, actual, expected and corrected states. They concluded that the applied computational method in their

Figure 8. Variations of various temperatures (K) versus net power output (KW) (Al-Hamdan & Ebadi, 2006)

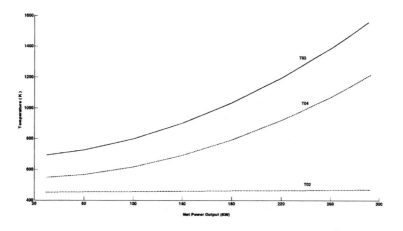

study could be adapted to other modelling, condition monitoring and diagnosis of gas turbines. Abdollahi and Vahedi developed a dynamic model of single-shaft micro turbine generation systems (Abdollahi & Vahedi, 2004). They provided a dynamic model for each component of the micro turbine including gas turbine, DC bridge rectifier, permanent magnet generator as well as power inverter. The models were implemented in SIMULINK/MATLAB. They showed that the models were suitable for dynamic analysis of micro turbines under different conditions, and recommended that the model could also be useful to study the effect of micro turbines on load sharing in power distribution network. A gas turbine fully-featured simulator was developed and implemented by Klang and Lindholm (Klang & Lindholm, 2005). They discussed simulator setup both technically and economically as well as chose a robust hardware solution based on the basic requirements. The simulator could be useful for testing the GT control system, trying out new concepts and training operators.

An aero-thermal model of gas turbines was presented by Camporeale et al. (2006). They modelled and simulated two different power plant gas turbines in SIMULINK environment of MATLAB. They also applied an object-oriented approach to develop a modular real time simulation code for gas turbines. The flexibility of the code allowed it to be adapted to any configuration of power plants. Figure 9 shows the diagram for the real-time simulation software interacted to hardware control devices (Camporeale et al., 2006).

Aguiar et al. investigated modelling and simulation of a natural gas based micro turbine using MATLAB (Aguiar et al., 2007). The main objective of the research was to present a technical and economical analysis of using *Micro Gas Turbines (MGTs)* for residential complex based on a daily simulation model. The results of the analysis could also be useful for the investors who are interested to predict the cost of investment, operation and maintenance of these turbines for power generation. A simplified desktop perfor-

Figure 9. The diagram for real-time simulation software interacted to hardware control devices (Camporeale, et al., 2006)

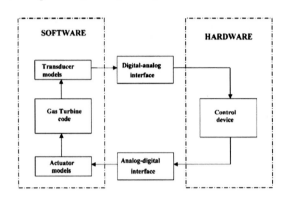

mance model of a typical single-shaft heavy-duty gas turbine in power generation systems was developed by Zhu et al. (2007). They built a model which could be accurate and robust to variations under different operational conditions. The researchers investigated a methodology for assessment and rapid analysis of the system alternations. The methodology could be implemented in a desktop computing environment. They applied sensitivity analysis to assess the model for a variety of fuels in terms of composition, moisture and carbon contents. The model could also be used to evaluate CO_2 emissions.

Development of single-shaft dynamic model for a combined-cycle plant was explored by Mantzaris and Vournas using SIMULINK/MATLAB (Mantzaris & Vournas, n.d.). They investigated stability of the turbine and its control system against overheat as well as changes in frequency and load. Yee et al. carried out a comparative analysis and overview of different existing models of power plant gas turbines (Yee & Milanovic, 2008). They identified, presented and discussed various kinds of GT models in terms of their application, accuracy and complexity. However, the study did not cover the black-box models of gas turbines.

A methodology was applied by Khosravy-el-Hossani et al. to determine the exhaust energy in the new edition of ASME PTC 22 which is about flow rate of flue gas (Khosravy-El-Hossani & Dorosti, 2009). They showed that this method could enhance the precision of ASME PTC 22 by more than one percent. The gas turbine performance test was improved based on the obtained operational data. A methodology was also suggested for calculation of GT performance estimation without measurement of input fuel components. The parameters of a single-shaft heavy-duty gas turbine were estimated using its operational data based on Rowen's model by Tavakoli et al. (2009). They applied simple physical laws and thermodynamic assumptions in order to derive the GT parameters. The study can be useful for educational purposes especially for the students and trainers who are interested in gas turbine dynamic studies.

Simple models of the systems for a power plant simulator were developed by Roldan-Villasana et al. based on the mass, momentum and energy principles (Roland-Villasana et al., 2010). The modelled systems were classified into seven main groups including water, steam, turbine, electric generator, auxiliaries, gas turbine and minimized auxiliaries. They concluded that the simulator could be very useful for training of operators. Yadav et al. applied graph networks approach to analyze and model a single-shaft open-cycle gas turbine (Yadav et al., 2010). They used graph theory and algorithms to identify pressure and temperature drops, work transfer rates, rate of heat and other system properties. Because of the similarities in the results from this approach with the results from conventional methods, it was suggested that the new technique could be used for optimization of GT process parameters. Shalan et al. employed a simple methodology to estimate parameters of a Rowen's model for heavy-duty single-shaft gas turbines (Shalan, 2011). Variety of simulated tests was performed using SIMU-LINK/MATLAB and the results were compared with and verified against the involved scientific articles in the literature.

4.2. White-Box Models of Aero Gas Turbines

Evans, Rees, and Hill examined a linear identification of fuel flow rate to shaft speed dynamics of a twin-shaft gas turbine which was a typical military Rolls Royce Spey engine (Evans et al., 1998). They studied direct estimation of s-domain models in frequency domain and showed that high-quality models of gas turbines could be achieved using frequency-domain techniques. They discussed that the technique might be used to model industrial systems, wherever a physical interpretation of the model is needed.

Arkov et al. employed four different system identification approaches to model a typical aircraft gas turbine using the obtained data from a twin-shaft Rolls Royce Spey engine (Arkov et al., 2000). The motivation behind their research was to minimize the cost and to improve the efficiency of gas turbine dynamical testing techniques. The four employed techniques by the researchers included "multi-sine and frequency-domain techniques for both linear and nonlinear models", "ambient noise excitation", "extended least-squares algorithms for finding time-varying linear models" and "multi-objective genetic programming for the selection of nonlinear model structures". A description of each technique and the relative merits of the approaches were also discussed in the study. Kim et al. developed a model for a single-spool turbojet engine using SIMULINK/MATLAB (Kim et al., 2000). The transient behavior and changes of different engine parameters could be predicted by the model based on variations of the fuel flow rate. The researchers considered different flight conditions in their simulation such as fuel cutoff. The simulation output was compared with another dynamic code for gas turbines and showed satisfactory results.

Evans et al. presented the linear multivariable model of a twin-shaft aero gas turbine (a typical Rolls Royce Spey military turbofan) using a frequency-domain identification technique (Evans et al., 2001). The technique was employed to estimate s-domain multivariable models directly from test data. The researchers examined the dynamic relationship between fuel flow rate and rotational speeds in the form of *Single-Input, Multi-Output (SIMO)*. The main advantage of the model was its capability to be directly compared with the linearized thermodynamic models of the GT. The output of the research showed that a second-order model could present the most suitable model and the best estimation of the engine. The techniques investigated in the research can be used to verify the linearized thermodynamic models of gas turbines. Figure 10 shows the Rolls Royce Spey engine modelled by Evans et al. (2001).

Arkov et al. discussed a life cycle support for dynamic modelling of aero engine gas turbines (Arkov et al., 2002). They investigated different mathematical models and their applications at life cycle stages of engine controllers and developed a unified information technology and a unified information space for creating and using GT mathematical models at the life cycle stages. Standard methodologies for system modelling

and appropriate software were employed for implementation of this new concept, and consequently performance enhancement of the control system.

5. BLACK-BOX (ANN-BASED) MODELS OF GAS TURBINES

One of the novel approaches for optimization of gas turbines is employing ANN-based identification and modelling technique. A neural network model is a group of interconnected artificial units (neurons) with linear or nonlinear transfer functions. Neurons are arranged in different layers including input layer, hidden layer(s) and output layer. The number of neurons and layers in an ANN model depends on the degree of complexity of the system dynamics. ANNs learn the relation between inputs and outputs of the system through an iterative process called training. Each input into the neuron has its own associated weight. Weights are adjustable numbers which are determined during training the network. Figure 11 shows a

Figure 10. A typical Rolls Royce Spey engine (Evans, et al., 2001)

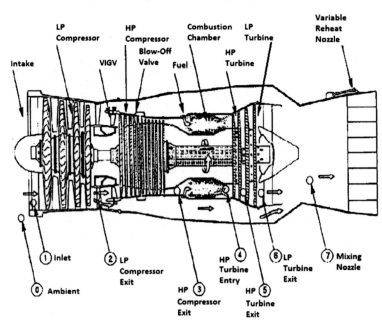

Figure 11. A simple structure of a typical artificial neural network (ANN) with input, hidden, and output layers (Wikimedia Commons, 2012)

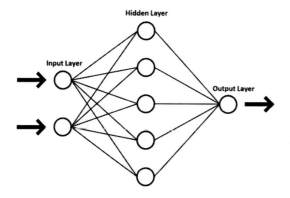

simple structure of a typical ANN with two inputs, one output and five neurons in the hidden layer (Wikimedia Commons, 2012).

ANN, as a data-driven model, has been considered as a suitable alternative to white-box models during the last few decades. ANN-based models can be created directly using the operational data from an actual GT or simulated data from *Original Equipment Manufacturers (OEMs)* performance. Simulated data may be used when operational data are not available. The obtained data should cover the whole operational range of the system. All transient data during start or stop process should be removed from the collected data before the modelling process.

ANN models for gas turbines can be created using different approaches based on the flexibility that ANNs provide. This flexibility is based on the number of neurons, number of hidden layers, values of the weights and biases, type of the activation function, structure of the network, training styles and algorithms as well as data structure. However, the best structure is the one which can predict dynamic behavior of the system as accurately as possible. Selecting the right parameters of GTs as inputs and outputs of a neural network is very important for making an accurate and reliable model. The availability of data for the selected parameters, system knowledge for identification of interconnections between different parameters,

and the objectives for making a model are basic factors in choosing appropriate inputs and outputs. Accuracy of the selected output parameters can be examined by sensitivity analysis. There are a considerable number of research sources regarding black-box system modelling and simulation of gas turbines in the literature. The following summarizes the most important studies which have been carried out so far in this area. As in white-box models, black-box (ANN-based) models can be categorized into power plant and aero gas turbine models.

5.1. Black-Box (ANN-Based) Models of Power Plant Gas Turbines

A single-shaft gas turbine design and off-design model was presented by Lazzaretto and Toffolo (2001). They used analytical method and feedforward neural network as two different approaches to model the system. Appropriate scaling techniques were employed to construct new maps for the gas turbine using the available generalized GT maps. The new maps were validated using the obtained experimental data. Ogaji et al. applied three different architectures of ANN for multi-sensor fault diagnosis of a stationary twin-shaft gas turbine using neural network toolbox in MATLAB (Ogaji et al., 2002). The results indicated that ANN could be used as a high-speed powerful tool for real-time control problems. Arriagata et al. applied ANN for fault diagnosis of a single-shaft industrial gas turbine (Arrigada et al., 2003). They obtained a comprehensive data set from ten faulty and one healthy engine conditions. The data were trained using feedforward *Multi-Layer Perceptron (MLP)* structure. The trained network was able to make a diagnosis about the gas turbine's condition when a new data set was presented to it. The results proved that ANN could identify the faults and generate warnings at early stages with high reliability. Figure 12 shows a schematic drawing of the ANN and the interpretation of the outputs in a graphical display (Arrigada et al., 2003).

Figure 12. A schematic drawing of the ANN model and the interpretation of the outputs in a graphical display (Arrigada, et al., 2003)

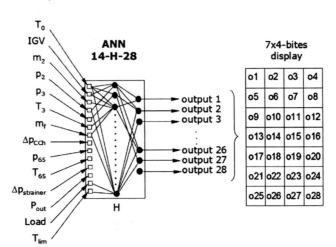

As it can be seen from the ANN architecture, the inputs correspond to the 14 measured parameters in the real engines, as well as the ones controlled by the operators and the control system. The parameters include ambient temperature, inlet guide vanes angle, mass flow rate, fuel flow rate, load, pressure, temperature, etc. As as it is shown in Figure 7, the desired outputs from the ANN are unique combinations of 28 binary numbers arranged in a graphical display. The training process of the ANN stopped when it showed the best performance based on a selected number of hidden neurons and weights for the network. The ANN can be named 14-H-28 according to its structure (Arrigada et al., 2003).

Basso et al. applied a *Nonlinear Autoregressive with Exogenous Inputs (NARX)* model to identify dynamics of a small heavy-duty power plant gas turbine (Basso et al., 2004). The objective was to make an accurate reduced-order nonlinear model using black-box identification techniques. They considered two operational modes for the gas turbine; when it was isolated from power network as a stand-alone unit and when it was connected to the power grid. They showed that in order to reduce the complexity and improve the simulating capability of the model, the ingredients should be chosen carefully.

Bettocchi et al. investigated artificial neural network model of a single-shaft gas turbine as an alternative to physical models (Bettocchi et al., 2004). They observed that ANN can be very useful for the real time simulation of GTs especially when there were not enough information about the system dynamics. In another effort, Bettocchi et al. developed a *Multiple-Input, Multiple Outputs (MIMO)* neural network approach for diagnosis of single-shaft gas turbine engines (Bettocchi et al., 2005). A NARX model was applied to model a power plant micro gas turbine and the related distribution system dynamics (Jurado, 2005). However, the nonlinear terms in the model were restricted to the second order. The resulting model was capable of modelling both low and high amplitude dynamics of MGTs. The quality of the model was examined by cross validation technique. The model was tested under different operational conditions and electrical disturbances. Simani et al. carried out a study to detect and isolate faults on a single-shaft industrial gas turbine prototype (Simani & Patton, 2008). They suggested exploiting an identified linear model in order to avoid nonlinear complexity of the system. For this purpose, black-box modelling and output estimation approaches were applied.

Magnus Fast et al. applied simulation data and ANN technique to examine condition-based maintenance of gas turbines (Fast et al., 2008). In another effort, Fast et al. used real data obtained from an industrial single-shaft gas turbine working under full load to develop a simple ANN model of the system with very high prediction accuracy (Fast et al., 2009a). A combination of ANN method and *Cumulative Sum (CUSMUS)* technique was utilized by Fast et al. for condition monitoring and detection of anomalies in GT performance (Fast et al., 2009b). To minimize the need for calibration of sensors and to decrease the percentage of shutdowns due to sensor failure, an ANN-based methodology was also developed for sensor validation in gas turbines by Fast et al. (Fast et al., 2009c). Application of ANN to diagnosis and condition monitoring of a combined heat and power plant was also discussed by Fast et al. (Fast & Palme, 2010). Fast applied different ANN approaches for gas turbine condition monitoring, sensor validation and diagnosis (Fast, 2010).

Yoru et al. examined application of ANN method to exergetic analysis of gas turbines which supplied both heat and power in a cogeneration system of a factory (Yoru et al., 2009). They compared the results of the ANN method with the exergy values from exergy analysis and showed that much closer exergetic results could be attained by using ANN method. Application of ANN and *Adaptive Network-Based Fuzzy Inference System (ANFIS)* to MGTs was investigated by Bartolini et al. (2011). They investigated the effects of changes of ambient conditions (temperature, pressure, humidity) and load on MGT's output power. The results indicated that ambient temperature variations had more effect on the output power than humidity and pressure. Besides, MGTs were less influenced by ambient conditions than load.

5.2. Black-Box (ANN-Based) Models of Aero Gas Turbines

A *Nonlinear Auto-Regressive Moving Average with Exogeneous Inputs (NARMAX)* model of an aircraft gas turbine was estimated by Chiras, Evans and Reesa (Chiras et al., 2001a). They employed nonparametric analysis in time and frequency domains to determine the order and nature of nonlinearity of the system. The researchers combined time-domain NARMAX modelling, time and frequency domain analysis, identification techniques and periodic test signals to improve GT nonlinear modelling. In another investigation, they applied a forward-regression orthogonal estimation algorithm to make a NARMAX model for a twin-shaft Rolls Royce Spey aircraft gas turbine (Chiras et al., 2001b). A nonlinear relationship between dynamics of shaft rotational speed and the fuel flow rate was also explored and discussed. To validate the model performance, the researchers examined static and dynamic behavior of the engine for small and large signal test. The results were satisfactory and could be matched with the results from another previously estimated model. In another effort, they used feedforward neural network to model the relationship between the fuel flow and shaft rotational speed dynamics for a Spey gas turbine engine (Chiras et al., 2002a). They validated performance of the nonlinear model against small and large signal engine tests. They also showed the necessity of using a nonlinear model for modelling high-amplitude dynamics of gas turbine engines. They also recommended a global nonlinear model of gas turbine dynamics using NARMAX Structures (Chiras et al., 2002b). They investigated both linear and nonlinear models of a twin-shaft Rolls Royce Spey gas turbine. Their suggestion for a global nonlinear model was based on the fact that linear models vary with operational points. They discussed a simple method for identification of a NARX model. The performance of this model was satisfactory for both small and high amplitude tests. However, due to inherent problems with discrete-time estimation and great variability of the model parameters, the physical interpretability of the model was lost.

Ruano et al. carried out a research regarding nonlinear identification of shaft-speed dynamics for a Rolls Royce Spey aircraft gas turbine (Ruanoa et al., 2003). They used two different approaches including NARX models and Neural Network (NN) models. The researchers realized that among the three different structures of NN including *Radial Basis Function (RBS)*, MLP and B-spline, the latter delivered the best results. They employed genetic programming tool for NARMAX and B-spline models to determine the model structure. Different configurations of *Back Propagation Neural Networks (BPNN)* were used by Torella, Gamma and Palmesano to study and simulate the effects of gas turbine air system on engine performance (Torella et al., 2003).

6. APPLICATIONS OF GT MODELS IN CONTROL SYSTEM DESIGN

Modelling and simulation of gas turbines can play a significant role in control areas. Research activities regarding applications of modelling in control systems can also be categorized into white-box and black-box (ANN-based) approaches.

6.1. White-Box Approach

A dynamic model for a twin-shaft gas turbine which was developed by Ricketts based on a generic methodology, were used for designing an appropriate adaptive controller (Ricketts, 1997). Pongrácz et al. used an input-output linearization method to design an adaptive reference tracking controller for a low-power gas turbine model (Pongracz et al., 2000). They discussed a third-order nonlinear state space model for a real low-power single-shaft gas turbine based on dynamic equations of the system. In their model, fuel mass flow rate and rotational speed were considered as input and output respectively. A linear adaptive controller with load torque estimation was also designed for the linearized model. According to the results of simulation, the required performance criteria were fulfilled by the controlled plant. The sufficient robustness of the system against the model parameter uncertainties and environmental disturbances were also investigated and approved.

Ashikaga et al. carried out a study to apply nonlinear control to gas turbines (Ashigaga et al., 2003). They reported two applications of nonlinear control. The first one was the GT starting control using the fuzzy control, and the other was the application of the optimizing method to *Variable Stator Vane (VSV)* control. The aim was to increase thermal efficiency and to decrease NO_x emission. However, the algorithms for solving optimization problems were complicated, time-consuming and too large to be installed easily in computers. Agüero et al. applied modifications in a heavy-duty power plant gas turbine control system (Aguero et al., 2002). One of the modifications limited speed deviations to the governor, which in its turn, limited power deviation over dispatch set point. Another modification could prevent non-desired unloading of turbine. According to the last modification, operators were allowed to adjust set points of dispatched power, grid frequency and required spinning reserve regulated by the dispatch center. The researchers investigated the turbine dynamic behavior before and after the modifications were made. Centeno et al. reviewed typical gas turbine dynamic models for power system stability studies (Centeno et al., 2002). They discussed main control loops including temperature and acceleration control loops, their applications and implementations. They also explained different issues which should be considered for modelling of temperature and acceleration control loops. The performance of the control loops were simulated against changes in gas turbine load. Figure 13 shows the block diagram of the basic temperature control loop for the GT model (Centeno et al., 2002).

6.2. Black-Box (ANN-Based) Approach

Investigation for the practical use of artificial neural networks to control complex and nonlinear systems was carried out by Nabney and Cressy (1996). They utilized multiple ANN controllers to maintain the level of thrust for aero gas turbines and to control system variables for a twin-shaft aircraft gas turbine engine in desirable and safe operational regions. The main idea behind the research was to minimize fuel consumption and to increase the engine life. They aimed to improve the performance of control system by using the capability of ANN in nonlinear mapping instead of using varieties of linear controllers. They used MLP architecture with a single hidden layer to train the networks. The researchers applied a reference model as an input to the ANN controller. The results showed that performance of the applied ANN controllers was better than conventional ones; but they could not track the reference models as closely as they expected.

Another effort was carried out by Dodd and Martin, more or less with the same goals (Dodd & Martin, 1997). They proposed an ANN-based adaptive technique to model and control an aero gas turbine engine and to maintain thrust at desired level while minimizing fuel consumption in the engine. They suggested a technique, which consequently could lead to maximizing thrust for a specified fuel, lowering the critical temperature of the turbine blades and increasing the engine life. In their research, a feedforward neural network with sigmoidal activation function was utilized to model the system. The simplicity and differentiability of the neural network helped the researchers to calculate necessary changes to controllable parameters of the engine and consequently to maintain the level of the thrust in a targeted point. Figure 14 shows the block diagram of the ANN model. The inputs correspond to fuel rate, final nozzle area and inlet guide vane angle. The only output is thrust (Dodd & Martin, 1997).

Figure 14. Block diagram of an ANN-based aero gas turbine model for system optimization consists of minimizing fuel while maintaining thrust (Dodd & Martin, 1997)

Figure 13. Block diagram of the basic temperature control loop for a gas turbine model (Centeno, et al., 2002)

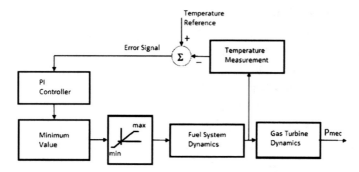

Mu and Rees investigated nonlinear modelling and control of a Rolls Royce Spey aircraft gas turbine (Mu & Rees, 2004). They used NARMAX and neural networks to indentify the engine dynamics under different operational conditions. An *Approximate Model Predictive Control (AMPC)* was applied in order to control shaft rotational speed. The results proved that the performance of AMPC as a global nonlinear controller was much better than gain-scheduling PID ones. AMPC showed optimal performance for both small and large random step changes as well as against disturbances and model mismatch. Mu, Rees and Liu, in another effort, examined two different approaches to design a global nonlinear controller for an aircraft gas turbine (Mu et al., 2004). They compared and discussed the properties of AMPC and *Nonlinear Model Predictive Control (NMPC)*. The results showed that the both controllers provided good performance for the whole operational range. However, AMPC showed better performance against disturbances and uncertainties. Besides, AMPC could be gained analytically, required less computational time and avoided the local minimum.

Spina and Venturini applied ANN to train operational data through different patterns in order to model and simulate a single-shaft gas turbine and its diagnostic system with a low computational and time effort (Spina & Venturini, 2007). Implementation of a *Model Predictive Control (MPC)* to a heavy-duty power plant gas turbine was investigated by Ghorbani et al. (Ghorbani et al., 2008). They modeled the system based on a mathematical procedure and *Autoregressive with Exogenous Input (ARX)* identification method. The goal was to design a controller which could adjust rotational speed of the shaft and exhaust gas temperature by the fuel flow rate and the position of IGV. The MPC controller showed superior performance to both PID controller and SpeedTronic control system. Using PID and ANN controllers for a heavy-duty gas turbine plant was investigated by Balamurugan et al. (2008). Their

work was based on the GT mathematical model already developed by Rowen. They applied *Ziegler Nichols's (ZN)* method to tune PID controller parameters. Besides, they trained an ANN controller using backpropagation method to control the speed of the GT. The simulation results showed that ANN controller performed better than the PID controller. Figure 15 shows a comparison of gas turbine plant response with PID and ANN controllers (Balamurugan et al., 2008).

7. CONCLUSION

Modelling and simulation of gas turbines plays a key role in manufacturing the most efficient, reliable and durable gas turbines. This fact has been a strong potential and motivation for scientists to keep carrying out significant research in this field. The outcome of the research, so far, has been very effective to evaluate and optimize performance of gas turbines before final design and manufacturing processes. Besides, GT models can also be used on industrial sites for optimization, condition monitoring, sensor validation, fault detection, trouble shooting, etc. There are different approaches and methodologies in modelling and control of gas turbines. Choosing the right method

Figure 15. A comparison of gas turbine plant response with PID and ANN controllers (Balamurugan, et al., 2008)

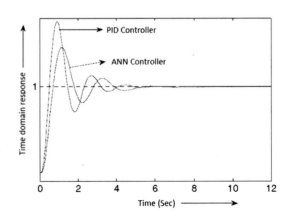

and creating the right model based on the required application depends on different factors such as GT type, GT configuration, modelling methods, model construction approaches and modelling objectives. By highlighting these factors, remarkable enhancements can be achieved in the process of modelling and control of gas turbines.

In this chapter, a brief overview of significant research activities in the area of modelling and simulation of gas turbines were briefly reviewed and discussed. Main white-box and black-box models and their applications in control systems were investigated for both aero and power plant gas turbines. Despite all the efforts carried out so far in the field of modelling, simulation and control of gas turbines using mathematical and experimental methods, there is still a great need for further system optimization. To approach an optimized model as closely as possible, researchers need to unfold the unknowns of complicated nonlinear dynamic behavior of these systems in order to minimize undesirable events such as unpredictable shutdowns, overheating and overspeed during operation. Further research and development activities can be carried out in the following areas:

1. Since it is desirable to design gas turbines with high performance, high reliability and cost effectiveness, an extensive effort still needs to be devoted towards understanding their complex natural dynamics and coupled parameters.

2. System disturbances arising from faults or from load fluctuations in power network of power plant gas turbines could drive GTs to instability. Exploring reaction of gas turbines to the system disturbances and environmental condition changes is still a challenging issue. Therefore, there is an increasing demand for accurate dynamic models, to investigate the system response to disturbances and to improve existing control system. Application of ANN as a fast and reliable method to stabilize the system against disturbances can be investigated further. In this case, dynamic behavior of the system can be predicted and controlled in the presence of a number of uncertainties, such as environmental conditions and load changes.

3. Approximating an ANN model with high generalization capabilities and robustness for industrial power plant gas turbines can be extensively investigated using operational data of real GTs and based on the flexibility that ANN provides for modelling of different types of systems. This flexibility is based on the varieties of structures of the network, training styles and algorithms, types of the activation function, number of neurons, number of hidden layers, values of the weights and biases as well as data structures. For this purpose, different ANN architectures can be explored for gas turbines using operational data in order to attain or customize the optimal model. The obtained optimal model should predict dynamic behavior of the system as accurately as possible. It can be used as a powerful tool in condition monitoring, trouble shooting and even maintenance of gas turbines.

4. A neural adaptive controller with superior control behavior and high adaptability needs to be designed for GT models. The controller should contribute towards high-performance, cost-effectiveness and high-reliability. The GT model can be used to predict the effect of controller changes on plant output, which consequently allows the updating of controller parameters. The goal is to maximize system robustness, output power and efficiency.

The upcoming efforts can lead to optimal models and control algorithms with minimal supervision and energy consumption. A methodology can be developed to identify system parameters and to predict dynamic behavior of the system as accurately as possible. This methodology can be applicable to a wide range of operational conditions. The future will bring advancements in technology that enables the development of optimized reliable gas turbines.

REFERENCES

Abdollahi, S. E., & Vahedi, A. (2004). *Dynamic modelling of micro-turbine generation systems using MATLAB/SIMULIN*. Tehran, Iran: Department of Electrical Engineering, Iran University of Science and Technology.

Aguero, J. L., Beroqui, M. C., & Di Pasquo, H. (2002). *Gas turbine control modifications for: Availability and limitation of spinning reserve and limitation of non-desired unloading*. Pluspetrol Energy, SA: Facultad de Ingeniería Universidad Nacional de La Plata, and Central Térmica Tucumán.

Aguiar, A. B. M., Pinto, J. O. P., & Nogueira, L. A. H. (2007). Modelling and simulation of natural gas micro-turbine for residential complexes. In *Proceedings of the World Congress on Engineering and Computer Science*. San Francisco, CA: WCECS.

Ailer, P., Santa, I., Szederkenyi, G., & Hangos, K. M. (2001). Nonlinear model-building of a low-power gas turbine. *Periodica Polytechnica Ser. Transp.*, *29*(1-2), 117–135.

Al-Hamdan, Q. Z., & Ebaid, M. S. Y. (2006). Modelling and simulation of a gas turbine engine for power generation. *Journal of Engineering for Gas Turbines and Power*, *128*, 302–311. doi:10.1115/1.2061287.

Arkov, V., Evans, C., Fleming, P. J., Hill, D. C., Norton, J. P., & Pratt, I. et al. (2000). System identification strategies applied to aircraft gas turbine engine. *Annual Reviews in Control*, *24*, 67–81.

Arkov, V., Kulikov, G., & Breikin, T. (2002). Life cycle support for dynamic modelling of gas turbines. In *Proceedings of the 15th Triennial World Congress*. Barcelona, Spain: World Congress.

Arrigada, J., Genrup, M., Loberg, A., & Assadi, M. (2003). Fault diagnosis system for an industrial gas turbine by means of neural networks. In *Proceedings of the International Gas Turbine Congress 2003*. Tokyo, Japan: IGTC.

Ashikaga, M., Kohno, Y., Higashi, M., Nagai, K., & Ryu, M. (2003). A study on applying nonlinear control to gas turbine systems. In *Proceedings of the International Gas Turbine Congress*. Tokyo, Japan: IGTC.

Balamurugan, S., Xavier, R. J., & Jeyakumar, A. E. (2008). ANN controller for heavy-duty gas turbine plant. *International Journal of Applied Engineering Research*, *3*(12), 1765–1771.

Bartolini, C. M., Caresana, F., Comodi, G., Pelagalli, L., Renzi, M., & Vagni, S. (2011). Application of artificial neural networks to micro gas turbines. *Energy Conversion and Management*, *52*, 781–788. doi:10.1016/j.enconman.2010.08.003.

Basso, M., Giarre, L., Groppi, S., & Zappa, G. (2004). NARX models of an industrial power plant gas turbine. *IEEE Transactions on Control Systems Technology*.

Bettocchi, R., Pinelli, M., Spina, P. R., & Venturini, M. (2005). Artificial intelligent for the diagnostics of gas turbine, part 1: Neural network approach. In *Proceedings of the ASME Turbo Expo 2005*. ASME.

Bettocchi, R., Pinelli, M., Spina, P. R., Venturini, M., & Burgio, M. (2004). Set up of a robust neural network for gas turbine simulation. In *Proceedings of the ASME Turbo Expo 2004*. Vienna, Austria: ASME.

Boyce, M. P. (2002). *Gas turbine engineering handbook* (2nd ed.). New York: Butterworth-Heinemann.

Burns, R. S. (2001). *Advanced control engineering*. New York: Butterworth-Heinemann Publications.

Camporeale, S. M., Fortunato, B., & Mastrovito, M. (2006). A modular code for real time dynamic simulation of gas turbines in SIMULINK. In *Proceedings of ASME,* (Vol. 128). ASME.

Centeno, P., Egido, I., Domingo, C., Fernandez, F., Rouco, L., & Gonzalez, M. (2002). *Review of gas turbine models for power system stability studies*. Madrid, Spain: Universidad Pontificia Comillas and Endesa Generación.

Chiras, N., Evans, C., & Rees, D. (2001a). Nonlinear gas turbine modelling using NARMAX structures. *IEEE Transactions on Instrumentation and Measurement, 50*(4), 893–898. doi:10.1109/19.948295.

Chiras, N., Evans, C., & Rees, D. (2001b). *Nonlinear modelling and validation of an aircraft gas turbine engine*. Wales, UK: School of Electronics, University of Glamorgan.

Chiras, N., Evans, C., & Rees, D. (2002a). *Nonlinear gas turbine modelling using feedforward neural networks*. Wales, UK: School of Electronics, University of Glamorgan. doi:10.1115/GT2002-30035.

Chiras, N., Evans, C., & Rees, D. (2002b). Global nonlinear modelling of gas turbine dynamics using NARMAX structures. *Journal of Engineering for Gas Turbines and Power, 124,* 817. doi:10.1115/1.1470483.

Clifton, D. (2006). *Condition monitoring of gas turbine engines*. St. Cross College.

Daenotes-Electronics Notes, Lectures, Theory and Projects for Engineering Students. (2012). Retrieved from http://www.daenotes.com

Dodd, N., & Martin, J. (1997, June). Using neural networks to optimize gas turbine aero engines. *Computing & Control Engineering Journal,* 129-135.

Evans, C., Chiras, N., Guillaume, P., & Rees, D. (2001). *Multivariable modelling of gas turbine dynamics*. Wales, UK: School of Electronics, University of Glamorgan.

Evans, C., Rees, D., & Hill, D. (1998). Frequency domain identification of gas turbine dynamics. *IEEE Transactions on Control Systems Technology, 6*(5), 651–662. doi:10.1109/87.709500.

Fast, M. (2010). *Artificial neural networks for gas turbine monitoring*. (Doctoral Thesis). Lund University, Lund, Sweden.

Fast, M., Assadi, M., & De, S. (2008). Condition based maintenance of gas turbines using simulation data and artificial neural network: A demonstration of feasibility. In *Proceedings of the ASME Turbo Expo 2008*. Berlin, Germany: ASME.

Fast, M., Assadi, M., & De, S. (2009a). Development and multi-utility of an ANN model for an industrial gas turbine. *Journal of Applied Energy, 86*(1), 9–17. doi:10.1016/j.apenergy.2008.03.018.

Fast, M., & Palme, T. (2010). Application of artificial neural network to the condition monitoring and diagnosis of a combined heat and power plant. *Journal of Energy, 35*(2), 114–1120.

Fast, M., Palme, T., & Genrup, M. (2009b). A novel approach for gas turbines monitoring combining CUSUM technique and artificial neural network. In *Proceedings of ASME Turbo Expo 2009*. Orlando, FL: ASME.

Fast, M., Palme, T., & Karlsson, A. (2009c). Gas turbines sensor validation through classification with artificial neural networks. In *Proceedings of ECOS 2009*. ECOS.

Ghorbani, H., Ghaffari, A., & Rahnama, M. (2008). Constrained model predictive control implementation for a heavy-duty gas turbine power plant. *WSEAS Transactions on Systems and Control, 6*(3), 507–516.

Giampaolo, T. (2009). *Gas turbine handbook – Principles and practice* (4th ed.). The Fairmont Press, Inc..

Habib, M. K. (2006). Mechatronics engineering the evolution, the needs and the challenges. In *Proceedings of the 32nd Annual Conference of IEEE Industrial Electronics Society (IECON 2006)*, (pp. 4510-4515). IEEE.

Habib, M. K. (2008). Interdisciplinary mechatronics: Problem solving, creative thinking and concurrent design synergy. *International Journal of Mechatronics and Manufacturing Systems, 1*(1), 264–269. doi:10.1504/IJMMS.2008.018272.

Hannett, L. N., Jee, G., & Fardanesh, B. (1995). A governer/turbine model for a twin-shaft combustion turbine. *IEEE Transactions on Power Systems, 10*(1). doi:10.1109/59.373935.

Jelali, M., & Kroll, A. (2004). *Hydraulic servosystems: Modelling, identification, and control.* Berlin: Springer Publications.

Jurado, F. (2005). Nonlinear modelling of microturbines using NARX structures on the distribution feeder. *Energy Conversion and Management, 46*, 385–401. doi:10.1016/j.enconman.2004.03.012.

Kaikko, J., Talonpoika, T., & Sarkomma, P. (2002). Gas turbine model for an on-line condition monitoring and diagnostic system. In *Proceeding of the Australasian Universities Power Engineering Conference (AUPEC2002)*. Melbourne, Australia: AUPEC.

Khosravi-el-Hossani, M., & Dorosti, Q. (2009). Improvement of gas turbine performance test in combine-cycle. World Academy of Science, Engineering and Technology, 58.

Kim, S. K., Pilidis, P., & Yin, J. (2000). *Gas turbine dynamic simulation using SIMULINK*. Society of Automation Engineers, Inc. doi:10.4271/2000-01-3647.

Klang, H., & Lindholm, A. (2005). *Modelling and simulation of a gas turbine.* (PhD Thesis). Linkopings University, Linkopings, Sweden.

Kulikov, G. G., & Thompson, H. A. (2004). *Dynamic modelling of gas turbines.* London: Springer-Verlag.

Lazzaretto, A., & Toffolo, A. (2001). Analytical and neural network models for gas turbine design and off-design simulation. *International Journal of Applied Thermodynamics, 4*(4), 173–182.

Lee, Y. K., Mavris, D. N., Volovoi, V. V., Yuan, M., & Fisher, T. (2010). A fault diagnosis method for industrial gas turbines using bayesian data analysis. *Journal of Engineering for Gas Turbines and Power, 132*, 041602–1. doi:10.1115/1.3204508.

Ljung, L., & Glad, T. (1994). *Modelling of dynamic systems.* Englewood Cliffs, NJ: PTR Prentice Hall.

MAN Diesel & Turbo Company. (n.d.). *THM gas turbine basic training.* Retrieved from http://www.mandieselturbo.com

Mantzaris, J., & Vournas, C. (2007). Modelling and stability of a single-shaft combined-cycle power plant. *International Journal of Thermodynamics, 10*(2), 71–78.

Mostafavi, M., Alaktiwi, A., & Agnew, B. (1998). Thermodynamic analysis of combined open-cycle twin-shaft gas turbine (Brayton cycle) and exhaust gas operated absorption refrigeration unit. *Applied Thermal Engineering, 18*, 847–856. doi:10.1016/S1359-4311(97)00105-1.

Mu, J., & Rees, D. (2004). Approximate model predictive control for gas turbine engines. In *Proceeding of the 2004 American Control Conference*, (pp. 5704-5709). Boston, MA: ACC.

Mu, J., Rees, D., & Liu, G. P. (2004). Advanced controller design for aircraft gas turbine engines. *Control Engineering Practice, 13,* 1001–1015. doi:10.1016/j.conengprac.2004.11.001.

Nabney, I. T., & Cressy, D. C. (1996). Neural network control of a gas turbine. *Neural Computing & Applications, 4,* 198–208. doi:10.1007/BF01413818.

Nagpal, M., Moshref, A., Morison, G. K., & Kundur, P. (2001). Experience with testing and modelling of gas turbines. In *Proceedings of the Power Engineering Society Winter Meeting.* IEEE.

Najjar, Y. S. H. (1994). Performance of single-cycle gas turbine engines in two modes of operation. *Energy Conversion and Management, 35*(5), 433–441. doi:10.1016/0196-8904(94)90101-5.

Norgaard, M., Ravn, O., Poulsen, N. K., & Hansen, L. K. (2000). *Neural networks for modelling and control of dynamic systems.* Berlin: Springer Publications. doi:10.1007/978-1-4471-0453-7.

Ogaji, S. O. T., Singh, R., & Probert, S. D. (2002). Multiple-sensor fault-diagnosis for a 2-shaft stationary gas turbine. *Applied Energy, 71,* 321–339. doi:10.1016/S0306-2619(02)00015-6.

Pongracz, B., Ailer, P., Hangos, K. M., & Szederkenyi, G. (2000). *Nonlinear reference tracking control of a gas turbine with load torque estimation.* Budapest, Hungary: Computer and Automation Research Institute, Hungarian Academy of Sciences.

Ricketts, B. E. (1997). Modelling of a gas turbine: A precursor to adaptive control. In *Proceedings of the IEE Colloquium on Adaptive Controllers in Practice '97* (Digest No: 1997/176). IEE.

Roland-Villasana, E. J., Vazquez, A., & Jimenez-Sanchez, V. M. (2010). Modelling of the simplified systems for a power plant simulator. In *Proceedings of the Fourth UKSim European Symposium on Computer Modelling And Simulation (EMS).* EMS.

Rowen, W. I. (1983). Simplified mathematical representations of heavy-duty gas turbines. *Journal of Engineering for Power, 105*(4). doi:10.1115/1.3227494.

Rowen, W. I. (1992, July/August). Simplified mathematical representations of single-shaft gas turbines in mechanical derive service. *Turbomachinery International.*

Ruanoa, A. E., Fleming, P. J., Teixeiraa, C., Vazquezc, K. R., & Fonsecaa, C. M. (2003). Non-linear identification of aircraft gas turbine dynamics. *Neurocomputing, 55,* 551–579. doi:10.1016/S0925-2312(03)00393-X.

Shalan, H. E. (2011). Parameter estimation and dynamic simulation of gas turbine model in combined-cycle power plants based on actual operational data. *Journal of American Science, 7*(5).

Simani, S., & Patton, R. J. (2008). Fault diagnosis of an industrial gas turbine prototype using a system identification approach. *Control Engineering Practice, 16,* 769–786. doi:10.1016/j.conengprac.2007.08.009.

Spina, P. R., & Venturini, M. (2007). Gas turbine modelling by using neural networks trained on field operating data. In *Proceedings of ECOS 2007.* Padova, Italy: ECOS.

Tavakoli, M. R. B., Vahidi, B., & Gawlik, W. (2009). An educational guide to extract the parameters of heavy-duty gas turbines model in dynamic studies based on operational data. *IEEE Transactions on Power Systems, 24*(3). doi:10.1109/TPWRS.2009.2021231.

Torella, G., Gamma, F., & Palmesano, G. (2003). Neural networks for the study of gas turbine engines air system. In *Proceedings of the International Gas Turbine Congress 2003.* Tokyo, Japan: IGTC.

Wikimedia Commons. (2012). Retrieved from http://commons.wikimedia.org

Yadav, N., Khan, I. A., & Grover, S. (2010). Modelling and analysis of simple open-cycle gas turbine using graph networks. In *Proceedings of the International Conference of Electrical and Electronics Engineering*. IEEE.

Yee, S. K., & Milanovic, J. V. (2008). Overview and comparative analysis of gas turbine models for system stability studies. *IEEE Transactions on Power Systems*, *23*(1). doi:10.1109/TP-WRS.2007.907384.

Yoru, Y., Karakoc, T. H., & Hepbasli, A. (2009). Application of artificial neural network (ANN) method to exergetic analyses of gas turbines. In *Proceedings of the International Symposium on Heat Transfer in Gas Turbine Systems*. Antalya, Turkey: IEEE.

Zhu, Y., Frey, H. C., & Asce, M. (2007). Simplified performance model of gas turbine combined-cycle systems. *Journal of Energy Engineering*. doi:10.1061/(ASCE)0733-9402(2007)133:2(82).

Chapter 15
Robotic CAM System Available for Both CL Data and NC Data

Fusaomi Nagata
Tokyo University of Science, Japan

Keigo Watanabe
Okayama University, Japan

Sho Yoshitake
Tokyo University of Science, Japan

Maki K. Habib
American University in Cairo, Egypt

ABSTRACT

This chapter describes the development of a robotic CAM system for an articulated industrial robot from the viewpoint of robotic servo controller. It is defined here that the CAM system includes an important function that allows an industrial robot to move along not only numerical control data (NC data) but also cutter location data (CL data) consisting of position and orientation components. A reverse post-processor is proposed for the robotic CAM system to online generate CL data from the NC data generated for a five-axis NC machine tool with a tilting head, and the transformation accuracy about orientation components in CL data is briefly evaluated. The developed CAM system has a high applicability to other industrial robots with an open architecture controller whose servo system is technically opened to end-users, and also works as a straightforward interface between a general CAD/CAM system and an industrial robot. The basic design of the robotic CAM system and the experimental result are presented, in which an industrial robot can move based on not only CL data but also NC data without any teaching.

1. INTRODUCTION

At the present stage, the relationship between CAD/CAM systems and industrial robots are not well established compared to NC machine tools that are widely spread in manufacturing industries. Generally, the main-processor of CAD/CAM system generates CL data according to each model's shape and machining conditions, then the post-processor produces suitable NC data according to an NC machine tool actually used. The controller of the NC machine tool sequentially deals with NC data and accurately controls the positions of main head and the angles of other axes. Thus, the CAM systems for NC machine tools are already established. On the other hand,

DOI: 10.4018/978-1-4666-4225-6.ch015

however, the CAM system for industrial robots has not been sufficiently considered and developed yet. A teaching pendant is generally used to obtain position and orientation data of the arm tip before an industrial robot works.

Nagata et al. (2001, 2006) developed a joystick teaching system for a polishing robot to safely obtain desirable orientation data of a sanding tool attached to the tip of robot arm. Maeda et al. (2002) proposed a simple teaching method for industrial robots by human demonstration. The proposed automated camera calibration enabled labor-saving teaching and compensated the absolute positional error of an industrial robot. Also, Kushida et al. (2001) proposed a method of force-free control for an industrial articulated robot arm. The control method was applied to the direct teaching of industrial articulated robot arms, in which the robot arm was directly moved by human force. Further, Sugita et al. (2003) developed two kinds of teaching support devices, i.e., a three-wire type and an arm type, for a deburring and finishing robot. The validity the proposed devices were verified through experiments by using an industrial robot. As for off-line teaching, Ahn and Lee (2000) proposed an off-line automatic teaching method using vision information for robotic assembly task. Also, CAD-based off-line teaching system was proposed by Neto et al. (2010), which allows users with basic CAD skills to generate robot programs off-line, without stopping the production by using a robot. Besides, Ge et al. (1993) showed a basic transformation from CAD data to position and orientation vectors for a polishing robot. As one of the pioneers about teaching-less industrial robotic system, Sugitani et al. (1996) developed welding robots which were successfully controlled by the teaching-less CAD/CAM system, in which there were 26 sets of arc welding robot for steel bridge panel fabrication.

This chapter describes the development of a robotic CAM system for an articulated industrial robot from the view point of robotic servo controller. It is defined here that the CAM system

includes an important function which allows an industrial robot to move along not only numerical control data (NC data) but also cutter location data (CL data) consisting of position and orientation components. In order to generate CL data from NC data, a reverse post-processor is proposed. In addition, the developed CAM system has a high applicability to other industrial robots with an open architecture controller whose servo system is technically opened to end-users. The developed robotic CAM system works as a straightforward interface between a general CAD/CAM system and an industrial robot.

Figure 1 shows an articulated-type industrial robot with an open architecture controller, that is used to evaluate the effectiveness of the proposed robotic CAM system. In Figure 2, the proposed robotic CAM system is shown in comparison with the conventional CAM process using an NC machine tool. Also, in Figure 3, the proposed robotic CAM system is compared with the conventional off-line teaching process for an industrial robot. As can be seen, the industrial robot with the proposed CAM system can work based on both NC data and CL data without conducting a teaching task and using a robot language.

In this chapter, a simple and straightforward CAM system without using any robot language is designed for a Cartesian-based servo system in

Figure 1. Articulated-type industrial robot with an open architecture controller

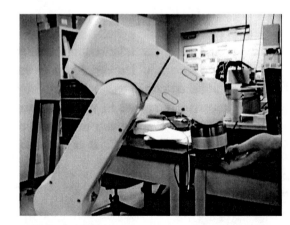

Figure 2. The proposed robotic CAM system compared with conventional CAM process using an NC machine tool

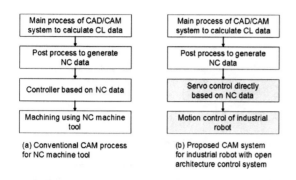

(a) Conventional CAM process for NC machine tool

(b) Proposed CAM system for industrial robot with open architecture control system

Figure 3. The proposed robotic CAM system compared with conventional off-line teaching process for an industrial robot

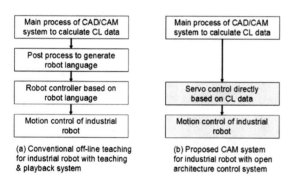

(a) Conventional off-line teaching for industrial robot with teaching & playback system

(b) Proposed CAM system for industrial robot with open architecture control system

order to enhance the relationship between a conventional CAD/CAM system and an industrial robot. The basic design of the robotic CAM system and the experimental result are presented. Further, the transformation accuracy about orientation components in CL data is briefly evaluated.

2. DESIRED TRAJECTORY

2.1. Main-Processor of General CAD/CAM System

Various kinds of CAD/CAM system such as Catia, Unigraphics, Pro/Engineer, etc. are widely used in manufacturing industries. The main-processor of each CAM can generate CL data consisting of position and orientation components along a 3D model. In this section, how to calculate desired position and orientation for robotic servo system is described in detail.

For example, robotic sanding task needs a desired trajectory so that the sanding tool attached to the tip of the robot arm can follow the object's surface, keeping contact with the surface from normal direction. If the object is fortunately designed by a CAD/CAM system and manufactured by an NC machine tool, then CL data can be referred as the desired trajectory consisting of position and orientation elements. It is important to show a guideline for raising the relationship between the main-processor of CAD/CAM system and an industrial robot, which is one of the roles of the CAM system proposed after this subsection.

2.2. Position and Orientation Components for Discrete-Time Control System

In order to realize non-taught operation, we have already proposed a generalized trajectory generator (Nagata et al., 2001) using CL data, that yields desired trajectory ${}^{w}\mathbf{r}(k)$ at the discrete time k given by

$$ {}^{w}\mathbf{r}(k) = [{}^{w}\mathbf{x}_{d}^{T}(k) \quad {}^{w}\mathbf{o}_{d}^{T}(k)]^{T} \qquad (1) $$

where the superscript w denotes the work coordinate system, ${}^{w}\mathbf{x}_{d}(k) = [{}^{w}x_{d}(k) \quad {}^{w}y_{d}(k) \quad {}^{w}z_{d}(k)]^{T}$ and ${}^{w}\mathbf{o}_{d}(k) = [{}^{w}o_{dx}(k) \quad {}^{w}o_{dy}(k) \quad {}^{w}o_{dz}(k)]^{T}$ are the position and orientation components, respectively. ${}^{w}\mathbf{o}_{d}(k)$ is the normal vector at the position ${}^{w}\mathbf{x}_{d}(k)$.. In the following, we explain in detail how to make ${}^{w}\mathbf{r}(k)$ using the CL data.

A target workpiece with curved surface is generally designed by a CAD/CAM system, so that CL data can be calculated by the main-processor. The CL data consist of sequential points along the model surface given by a zigzag path or a whirl path. In this approach, the desired trajectory $^w\mathbf{r}(k)$ is generated along the CL data. The CL data are usually calculated with a linear approximation along the model surface. The i-th step is written by

$$\mathbf{cl}(i) = [p_x(i)\ p_y(i)\ p_z(i)\ n_x(i)\ n_y(i)\ n_z(i)]^T \qquad (2)$$

$$\{n_x(i)\}^2 + \{n_y(i)\}^2 + \{n_z(i)\}^2 = 1 \qquad (3)$$

w h e r e $\mathbf{p}(i) = [p_x(i)\ p_y(i)\ p_z(i)]^T$ a n d $\mathbf{n}(i) = [n_x(i)\ n_y(i)\ n_z(i)]^T$ are position and orientation vectors viewed from the origin ^wO, respectively. $^w\mathbf{r}(k)$ is obtained by using both linear equations and a tangential velocity scalar v_t called the feed rate.

A relation between $\mathbf{cl}(i)$ and $^w\mathbf{r}(k)$ is shown in Figure 4. In this case, assuming $^w\mathbf{r}(k) \in [\mathbf{cl}(i-1), \mathbf{cl}(i)]$ we obtain $^w\mathbf{r}(k)$ through the following procedure. First, a direction vector $\mathbf{t}(i) = [t_x(i)\ t_y(i)\ t_z(i)]^T$ is given by

$$\mathbf{t}(i) = \mathbf{p}(i) - \mathbf{p}(i-1) \qquad (4)$$

so that v_t is decomposed into x-, y- and z-components in work coordinate system as written by

$$v_{tj} = v_t \frac{t_j(i)}{\|\mathbf{t}(i)\|} \qquad (j = x, y, z) \qquad (5)$$

Figure 4. Relation between CL data $\mathbf{cl}(i)$ *and desired trajectory* $^w\mathbf{r}(k)$*, in which wO is the origin of work coordinate system*

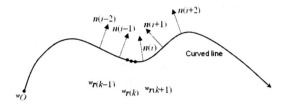

Using a sampling width Δt, each component of the desired position $^w\mathbf{x}_d(k)$ is represented by

$$^w x_d(k) = {}^w x_d(k-1) + v_{tx}\Delta t \qquad (6)$$

$$^w y_d(k) = {}^w y_d(k-1) + v_{ty}\Delta t \qquad (7)$$

$$^w z_d(k) = {}^w z_d(k-1) + v_{tz}\Delta t \qquad (8)$$

Next, how to calculate the desired orientation $^w\mathbf{o}_d(k)$ is considered. By using the orientation components of two adjacent steps in CL data, a rotational direction vector

$$\mathbf{t}_r(i) = [t_{rx}(i)\ t_{ry}(i)\ t_{rz}(i)]^T$$

is defined as

$$\mathbf{t}_r(i) = \mathbf{n}(i) - \mathbf{n}(i-1) \qquad (9)$$

Each component of desired orientation vector can be linearly calculated with $\mathbf{t}_r(i)$ as

$$^w o_{dj}(k)$$
$$= n_j(i-1) + t_{rj}(i)\frac{\|\mathbf{x}_d(k) - \mathbf{p}(i-1)\|}{\|\mathbf{t}(i)\|} \qquad (10)$$
$$(j = x, y, z)$$

$^w\mathbf{x}_d(k)$ and $^w\mathbf{o}_d(k)$ shown above are directly obtained from the CL data without either any conventional complicated teaching process or recently proposed off-line teaching methods. The desired position and orientation in the discrete time domain are very important to control the tip of an industrial robot in real time, i.e., to design a feedback control system.

If the linear approximation is applied when CL data are generated by the main-processor of a CAD/CAM system, the CL data forming a curved line are composed of continuous minute lines such as $\mathbf{p}(i) - \mathbf{p}(i-1)$ and $\mathbf{p}(i+1) - \mathbf{p}(i)$ shown in Figure 5. In this case, it should be noted that each position vector in CL data such as $\mathbf{p}(i)$ and $\mathbf{p}(i+1)$ have to be carefully dealt with in order to be accurately followed along the CL data. For example, ${}^{w}\mathbf{x}_{d}(k+1)$, ${}^{w}\mathbf{x}_{d}(k+5)$ and ${}^{w}\mathbf{x}_{d}(k+9)$ are not calculated by using Equations (6), (7) and (8) but have to be directly set with $\mathbf{p}(i)$, $\mathbf{p}(i+1)$ and $\mathbf{p}(i+2)$, respectively, just before the feed direction changes. $\mathbf{p}(i) - {}^{w}\mathbf{x}_{d}(k)$, $\mathbf{p}(i+1) - {}^{w}\mathbf{x}_{d}(k+4)$ and $\mathbf{p}(i+2) - {}^{w}\mathbf{x}_{d}(k+8)$ are called the fraction vectors.

3. IMPLEMENTATION TO AN INDUSTRIAL ROBOT

3.1. Communication with an Open Architecture Controller of an Industrial Robot

A Windows PC and an open architecture controller of RV1A are connected with Ethernet as shown in Figure 6. The servo system in Cartesian coordinate system of the RV1A is technically opened to users, so that absolute coordinate vectors of position and orientation can be given to the reference of the servo system. The servo rate of the

robot is fixed to 7.1 ms by the robot maker. Figure 7 illustrates the communication scheme by using UDP packet, in which sampling period is set to 10 ms. The data size in a UDP packet is 196 bytes. The packet transmitted by "sendto()" includes values of desired position $\mathbf{X}_{d}(k) = [X_{d}(k)\,Y_{d}(k)\,Z_{d}(k)]^{T}$ (mm) and desired orientation $\mathbf{O}_{d}(k) = [\varphi_{d}(k)\,\theta_{d}(k)\,\psi_{d}(k)]^{T}$ (rad) in

Figure 6. Block diagram of communication system by using UDP packet between a PC and an industrial robot RV1A

Figure 7. Communication scheme implemented in the PC by using Windows socket functions for UDP, in which the sampling period Δt is set to 10 ms

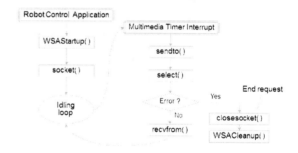

Figure 5. Relation between position component $\mathbf{p}(i)$ in CL data and desired position ${}^{w}\mathbf{x}_{d}(k)$, in which $\mathbf{p}(i) - {}^{w}\mathbf{x}_{d}(k)$, $\mathbf{p}(i+1) - {}^{w}\mathbf{x}_{d}(k+4)$, and $\mathbf{p}(i+2) - {}^{w}\mathbf{x}_{d}(k+8)$ are called the fraction vectors

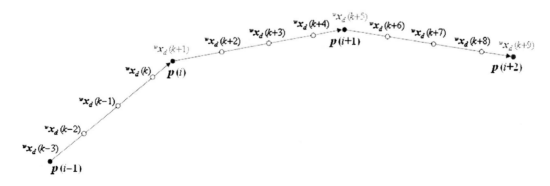

robot absolute coordinate system. $\varphi_d(k)$, $\theta_d(k)$ and $\psi_d(k)$ are the rotational angles around x-, y- and z-axes, respectively, which are called X-Y-Z fixed angles or roll, pitch and yaw angles. The desired position and the desired orientation are set in a UDP packet as the reference of arm tip for the Cartesian servo system. Also, the packet received by "recvfrom()" includes values of current position $\mathbf{X}(k) = [X(k)\,Y(k)\,Z(k)]^T$ (mm) and current orientation $\mathbf{O}(k) = [\varphi(k)\,\theta(k)\,\psi(k)]^T$ (rad) in robot absolute coordinate system, which can be used for feedback quantity.

3.2. Desired Position and Orientation Based on CL Data

How to make $\mathbf{X}_d(k)$ and $\mathbf{O}_d(k)$ based on the position and the orientation given by Equation (1) is discussed in detail by using an example shown in Figure 4, in which the tip of robot arm follows the trajectory keeping the orientation along normal direction to the surface. Each components of $\mathbf{X}_d(k)$ is represented with the initial position $\mathbf{X}_d(0) = [X_d(0)\,Y_d(0)\,Z_d(0)]^T$ as

$$X_d(k) = X_d(0) + {}^w x_d(k) \tag{11}$$

$$Y_d(k) = Y_d(0) + {}^w y_d(k) \tag{12}$$

$$Z_d(k) = Z_d(0) + {}^w z_d(k) \tag{13}$$

where $\mathbf{X}_d(0)$ means the coordinate values of ${}^wO$ in robot absolute coordinate system.

Next, we consider the rotational components. Generally, rotational matrices R_X, R_Y and R_Z around x-, y- and z-axes are respectively given by

$$R_X = \begin{pmatrix} 1 & 0 & 0 \\ 0 & \cos\varphi(k) & -\sin\varphi(k) \\ 0 & \sin\varphi(k) & \cos\varphi(k) \end{pmatrix} \tag{14}$$

$$R_Y = \begin{pmatrix} \cos\theta(k) & 0 & \sin\theta(k) \\ 0 & 1 & 0 \\ -\sin\theta(k) & 0 & \cos\theta(k) \end{pmatrix} \tag{15}$$

$$R_Z = \begin{pmatrix} \cos\psi(k) & -\sin\psi(k) & 0 \\ \sin\psi(k) & \cos\psi(k) & 0 \\ 0 & 0 & 1 \end{pmatrix} \tag{16}$$

Accordingly, the rotational matrix $R_Z R_Y R_X$ with a roll angle $\varphi(k)$, a pitch angle $\theta(k)$, and a yaw angle $\psi(k)$ is given by

$$R_Z R_Y R_X =$$
$$\begin{pmatrix} C_\theta C_\psi & S_\varphi S_\theta C_\psi - C_\varphi S_\psi & C_\varphi S_\theta C_\psi + S_\varphi S_\psi \\ C_\theta S_\psi & S_\varphi S_\theta S_\psi + C_\varphi C_\psi & C_\varphi S_\theta S_\psi - S_\varphi C_\psi \\ -S_\theta & S_\varphi C_\theta & C_\varphi C_\theta \end{pmatrix} \tag{17}$$

where, for example, S_θ and C_θ means $\sin\theta(k)$ and $\cos\theta(k)$, respectively. Figure 8 illustrates the trajectory following control along CL data shown in Figure 4. As can be seen, the direction of ${}^w\mathbf{o}_d(k)$ is just inverse to the one of the arm tip. When referring the orientation components in CL data, the arm tip can be determined only with the roll

Figure 8. Orientation control of arm tip based on CL data shown in Figure 4, in which the initial orientation is given with $[\varphi(k)\,\theta(k)\,\psi(k)]^T = [0\ 0\ 0]^T$

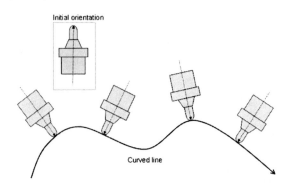

angle and the pitch angle except for the yaw angle. That means $\psi(k)$ can be always fixed to 0 (rad). Thus, the roll angle $\varphi(k)$ and the pitch angle $\theta(k)$ at the discrete time k are simply obtained by solving

$$\begin{pmatrix} -{}^{w}o_{dx}(k) \\ -{}^{w}o_{dy}(k) \\ -{}^{w}o_{dz}(k) \end{pmatrix} = \begin{pmatrix} C_{\theta} & S_{\varphi}S_{\theta} & C_{\varphi}S_{\theta} \\ 0 & C_{\varphi} & -S_{\varphi} \\ -S_{\theta} & S_{\varphi}C_{\theta} & C_{\varphi}C_{\theta} \end{pmatrix} \begin{pmatrix} 0 \\ 0 \\ 1 \end{pmatrix} \qquad (18)$$

where $(0\ 0\ 1)^{T}$ is the initial orientation vector in robot absolute coordinate system as shown in Figure 8. Finally, $\varphi(k)$ and $\theta(k)$ are calculated by using inverse trigonometric functions as

$$\varphi(k) = \text{atan2}\{-{}^{w}o_{dx}(k), -{}^{w}o_{dz}(k)\} \qquad (19)$$

$$\theta(k) = \text{asin}\{{}^{w}o_{dy}(k)\} \qquad (20)$$

If the desired yaw angle $\psi_{d}(k)$ is fixed to 0, then the desired roll angle $\varphi_{d}(k)$ and pitch angle $\theta_{d}(k)$ can be calculated with Equations (19) and (20), respectively.

4. EXPERIMENT

In this section, an experiment of trajectory following control based on CL data is conducted to evaluate the effectiveness of the proposed CAM system. Figure 9 shows the desired trajectory generated by using the main-processor of 3D CAD/CAM Pro/Engineer, which consists of position and orientation components written by multi−lined "GOTO/" statements. It is tried that the arm tip is controlled so as to follow the position and orientation. The value of feed rate, i.e., tangent velocity, is set to 1 mm/s; the desired values composed of $\mathbf{X}_{d}(k)$ and $\mathbf{O}_{d}(k)$ are given to the references of the servo controller in RV1A every sampling period through UDP packets.

Figure 9. CL data $\mathbf{cl}(i) = [\mathbf{p}^{T}(i)\ \mathbf{n}^{T}(i)]^{T}$ *consisting of position and orientation components, which is used for desired trajectory of the tip of robot arm*

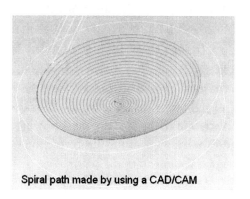

Spiral path made by using a CAD/CAM

When the desired position $\mathbf{X}_{d}(k)$ and orientation $\mathbf{O}_{d}(k)$ is transmitted to the robotic servo system in RV1A, the roll angle $\varphi_{d}(k)$ has to be given within the range from $-\pi$ to π. Figure 10 shows four examples of roll angle around x-axis, in which (a) is the initial angle represented by

Figure 10. Roll angles around x-axis, in which (a) is the initial angle represented by $\phi(k) = 0$, (b) is the case of $-\pi \le \phi(k) < 0$, (c) is the case of $\phi(k) = \pm \pi$, and (d) is the case of $0 < \phi(k) \le \pi$

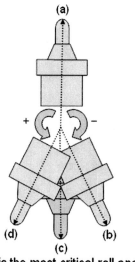

(c) is the most critical roll angle.

$\phi(k) = 0$, (b) is the case of $-\pi \leq \phi(k) < 0$, (c) is the case of $\phi(k) = \pm \pi$, and (d) is the case of $0 < \phi(k) \leq \pi$. It should be noted how to calculate the desired roll angle $\varphi_d(k)$ for an manipulated value if the orientation changes as (b)→(c)→(d) or (d)→(c)→(b), i.e., in passing through the situation (c). In such cases, the sign of $\varphi_d(k)$ has to be suddenly changed as shown in Figure 11, i.e., (c) is the most critical orientation from the viewpoint of roll angle. In order to smoothly control the orientation of the arm tip, the following control rules were applied for the correction of the roll angle (Yoshitake et al., 2012).

if $\phi_d(k) > 0$ and $\phi(k) < 0$, then $\Delta\phi(k) = K_p\{\phi_d(k) - \phi(k) - 2\pi\}$ (21)

else if $\phi_d(k) < 0$ and $\phi(k) > 0$, then $\Delta\phi(k) = K_p\{\phi_d(k) - \phi(k) + 2\pi\}$ (22)

else if $\sin\{\phi_d(k)\} = \sin\{\phi(k)\}$, then $\Delta\phi(k) = K_p\{\phi_d(k) - \phi(k)\}$ (23)

where $\varphi_d(k)$ and $\varphi(k)$ are the desired roll angle transmitted to the robotic servo controller from the PC and actual roll angle transmitted from the robotic servo controller to the PC, respectively. $\Delta\varphi(k)$ is the output of proportional control action with a gain K_p. In fact, the desired roll angle $\varphi_d(k+1)$ can be generated as

$$\phi_d(k+1) = \phi(k) + \Delta\phi(k) \qquad (24)$$

It was confirmed from an experiment that desirable control results of position and orientation could be obtained as shown in Figure 12. The arm tip could gradually move up from the bottom center along the desired spiral path shown in Figure 9. In this case, the orientation of the arm

Figure 11. Initial part of desired roll angle $\phi_d(k)$ calculated from the desired trajectory shown in Figure 9

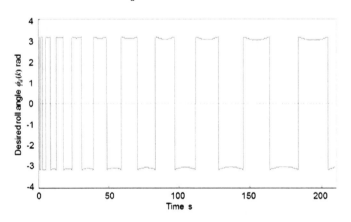

Figure 12. 3D view illustrated in robot absolute coordinate system, in which actual controlled results X(k), Y(k), and Z(k) in the initial range are plotted

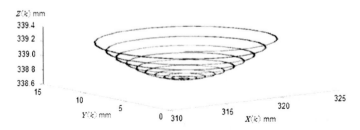

tip was simultaneously controlled so as to be normal direction to the surface. As an example, position error $X_d(k) - X(k)$ in x-direction is shown in Figure 13. The values of the errors are under the neighborhood of 0.15 mm that is well-known as the repetitive position accuracy officially announced by industrial robots' maker. Also, Figure 14 shows the position error $X_d(k) - X(k+1)$. It is observed that the one sampling period delayed x-directional position $X(k+1)$ could accurately follow the reference $X_d(k)$. The position error $X_d(k) - X(k)$ shown in Figure 13 is the typical problem concerning the delay of response when a feedback control law is applied to a dynamic system.

Thus, it was successfully demonstrated that the proposed CAM system allows the tip of the robot arm to desirably follow the desired trajectory given by multi-axis CL data without any complicated teaching tasks.

5. ROBOTIC CAM SYSTEM EXTENDED FOR DEALING WITH NC DATA

A robotic CAM system based on CL data and the implementation to an industrial robot have been introduced up to the previous section. Generally, the main processor of a CAD/CAM system calculates CL data, then the post-processor

Figure 13. Initial range of position error $X_d(k) - X(k)$ in x-direction

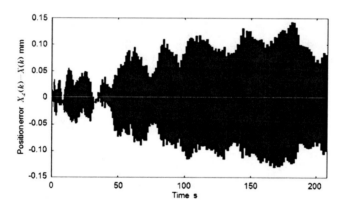

Figure 14. Initial range of one sampling period delayed position error $X_d(k) - X(k+1)$ in x-direction

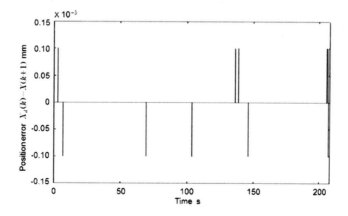

Figure 15. Five-axis NC machine tool with a tiling head, which is used in the advanced woodworking industry

Figure 16. In the proposed reverse post-processor, tilting and rotation angles have to be converted to a normal direction vector

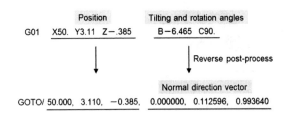

transforms the CL data to the corresponding NC data according to various types of NC machine tools actually used. In this section, the proposed robotic CAM system is extended for dealing with NC data post-processed for a five-axis NC machine tool with a tilting head (Nagata et al., 2009) as shown in Figure 15. If the NC data can be reversely transformed to the original CL data, it means that the proposed robotic CAM system available for CL data allows the industrial robot to be also controlled based on the NC data. We call this transformation (NC data → CL data) the reverse post-process as shown in Figure 16.

In the proposed reverse post-processing shown in Figure 17, the position components have only to be directly extracted. On the other hand, the orientation components have to be dealt with considering the geometric configuration, e.g., the tilt angle $-90 \le B(i) \le 90$ and the turn angle $0 \le C(i) \le 360$ at the i-th step in NC data can be transformed to the corresponding normal vector $\mathbf{n}_r(i) = [n_{rx}(i) \ n_{ry}(i) \ n_{rz}(i)]^T$ in CL data by calculating:

$$n_{rx} = \sin\left\{\frac{B(i)}{180}\pi\right\}\cos\left\{\frac{C(i)}{180}\pi\right\} \qquad (25)$$

Figure 17. Post-process means CL data → NC data, also proposed reverse post-processor conducts NC data → CL data

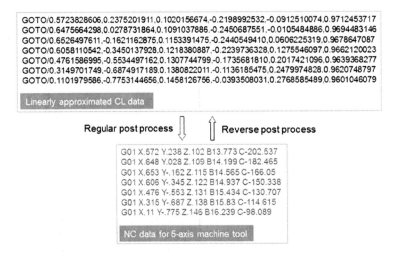

$$n_{ry} = \sin\left\{\frac{B(i)}{180}\pi\right\}\sin\left\{\frac{C(i)}{180}\pi\right\} \qquad (26)$$

$$n_{rz} = \cos\left\{\frac{B(i)}{180}\pi\right\} \qquad (27)$$

The accuracy of the transformed $\mathbf{n}_r(i)$ to the original $\mathbf{n}(i)$ was examined by using the case shown in Figure 17, so that it was confirmed that the errors were relatively small, i.e., $\max_i \left| n_{rx}(i) - n_x(i) \right| = 0.0000089$, $\max_i \left| n_{ry}(i) - n_y(i) \right| = 0.0000085$ and $\max_i \left| n_{rz}(i) - n_z(i) \right| = 0.0000055$. Here, $\mathbf{n}(i)$ is the normal vector included in the original CL data in Figure 17, which is generated from the main-processor of 3D CAD/CAM Pro/Engineer. Also, $B(i)$ and $C(i)$ included in NC data in Figure 17 are the tilt angle and the turn angle, respectively, which are calculated by the post-processor of Pro/Engineer. Further, $\mathbf{n}_r(i)$ is the normal vector calculated by using the proposed reverse post-processor given by Equations (25), (26). and (27).

6. CONCLUSION

A CAM system for an articulated-type industrial robot RV1A has been proposed in order to enhance the relationship between a design tool such as CAD/CAM and industrial robots spread to industrial manufacturing fields. The CAM system required for an industrial robot RV1A was realized as an integrated system including not only conventional main-processor of CAM but also robotic servo system and kinematics. Further, the reverse post-processor has been proposed so that the arm tip of the industrial robot can be flexibly controlled based on both CL data and NC data. Finally, an example of transformation accuracy about the orientation components in CL data has been numerically shown.

REFERENCES

Ahn, C. K., & Lee, M. C. (2000). An off-line automatic teaching by vision information for robotic assembly task. In *Proceedings of IEEE International Conference on Industrial Electronics, Control and Instrumentation* (pp. 2171–2176). IEEE.

Ge, D. F., Takeuchi, Y., & Asakawa, N. (1993). Automation of polishing work by an industrial robot – 2nd report, automatic generation of collision- free polishing path. *Transactions of the Japan Society of Mechanical Engineers, 59*(561), 1574–1580. doi:10.1299/kikaic.59.1574.

Kushida, D., Nakamura, M., Goto, S., & Kyura, N. (2001). Human direct teaching of industrial articulated robot arms based on force-free control. *Artificial Life and Robotics, 5*(1), 26–32. doi:10.1007/BF02481317.

Maeda, Y., Ishido, N., Kikuchi, H., & Arai, T. (2002). Teaching of grasp/graspless manipulation for industrial robots by human demonstration. In *Proceedings IEEE/RSJ International Conference on Intelligent Robots and Systems* (pp. 1523–1528). IEEE.

Nagata, F., Kusumoto, Y., & Watanabe, K. (2009). Intelligent machining system for the artistic design of wooden paint rollers. *Robotics and Computer-integrated Manufacturing, 25*(3), 680–688. doi:10.1016/j.rcim.2008.05.001.

Nagata, F., Watanabe, K., & Izumi, K. (2001). Furniture polishing robot using a trajectory generator based on cutter location data. In *Proceedings of 2001 IEEE International Conference on Robotics and Automation* (pp. 319–324). IEEE.

Nagata, F., Watanabe, K., & Kiguchi, K. (2006). Joystick teaching system for industrial robots using fuzzy compliance control. In Industrial Robotics: Theory, Modelling and Control (pp. 799–812). INTECH.

Neto, P., Pires, J. N., & Moreira, A. P. (2010). CAD based off-line robot programming. In *Proceedings of IEEE International Conference on Robotics Automation and Mechatronics* (pp. 516–521). IEEE.

Sugita, S., Itaya, T., & Takeuchi, Y. (2003). Development of robot teaching support devices to automate deburring and finishing works in casting. *International Journal of Advanced Manufacturing Technology, 23*(3/4), 183–189.

Sugitani, Y., Kanjo, Y., & Murayama, M. (1996). Systemization with CAD/CAM welding robots for bridge fabrication. In *Proceedings of 4th International Workshop on Advanced Motion Control* (pp. 80–85). IEEE.

Yoshitake, S., Nagata, F., Otsuka, A., Watanabe, K., & Habib, M. K. (2012). Proposal and implementation of CAM system for industrial robot RV1A. In *Proceedings of the 17th International Symposium on Artificial Life and Robotics* (pp. 158–161). IEEE.

KEY TERMS AND DEFINITIONS

CAD/CAM: Computer Aided Design and Computer Aided Manufacturing. Here, CAD/CAM is used to make a desired trajectory of the robot arm without a robotic teaching process.

CL Data: Cutter location data are generated from the main-processor of CAM, which are written with multi-lined "GOTO" statements including position and orientation vectors.

Industrial Robot: Articulated-type industrial robot RV1A with six degree-of-freedoms. The servo system and communication interface are technically opened to users.

NC Data: Numerical control data according to the type of machine tool are converted from CL data. The tool for this process is called the post-processor.

Reverse Post-Processor: Reverse post-processor is the online conversion tool from NC data to CL data in the proposed robotic CAM system.

Robotic Teaching: Robotic teaching is the well-known process to make a desired trajectory by using a teaching pendant. The robotic teaching is a time-consuming task.

Robot Language: An industrial robot has an original robot language provided by the robot maker to describe a program for a robotic application. The drawback is that it is not well standardized among various types of industrial robots.

Servo Controller: Six servo motors are built in the corresponding six joint of the industrial robot RV1A. The joint angles can be controlled to follow reference values by the servo controller.

Chapter 16
Robotic Grippers, Grasping, and Grasp Planning

Seyed Javad Mousavi
Tarbiat Modares University, Iran

Ellips Masehian
Tarbiat Modares University, Iran

ABSTRACT

Utilizing robotic hands for manipulating objects and assembly requires one to deal with problems like immobility, grasp planning, and regrasp planning. This chapter integrates some essential subjects on robotic grasping: the first section presents a concise taxonomy of robotic grippers and hands. Then the basic concepts of grasping are provided, including immobility, form-closure, and force-closure, 2D and 3D grasping, and Coulomb friction. Next, the principles of grasp planning, measures of grasping quality, pre-grasp, stable grasps, and regrasp planning are presented. The chapter presents comparisons for robotic grippers, a new classification of measures of grasp quality, and a new categorization of regrasp planning approaches.

INTRODUCTION

Mechatronics Engineering concerns with a wide spectrum of technologies, puts forth new concepts in modern product design, and provides methodologies not only for producing high-quality products, but also for their maintenance and application. In other words, the whole period of the product lifecycle is considered. Mechatronics can be viewed as a width synergy of the techniques and technologies that deal with precise and intelligent mechanisms, smart sensors, moving devices (e.g., robots, AGVs, etc.), real-time control, biomimetics, etc. In this field, engineers encounter with a wide range of systems, devices, tools, robots, and information (Habib, 2008). Therefore, when the product lifecycle is concerned, the role of robots in reducing production costs, improving quality, and increasing production speed become more significant.

DOI: 10.4018/978-1-4666-4225-6.ch016

A vital part of any industrial robot is its end-effector, which is a gripper, hand or tool devised for performing the task entrusted to the robot. Accomplishing a given task successfully depends on correct and effective planning of the end-effector movements, and that's why robot and tool path planning has been and currently is on the focus of many robotics researchers. Specifically, when the end-effector is in the form of a human hand, such a planning becomes a challenge since it must be able to stably hold and manipulate objects and parts.

Performing tasks like automated grasping, manipulating, and placing of parts by robotic hands requires advanced skills in object identification, pose analysis, motion planning, effective and stable grasp calculation, and precise execution. In this regard, planning reliable and effective grasping for robot end-effectors like grippers and hands becomes indispensable.

Utilizing robotic hands for manipulating objects and assembly parts needs dealing with problems like immobility, grasp planning, and regrasp planning. Immobility is a state in which a series of kinematic constraints prevent an object from translation along orthogonal axes or rotation about any arbitrary axis. Grasping addresses whether or not an object is being held in form-closure or force-closure, and grasp planning specifies how form- and force-closure grasps can be done, defines an objective function, and optimizes the grasping based on that function. A problem may occur when the position of an object in a feasible grasping does not match with its final and required placement position, and that is when an additional grasping is needed from another direction or position. This is called regrasping, and planning for it is known as regrasp planning, which can be done in various ways.

This chapter integrates some essential subjects on robotic grasping: the first section presents a concise taxonomy of robotic grippers and hands. Then the basic concepts of grasping are provided, including immobility, form-closure and force-closure, 2D and 3D grasping, and Coulomb friction.

Next the principles of grasp planning, measures of grasping quality, pre-grasp, stable grasps and regrasp planning are presented.

The purpose in writing this chapter has been to present a unified treatise on the principles and state of the art of grasping and grasp planning as a guide to the interested researchers who wish to explore this vibrant research field.

ROBOTIC GRIPPERS

A robotic gripper is a type of end-effector that comes into contact with the workpiece and hold or manipulates it. The varieties of robotic grippers can be divided into four main categories: mechanical grippers, vacuum and magnetic grippers, universal grippers, and multi-fingered hands. End-effectors are inspired from either tools with extended capabilities, or human hand and fingers, although there have been some new creations with novel mechanisms. In the following each category is explained in brief (Praveen, 2013).

1. **Mechanical Grippers:** Based on their mechanisms, mechanical grippers exist in the following main types:
   a. **Linkage Grippers:** In this type no gears and racks are used and their movement is caused by inward and outward connection of joints and junctions. The outer jaw opens and closes as a result of the movement of internal joints (Figure 1a).
   b. **Gear and Rack Grippers:** In this type the movement of the gripper's jaw is caused by motion of a gear which in turn moves the racks being in contact with it (Figure 1b).
   c. **Cam-Actuated Grippers:** In this type the motion of cams produce translational movements of links in contact with them, and results in gripping. A variety of cam profiles can be used for

Figure 1. (a) Linkage grippers; (b) gear and rack grippers

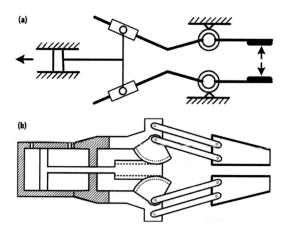

producing constant velocities in the form of circles and harmonic curves (Figure 2a).

d. **Screw-Driven Grippers:** In this type rotating a screw leads to movements in the links attached to it, causing a

Figure 2. (a) Cam-actuated grippers; (b) screw-driven grippers; (c) rope and pulley gripper

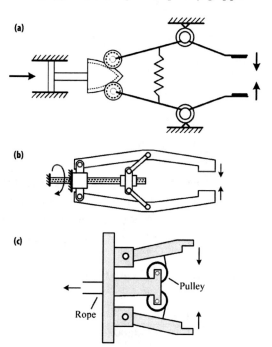

translation of gripper's jaws. The rotation of the screw can be produced by an electric motor (Figure 2b).

e. **Rope and Pulley Grippers:** In this type rotating a pulley by a motor leads to winding and unwinding of a rope around it which moves a pair of gripper's jaws (Figure 2c).

2. **Vacuum and Magnetic Grippers**: These grippers operate based on applying distributed forces on the object and have two main classes:

    a. **Vacuum Grippers:** This gripper is used for grasping flat surfaces and nonferrous objects. It can be made of standard rubber cups and pads, which by the way are inappropriate for curved and holed surfaces (Figure 3).

    b. **Magnetic Grippers:** This gripper is suitable for grasping ferrous materials like steel plates. Its magnetic head is made of a ferromagnetic core and a coil (Figure 4a).

3. **Universal Grippers**: These types of grippers have simple structures different from previous grippers, and can be divided into four classes:

    a. **Inflatable Grippers:** These grippers are used for grasping tiny, delicate, or irregular-shaped objects without exerting a concentrated load. Initially, the air vent of the gripper is open and the air in a bag (container) creates pressure on the object. When the covered surface grows large enough the internal pressure in the bag causes it to be hardened and rigid and consequently the object is grasped (Figure 4b).

    b. **Soft Grippers:** These types of grippers consist of small links and a pulley that are connected together with a pair of cords. The gripper assumes the outer shape of the object and holds it with uniform pressure (Figure 5a).

Figure 3. Vacuum grippers

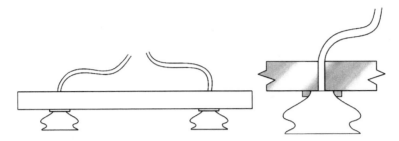

Figure 4. (a) Form-adaptable magnetic gripper; (b) inflatable gripper

Figure 5. (a) Soft gripper; (b) three-fingered gripper (triple jaw)

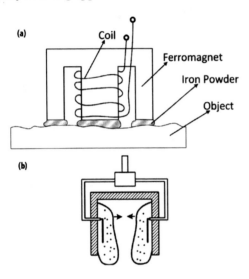

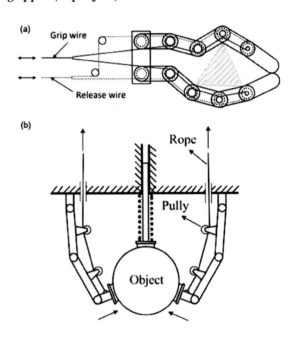

c.   **Three-Fingered Grippers (Triple Jaw):** The grabbing task by this gripper is usually performed by movements of two fingers and the third finger acts as a support. Fingers can exert beat, bite, and parallel jaw movements, and can grasp and ungrasp an object (Figure 5b).

d.   **Formless Grippers:** These kinds of grippers were developed at Cornell University using a rubber bag filled of sand, powder, or coffee ground. The gripper has a simple design and can easily adapt with a variety of object shapes and can firmly grab them, i.e., for sphere-like objects that are larger

than half the size of the gripper, or for thin disks (Brown Eric et al., 2010). The holding force is influenced by the object's shape. Indeed this is a new approach for grasping objects using amorphous powder rather than with fingers, and enables grasping objects with complex shapes (Figure 6a).

4.   **Multi-Fingered Hands:** This kind of end-effectors can perform more tasks than ordinary grippers, and are designed for complex operations like in-hand manipulation. In this kind of end-effectors it has been tried to

imitate human hand attributes like degrees of freedom and forces exerting on objects. These end-effectors are called *Robotic Hands* and are the main gripper type discussed in the rest of this chapter. The internal mechanism of these-end effectors is a combination of the systems used in Three-fingered Grippers, Soft Grippers, Rope and Pulley Grippers, Screw-driven Grippers, Cam-actuated Grippers, Gear and Rack Grippers, and Linkage Grippers. There are two main classes:

a. **Multi-Fingered Hand with Parallel Fingers:** This kind of grippers usually has three or four identical fingers installed equidistantly on the circumference of a circle and appropriate for fingertip-grasping and in-hand manipulation (Figure 6b).

b. **Multi-Fingered Hand with Thumb:** This kind of grippers has a row of two, three or four fingers and a finger (thumb) located against them that operate like a human hand for grasping objects. The gripper can do both envelope- and fingertip-grasping (Figure 6c).

In Table 1 we provide a comparison featuring the characteristics of the prior four main classes of robotic grippers. After having introduced the major types of robotic grippers, principles of grasping are presented in the next section. In the rest of the chapter we will consider multi-fingered robotic hands as the means for grasping and re-grasping objects.

GRASPING

In grasping of objects, whether the object is moving or fixed relative to an external observer coordinate system, it is important to maintain equilibrium of forces at different positions. This means that in envelope grasping it is intended to make the relative movements of the object and the hand equal to zero, and in fingertip grasping, the total forces and moments on the center of mass of the object must be zero as well. Grasping discussions always come with Fixturing and Immobility issues, so in this section the concept of immobility and its principles are mentioned before going through grasping. We start with two-dimensional immobility and then discuss three-dimensional cases, which are more complex. In two-dimensional immobility, the object must stay stable under maximal external forces exerted on it.

Immobility

The principles of immobility can be summarized in three inquiries: (1) Does the arrangement of the fingers immobilize the object? (2) How many fingers are necessary or sufficient for immobiliz-

Figure 6. (a) Adaptive and soft gripper made from coffee ground and rubber bag (Brown, et al., 2013); (b) multi-fingered hand with parallel fingers (Hasegawa & Morooka Laboratory, 2012); (c) Utah/MIT hand (Computer History Museum, 2012)

Table 1. Comparison between different features of robotic grippers

Gripper Type	Grasping in Form-Closure	Ability of ...				Computational Complexity of Grasp Planning
		Fine-tuning the grasp span	Gripping delicate objects	Regrasping	In-hand manipulation	
Mechanical	No	Only for screw-driven grippers	No	Yes (only through an intermediate state)	No	Low (except for objects with complex shapes)
Vacuum and Magnetic	No	N/A	Vacuum Grippers (only flat surfaces)	No	No	Very Low
Universal	Yes	No	Yes (except for three-fingered grippers)	Yes (only through an intermediate state)	No	Low
Multi-fingered	Yes	Yes	Yes	Yes	Yes	High

ing an object? (3)Where should be the positions of fingers for immobilizing the object?

For answering the previous questions there are two analytic approaches: based on velocities and movements, and based on forces and moments. Consequently, two kinds of immobility analyses are proposed: (a) The *Form-closure analysis*, that analyzes instant velocities, and (b) The *Force-closure analysis*, that analyzes the forces and moments. Creating form-closure means to limit infinitesimal movements (displacements) of an object through contacts. A direct method for computing proper contacts is to analyze the object in its own 2D configuration, since the geometric nature of such an approach allows it to be converted from the problem of finding a form-closure configuration into a problem of geometric search (it is noted that the focus is on rigid bodies). By displacement it is meant a mere change in configuration of the object without

changing the distance between any pair of points on it and the (left- or right-) handedness of the shape. For instance, in Figure 7, parts (b) and (c) are not displacements.

Rotation and Translation are two kinds of displacements: Rotation is a displacement that has at least one fixed point, which can be internal or external. In Translation all points move equally along parallel lines. Figure 8 depicts some examples of Rotation and Translation. Any general displacement can be expressed as a combination of translations and rotations (around point O) in $\Re^d$ space.

For every object an attached coordinates frame (system) can be considered that rotates and translates with it. Then the position and orientation of the object can be defined by the position and orientation of that coordinates frame relative to another coordinates frame. For example in Figure 9(a) the position and orientation of the object can

Figure 7. (a) Rotation about an axis perpendicular to the plane; (b) changing size; (c) rotation about an axis lying on the plane (flip)

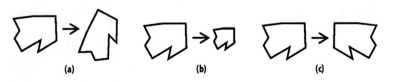

Figure 8. (a) Rotation about a point on boundary; (b) rotation about an internal point; (c) rotation about an external point; (d) translation

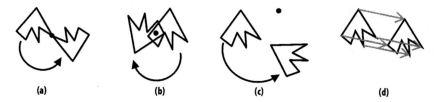

be determined by finding the position and orientation of frame O' relative to frame O. In $\mathfrak{R}^2$ space, a translation can be considered as a rotation around an arbitrary point in infinity. As shown in Figures 9(b) and (c), by moving the center of rotation C to infinity, the rotation converts into a translation. Accordingly, it is possible to completely represent the velocity and direction of a motion by centers of rotation and plus and minus signs of rotation. These centers are also called 'velocity centers'.

In presence of only one contact point (Figure 10a) and by drawing a line passing through the contact point and perpendicular to the contacted edge, on the right hand side of the line only negative centers of rotation, and on the left hand side only positive centers of rotation exist. The minus sign is used for clockwise (negative) rotations and the plus sign for counterclockwise (positive) rotations. Consider a polygonal object that is being held with four frictionless spot fingers that try to limit its movement. In Figure 10(b) the fingers limit movements of the object at different directions, and so the solution to the immobility problem is to remove negative and positive centers of rotation from the whole space (van der Stappen, Wentink, & Overmars, 1999; Czyzowicz, Stojmenovic, & Urrutia, 1999).

In grasping a polygonal object, the constraints imposed on it by contact points must prevent any movement. For this purpose, all clockwise and counterclockwise velocity centers should be removed. A simple way to do that is to draw an inward vector at any contact point and normal to the contacted edge, and then to mark left half-

spaces of all vectors. If the whole space becomes marked then there is no clockwise center of rotation (Figure 11). This procedure is repeated for

Figure 9. (a) The coordinates frame O is fixed and the frame O' moves relative to it; (b) a simple rotation; (c) a rotation about a point in infinity

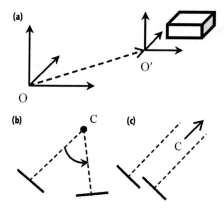

Figure 10. (a) Centers of rotation in presence of one contact point; (b) an immobilized polygonal part held with spot fingers

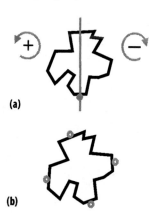

counterclockwise centers of rotation, which results in marking right half-spaces of all normal vectors (van der Stappen, Wentink, & Overmars, 2000). The grasping will be form-closure if there are no clockwise and counterclockwise centers of rotation.

Four fingers are necessary and sufficient for a form-closure grasp in the plane. Sometimes, however, the sufficient condition precedes the necessary condition, as in Figure 12(a) where grasping of a triangular part in form-closure is possible by just three fingers. In this case two fingers are superimposed, and the four-finger condition is sufficient but not necessary (Overmars, Rao, Schwarzkopf, & Wentink, 1995). Such cases occur when polygons do not have parallel edges. Therefore, many 2D polygonal parts can be immobilized in form-closure if the extended contacted edges form a triangle, as shown in Figure 12(b) (Czyzowicz et al., 1999). For immobilizing a rectangular part (and any arbitrary polygonal shape that has some parallel edges) as in Figure 12(c) four fingers are necessary (van der Stappen et al., 2000). In summary, using four fingers is a necessary condition for immobilizing an arbitrary polygonal object in 2D, unless the fingers are not placed on parallel edges and the extended touched edges form a triangle.

After introducing the principles of immobility, in the following the form-closure grasp is discussed in more detail.

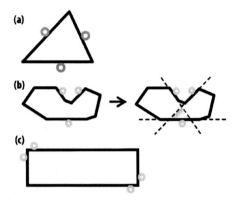

Figure 12. (a) Three-finger grasping; (b) grasping a concave part with three fingers; (c) immobilizing a rectangle (more generally, a parallelogram)

(a)

(b)

(c)

Form-Closure and Force-Closure

As stated previously, a form-closure grasp limits infinitesimal displacements of an object through contacts, by not necessarily exerting forces on it. The form-closure problem can be expressed in mathematical form such that the exerted forces and moments on an object in the equilibrium state are equal to zero:

$$\sum_{i=1}^{l} a_i\, \hat{n}(r_i) = 0 \tag{1}$$

$$\sum_{i=1}^{l} \vec{r}_i \times a_i\, \hat{n}(r_i) = 0 \tag{2}$$

Figure 11. Removing clockwise centers of rotation

in which l is the number of contacted fingers, $\vec{r}_i$ is a point on the boundary of the object being contacted with the i-th finger, $\hat{n}(r_i)$ is the surface inward normal in $\vec{r}_i$, and a_i is the size of the perpendicular force on the object at the i-th contact. Equation (1) stipulates equilibrium of forces and Equation (2) the equilibrium of moments. Now if external forces or moments are exerted on the object, the equilibrium state equations become:

$$F - \sum_{i=1}^{l} a_i \,\hat{n}(r_i) = 0$$

$$M - \sum_{i=1}^{l} \vec{r}_i \times a_i \,\hat{n}(r_i) = 0 \qquad (3)$$

where F is the external force and M is the external moment (Zuo & Sun, 1991). Therefore, we can equilibrate and immobilize objects with forces and moments of fingers and external forces and moments.

Another way is to decompose the force exerted on the object into its tangent and perpendicular components. If $\vec{f}_i$ is the force vector applied by the i-th finger, it can be decomposed as follows:

$$\vec{f}_i = \vec{f}_{ni} + \vec{f}_{ti} = a_i\hat{n}_i + d_i\hat{t}_i \qquad i = 1,2,3 \qquad (4)$$

in which $\hat{n}_i$ and $\hat{t}_i$ are respectively the normal and tangent unit vectors of the force exerted by the i-th contact point(as shown in Figure 13a), and d_i and a_i are the tangent and normal force components, respectively. Since $\hat{n}$ is perpendicular to $\hat{t}$, we have $\hat{n}_i = [n_{ix} + n_{iy}]^T$ and $\hat{t}_i = [n_{iy} - n_{ix}]^T$. Also, the squeeze and friction constraints are:

$$a_i > 0 \quad \text{and} \quad |d_i| < \mu a_i \,, \qquad i = 1,2,3 \qquad (5)$$

In (5), the first inequality is a prerequisite for satisfying the second inequality, so it follows that $|a_i| = a_i$.

Figure 13. Grasping by (a) nonparallel and (b) parallel contact normals

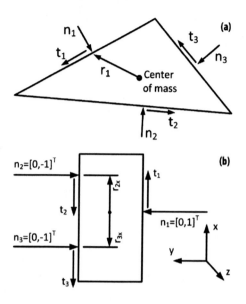

When grasping a planar object by three fingers, let F' and M' be the external force and moment at the object's center of mass on the plane. Then the relation between the force and moment at the object's center of mass (F'_x, F'_y, M'_z) with normal and tangent components of the contact forces can be shown as, $[F'_x, F'_y, M'_z]^T = [W][d_1, d_2, d_3, a_1, a_2, a_3]^T$, in which

$$[W] = \left[[W_t], [W_n]\right] = \begin{bmatrix} \hat{t}_1 & \hat{t}_2 & \hat{t}_3 & \hat{n}_1 & \hat{n}_2 & \hat{n}_3 \\ m_{t_1} & m_{t_2} & m_{t_3} & m_{n_1} & m_{n_2} & m_{n_3} \end{bmatrix}$$

$$= \left[w_1, w_2, w_3, w_4, w_5, w_6\right] \qquad (6)$$

$$m_{ti} = \vec{r}_i \otimes \hat{t}_i \quad \text{and} \quad m_{ni} = \vec{r}_i \otimes \hat{n}_i \,, \quad i = 1,2,3$$

in which r_i is the distance of the i-th contact to the center of mass, m_{ti} and m_{ni} are the moments exerted on the center of mass perpendicular to the tangent and normal unit vectors at the i-th contact (Figure 13a). The operator $\otimes$ is the cross product in 2D; for example, $\vec{r}_i \otimes \hat{n}_i = r_{ix} n_{iy} - r_{iy} n_{ix}$. The $[W_t]_{3\times3}$ and $[W_n]_{3\times3}$ are the tangent and normal contact matrices, respectively, and $[W]_{3\times6}$ is called a *Wrench*.

In order to create the sufficient condition for form closure grasp, the inequality (7) must be satisfied, which is based on the friction constraint $|d_{hi}| \le \mu a_{hi}$ related to the contacts between the object and the fingers:

$$|(w_3 \times w_2) \cdot (k_1 w_4 + k_2 w_5 + k_3 w_6)| \le \mu k_1 (w_1 \times w_2) \cdot w_3$$
$$|(w_1 \times w_3) \cdot (k_1 w_4 + k_2 w_5 + k_3 w_6)| \le \mu k_2 (w_1 \times w_2) \cdot w_3 \quad (7)$$
$$|(w_2 \times w_1) \cdot (k_1 w_4 + k_2 w_5 + k_3 w_6)| \le \mu k_3 (w_1 \times w_2) \cdot w_3$$

Apart from the obvious solution $k_1 = k_2 = k_3 = 0$, satisfying the inequalities (7) and the conditions $a_i > 0$ yields a possible region for nonzero k_1, k_2, and k_3. If the possible region is null, then the fingers in the contact points cannot hold the object firmly, and so the contact configuration is not form-closure (Chen & Trinkle, 1993). For $i = 1, 2, 3$, we have

$$k_i > \max\left\{\left(|d_{pi}| - \mu a_{pi}\right) / \delta_i\right\}, \delta_i = \mu a_{hi} - |d_{hi}| > 0.$$

A special case is when the contacted faces are mutually parallel. Because there are three contacts points, one of the internal contact normals should be in the opposite direction of the others (Figure 13b). The coordinate system is attached to the center of mass and the direction of its x and y axes are along the tangent and normal vectors, respectively. In Figure 13(b), the contact forces are able to produce any force at the y direction and any moment perpendicular to the object plane.

Grasping in 3D

For grasping non-planar objects we should decompose the forces differently. The forces and moments (F_A, M_A) and (F_B, M_B) exerted on points A and B in the object, respectively, will have the same effect if $F_A = F_B$. Then the moments will be related as:

$$M_B = M_A + r_{AB} \times F_A \quad (8)$$

in which r_{AB} is the distance between A and B. So if a set of forces create a pure moment in the point A (i.e., no force), based on (8), all points in the object will experience the following moment:

$$F_A = 0 \quad \Rightarrow \quad M_B = M_A \quad (9)$$

This principle can be used for grasping planar objects. Grasping a curved planar object with three contact points is equivalent to grasping a triangular planar object with edges tangent to the curved object at contact points. In the same way, grasping a 3D curved object is like grasping a tetrahedron with faces tangent to contact points.

Now we will analyze contact forces in grasping 3D objects. Any contact force can be decomposed along normal unit vectors $(\hat{n}_i)$ and tangent unit vectors $(\hat{t}_i, \hat{o}_i)$:

$$\vec{f}_i = \vec{f}_{ni} + \vec{f}_{ti} = a_i \hat{n}_i + d_i \hat{t}_i + c_i \hat{o}_i \quad (10)$$

in which $\hat{t}_i$ and $\hat{o}_i$ are orthonormal unit vectors defining the contact plane. The relation between the desired force and moment at the center of mass and the tangent and normal forces is:

$$\begin{bmatrix} F' \\ M' \end{bmatrix}_{6 \times 1} = \begin{bmatrix} [W_t]_{6 \times l} & [W_o]_{6 \times l} & [W_n]_{6 \times l} \end{bmatrix} \begin{bmatrix} [D] \\ [C] \\ [A] \end{bmatrix}_{3l \times 1} \quad (11)$$

in which l is the number of contact points, $[W_n]$ is the matrix of normal contact forces, and $[W_o]$ and $[W_t]$ are the matrices of tangent contact forces. In other words, $[W_t]$, $[W_o]$, and $[W_n]$ are the forces and moments (wrenches) produced by $\hat{t}_i$, $\hat{o}_i$ and $\hat{n}_i$, respectively, defined by:

$$[W_{ti}] = \begin{bmatrix} \hat{t}_i \\ \vec{r}_i \times \hat{t}_i \end{bmatrix}_{6 \times 1}, \quad [W_{oi}] = \begin{bmatrix} \hat{o}_i \\ \vec{r}_i \times \hat{o}'_i \end{bmatrix}_{6 \times 1},$$

$$[W_{ni}] = \begin{bmatrix} \hat{n}_i \\ \vec{r}_i \times \hat{n}_i \end{bmatrix}_{6 \times 1}, \quad [D] = [d_1, d_2, d_3, \cdots, d_l], \quad (12)$$

$$[C] = [c_1, c_2, c_3, \cdots, c_l], [A] = [a_1, a_2, a_3, \cdots, a_l].$$

The vector $\vec{r}_i$ denotes the distance of i-th contact point to the center of mass. In order to maintain contact between the hand and the object, the squeeze and friction constrains should be satisfied, expressed as a relation between normal and tangent forces for each contact point (finger):

$$a_i > 0, \quad \sqrt{c_i^2 + d_i^2} \le \mu a_i \quad \text{for} \quad i = 1, 2, 3, \cdots, l. \quad (13)$$

A proper and computationally efficient alternative for (13) is:

$$a_i > 0, \quad \left| c_i^2 \right| + \left| d_i^2 \right| \le \mu a_i \quad \text{for} \quad i = 1, 2, 3, \cdots, l. \quad (14)$$

Using Equation (11) the general solution for the values of contact forces is:

$$
\begin{aligned}
&\left[[D]^T_{l \times 1}, \ [C]^T_{l \times 1}, \ [A]^T_{l \times 1} \right] \\
&= \begin{bmatrix} [d_{p1} & d_{p2} & \cdots & d_{pl}], \\ [c_{p1} & c_{p2} & \cdots & c_{pl}], \\ [a_{p1} & a_{p2} & \cdots & a_{pl}] \end{bmatrix} \\
&+ k \begin{bmatrix} [d_{h1} & d_{h2} & \cdots & d_{hl}], \\ [c_{h1} & c_{h2} & \cdots & c_{hl}], \\ [a_{h1} & a_{h2} & \cdots & a_{hl}] \end{bmatrix}
\end{aligned} \quad (15)
$$

respectively. A particular solution may be obtained by setting $3l - 6$ elements of d_i, c_i, and a_i to zero and solving the remaining 6 elements which is equivalent to a column of $[W]$. The answer is linearly independent. The form-closure is related to the second part of Equation (15). We should define the values of d_{hi}, c_{hi}, and a_{hi} for satisfying the squeeze and friction constraints.

If the forces are centralized and $k = 1$ and $l = 4$, the moment produced by normal contact forces will be:

$$
\begin{aligned}
m_n &= \sum_{i=1}^{4} m'_{ni} = (\hat{n}_3 \times \hat{n}_2 \cdot \hat{n}_4)[r_1 \times \hat{n}_1] \\
&+ (\hat{n}_1 \times \hat{n}_3 \cdot \hat{n}_4)[r_2 \times \hat{n}_2] + (\hat{n}_2 \times \hat{n}_1 \cdot \hat{n}_4)[r_3 \times \hat{n}_3] \\
&+ (\hat{n}_1 \times \hat{n}_2 \cdot \hat{n}_3)[r_4 \times \hat{n}_4]
\end{aligned}
$$

(16)

As said before, grasping a 3D curved object is like grasping a tetrahedron with faces tangent to contact points. Let's consider a regular tetrahedron shown in Figure 14, which its geometrical features can be obtained from the equations of its constituting planes. Four special sets of contact forces should be investigated such that each set produces just a moment on the center of mass, and not a linear force. For each set of contact forces, $ka_{hi} > 0$ should be satisfied, and $\mu |a_{hi}| > |d_{hi}| + |c_{hi}|$ for $i = 1, 2, ..., l$ and $k \in \mathfrak{R}$.

These four different sets of forces produce distinct pairs of tangent and normal forces which can be computed by geometrical properties of the tetrahedron and equations of the sets of forces. As a result, four sets of equations like those in (7) are formed, which yield values for k_1, k_2, k_3, and k_4 (Yu-che and Walker, 1994).

We conclude this subsection with a comparison between the geometric method of creating immobility and form- and force-closed grasping. In the

Figure 14. A regular tetrahedron

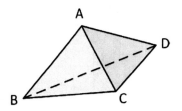

geometric method there is no need to exert forces by the fingers and it is sufficient to determine the geometry of the object and the contact points. Thus, following a simple calculation, it is possible to determine whether the centers of rotation are eliminated from the whole space or not. Also, this method can be easily generalized to 3D. On the other hand, in form- and force-closed grasping, in addition to the object's geometry and contact points, the force at each finger is also needed, which means more information and calculations as well. But since the fingers normally do not stay fixed on the object and external moments and forces like gravity influence the object, we must ensure that the object can withstand the exerted forces and moments, and hence the form- and force-closure method is more efficient, being extendable to 3D as well. Nevertheless, the geometric method can be used for preprocessing purposes considering its low computational burden.

First and Second Order Analyses

In this section we investigate the movement of object β that is held at l contact points by $A_1, A_2, ..., A_l$ frictionless fingers, and analyze it in the configuration space of β, as depicted in Figure 15. Points are denoted by r and x in the space $\Re$ and physical space ε, respectively.

The form-closure and force-closure concepts are usually used for multi-finger grasping, fixture planning and contacting object kinematics. Un-

Figure 15. Obstacle configuration space

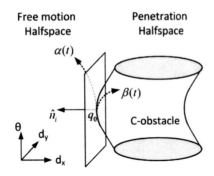

fortunately, the distinction between the two terms is somewhat vague. Their main difference lies in the way of modeling the contacts between the object and fingers: in form-closure, the contact between β and A_i is modeled as a constraint on the movement of β. From the configuration space point of view, a grasping is form-closure or immobilized if the configuration of β is completely bounded by the configurations of fingers. The concept of i-th mobility and i-th mobility index provide a precise definition for the form-closure.

Let β be in contact with k fingers in configuration q_0 (Figure 15). Then, β is in the 1st form-closure if it is in the 1st immobility by fingers; that is, if the free 1st half-spaces $M_i^1(q_0)$ ($\forall i = 1, 2, ..., l$) contain only the origin and $\dot{q} = 0$ in the space $T_{q_0} \times \Re^m$, in which

$$M_i^1(q_0) = \{\dot{q} \in T_{q_0} \times \Re^m : \hat{n}_i(q_0) \cdot \dot{q} \geq 0\}$$ and $\hat{n}_i(q_0)$ shows the outward pointing unit normal, and $T_{q_0} \times \Re^m$ describes the dimension of the configuration space (one rotational and m translational dimensions).

Moreover, β is in the 2nd form-closure if it is in the 2nd immobility; that is, if the free 2nd motion sets $M_i^2(q_0)$ ($\forall i = 1, 2, ..., l$) contain only zero velocity and acceleration vectors, in which

$$M_i^2(q_0) = \{(\dot{q}, \ddot{q}) | \hat{n}_i(q_0) \cdot \dot{q} = 0, \wedge \dot{q}^T[D\hat{n}_i(q_0)]\dot{q} + \hat{n}_i(q_0) \cdot \ddot{q} \geq 0\}$$ and $D\hat{n}_i(q_0) = D^2 d_i(q_0)$ formulates the curvature of obstacle configuration space surface at q_0.

If β is under equilibrium grasped by l frictionless fingers and each finger exerts nonzero forces, then the grasp is in 1st force-closure if and only if it is 1st form-closure. If β is under equilibrium grasped by l frictionless fingers and all fingers are essential to maintain the grasp, then the grasp is in 2nd force-closure if and only if it is 2nd form-closure. A proof for the prior is provided in (Rimon & Burdick, 1996). Figure 16 shows the hierarchy of form closure grasps.

Figure 16. Hierarchy of form-closure grasps

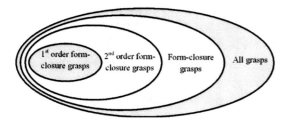

The grasps shown in Figures 17(a) and 17(b) are respectively 1st and 2nd form-closure grasps, while 17(c) is not form-closure (Krut, Bégoc, Dombre, & Pierrot, 2010).

We can conclude that by checking whether the origin of the operational space is in the convex hull of contact normal vectors (into the object), the 1st order form-closure analysis can be done (Krut et al., 2010). If we call this convex hull *P*, then:

1. A grasp is not form-closure if and only if the origin of the operational space is completely out of the convex hull *P*.
2. A grasp is 1st form-closure if the origin of operational space is completely in the convex hull *P*.
3. A grasp is unknown form-closure if the origin of the operational space is on the boundary of the convex hull *P*.

Since in the next section the grasp planning will be discussed, it is important to determine Nguyen Regions which guide the placement of fingers. This, in turn, is based on the *Coulomb friction* and *friction cone* concepts, which are explained next.

Coulomb Friction

Under the Coulomb friction, a contact force is bounded if it lies within a cone with its apex at the contact point and centered at the internal normal vector by half angle α. The tangent of the angle α is the friction coefficient. As shown in Figure 18 the *friction cone* at C_1 is bounded by vectors n_{11} and n_{12}, and any force f_1 is a non-negative combination of these two vectors. A *wrench* is defined as the combination of a force f and a moment m in the form of $w = [f, m]^T \in \Re^k$.

In planar mechanics $k = 3$ and in solid mechanics $k = 6$. A set of n wrenches $w_1, ..., w_n$ is said to be in *equilibrium* when the convex hull of points $w_1, ..., w_n$ in $\Re^k$ includes the origin.

A grasp is force-closure when it can resist arbitrary forces and moments. Force-closure implies an equilibrium but there can be an equilibrium wrench system that is not force-closure, as shown in Figure 19(a). Non-marginal grasping of three fingers is also force-closure (Li, Liu, & Cai, 2003).

The previous discussion is about a stable state in which an object is being contacted by fingers. In this case, the stability can be analyzed considering the object's geometric or physical properties. Now for creating stable grasps, based on geometric analysis of immobility and mathematic analy-

Figure 17. Fingers in (a) prevent rotation, and in (b) and (c) do not prevent rotation. In (b) the object will stop after infinitesimal rotations, but in (c) it will escape freely

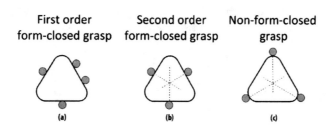

Figure 18. Coulomb friction

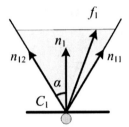

sis of form-closure and force-closure, as well as the concept of Coulomb friction, we can find a way of grasping through which immobility of the object is guaranteed. This is described in the next section.

Grasp Planning

After identifying states of stable grasps through immobility analysis, we should find a way to attain this state regarding the object's physical and geometrical properties. Due to the variety in final stable grasp configurations, and delays and

uncertainties in reaching stability, planning for a stable grasp is an optimization problem, which has been investigated in a number of researches and for which some algorithms have been developed. However, before dealing with grasp planning methods, it is essential to be familiar with Nguyen Regions.

Generally, a desired grasp is the one that needs minimum precision, which means that the fingers can independently locate on large regions of the object's surface rather than on single points. These regions are called *independent contact regions*. Figure 20 shows various grasp configurations of a multi-fingered robotic hand that performs alternative stable grasps of an object by selecting different sets of contact points inside the independent contact regions.

If an object with almost parallel edges is grasped by two fingers, the line segment connecting the two contact points must lie inside the two friction cones generated by the contact points (as in Figure 21a) (Nguyen, 1986). Here the tangent of angle φ is equal to the coefficient of friction.

Figure 19. Three finger grasping: (a) equilibrium grasp without force-closure; (b) non-marginal equilibrium and hence force closure grasp; (c) non-equilibrium grasp

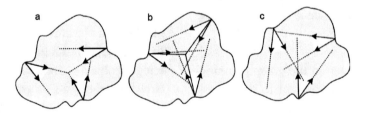

Figure 20. Four independent contact regions and various stable grasps (Krug, Dimitar, Krzysztof, & Boyko, 2010)

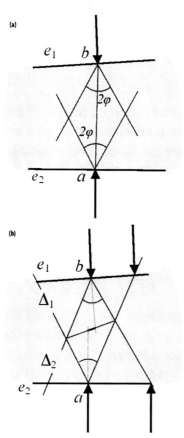

Figure 21. (a) Grasping by fingers with friction; (b) intersection of friction cones

A well-known algorithm has been developed by V.-D. Nguyen in 1986 for determining the set of optimal grasps (independent contact regions) on the perimeter of a planar object with almost parallel edges e_1 and e_2 grasped by two fingers (Nguyen, 1986):

1.   Find the friction cone that intersects the whole edge e_1 and a small portion or a single point of the edge e_2. In Figure 21(b), the cone includes the edge e_1 and the vertex a, and is denoted by the triangle Δ_1. So is created the triangle Δ_2 that includes the edge e_2 and vertex b.
2.   Find the "trade-off region" by intersecting the two triangles Δ_1 and Δ_2. The edges of the cones intersect at two points, and the

line-segment connecting these two points is called the tread-off region (the dotted line in Figure 21b). Note that in 3D the trade-off region will be ellipse-shaped.

3.   The bisectors of e_1 and e_2 intersect the tread-off region and identify a sub-segment on it. Depending on the direction of the cones formed by the two edges, a point or one of the endpoints of this sub-segment is the optimal vertex, which its projection on the two edges e_1 and e_2 give the best contact points. If no intersection exists, then the optimal vertex is a point on the trade-off region that is nearest to the bisector.

If the trade-off region is projected on the left and right sides of the contact points a and b on the edges e_1 and e_2, then the obtained intervals denote the independent contact regions. For instance, the intervals between the two arrows on edges in Figure 21(b) show right projections of the trade-off region on the edges. For any pair of contact points within the regions on e_1 and e_2, their connecting line remains in both of the two friction cones and so a stable grasp is secured.

For finding the optimal grasp set for three or four edges, a double-side cone can be fitted between two parallelograms and the procedure repeated for every two parallel edges. The Nguyen's method has been generalized for grasping of 3D objects in (Krug et al., 2010).

Measures of Grasp Quality

In order to optimize a certain grasp, different measures of grasp quality have been developed and used as fitness functions of grasp planning optimization problems. Many of the measures relate to assuring stability of the grasp. After reviewing a large a number of relevant works, we have identified and categorized some important measures of grasp quality as presented next. Some representative references where the measures have been implemented are also indicated.

- **System Vibration Measure:** During grasping an object the system undergoes a vibration that attenuates with a time constant. The less is the duration of vibration, the faster the stability will be achieved. So the grasping matrix (explained in Definitions 1-4) should be chosen in a way that the time constant is minimized and the vibration disappears very fast in the system (Guo, Gruver, & Jin, 1992).

- **Contact Precision Measure:** Due to limitations and imprecision in finger joints and their control, there are always differences between the real grasp contact points and the desired contact points calculated by analytical methods. Minimizing these differences can help optimizing the grasp (Zhang & Gu, 2012).

- **Contact Triangle Measure:** This measure pertains to grasping by three fingers and refers to the 'contact' triangle formed by the three contact points. Two sub-measures can be defined in this class: (1) by maximizing the area of the contact triangle the dynamic stability can be maximized, and (2) by minimizing the distance between the triangle's centroid and the object's center of mass, the object's reaction against gravity forces and moments can be optimized (Prado & Suarez, 2005).

- **Equable Forces Measure:** In the absence of external turbulences, magnitudes of forces exerted by the fingertips should be as equal as possible. That will extend the possibility of modifying fingertip forces when trying to maintain the grasp stability in the presence of external turbulences (Prado & Suarez, 2005).

- **Robust Wrenches Measure:** In a stable grasp, in the absence of external wrenches, the wrenches exerted by the fingertips are in different directions and are in balance. If these wrenches are plotted in $\mathfrak{R}^6$ space (re-

fer to (6)) from the origin of the coordinate system, a convex hull can be constructed for the endpoints of the wrenches. In order to maximally withstand external wrenches, the origin must lie inside that convex hull (by the way, this fact is being used as a force-closure test). Maximizing the distance of the origin from the inner boundaries of the convex hull leads to the most robust and stable grasp that can tolerate an external wrench with the worst direction (Phoka, Niparnan, & Sudsang, 2006).

The characteristics of the previous measures of grasp quality are summarized in Table 2, by which it is possible to determine proper measures for optimizing grasps by a specific end-effector.

Pre-Grasping

For a more stable grasp, a preliminary step can be considered at the beginning of the grasping operation in which the hand prepares for grasping. This can be done in various methods, two of which, namely, eigengrasp and pre-grasp measure methods, are described here.

In the eigengrasp (a term equivalent to "hand synergies" in human hand) method, the grasp space is constructed based on eigengrasps. For a hand with d degrees of freedom, an eigengrasp is a d-dimensional vector in the d-dimensional grasp space. A linear combination of the eigengrasps can define a region of hand postures that are suitable for performing a grasp. Using the eigengrasps, proper pre-shaping and a preliminary path can be calculated for the hand. The method has two stages. First, like human hand, the robotic hand takes an appropriate form prior to real grasping, and then approaches the object along the preliminary path till the fingers contact the object and the grasping is accomplished. Using eigengrasps reduces the dimension of the search space but for a good grasp sufficient information is needed (Ciocarlie, 2010).

Table 2. Comparison of measures of grasp quality

Measure	Capability of direct calculation	Appropriate end-effector	Number of fingers	Complexity of detecting source of errors
System Vibration	Indirect, only through solving the system equation	Mechanical grippers, Triple-jaws, Multi-fingered hands	unlimited	High
Contact Precision	Direct, through joint angles and the object's position relative to the palm	Multi-fingered hands	unlimited	Medium
Contact Triangle	Direct, by calculating the positions of fingertips relative to an origin	Multi-fingered Hands	three	Medium
Equable Forces	Direct, by using sensors at fingertips	Multi-fingered hands	three	Low
Robust Wrenches	Direct, by measuring fingertip forces and their distance to the center of mass	Mechanical grippers, Triple-jaws, Multi-fingered hands	unlimited	Medium to high

In the pre-grasp measure method, some measures of pre-grasp quality are used as the fitness function of an optimization problem, which is best solved by metaheuristic methods for the following reasons:

- The hand's workspace is too large to be searched effectively and efficiently.
- The configuration space is generally discontinuous due to obstacles, kinematic constraints, or unfavorable hand postures and positions.
- The measures of grasp quality are often nonlinear.

In this method the most important issue is defining a proper measure. For example, the sum of distances between the desired and actual positions of fingertips (i.e., $Q = \sum_{\text{all contacts}} \delta_i$) can be used as a fitness function. This optimization problem has been solved using the Genetic Algorithms by representing a hand posture as a chromosome (Zhang & Gu, 2012).

The next two subsections describe two major methods for calculating (planning) a stable grasp; namely, the Stiffness Matrix and Friction Cones methods. The first one is based on forces, and the second one is based on the geometry of friction cones.

Calculating Stable Grasps using Stiffness Matrix

In order for an object to be grasped by a multi-fingered hand and safely carried by a robotic arm, two procedures must be carried out:

- Detection of a stable and appropriate grasp.
- Cooperative operation of the hand and the arm for manipulating the object.

The procedure of detecting stable grasps is called *grasp planning*. It is comprised of selecting an optimal grasp posture G by which the grasp is tangentially stable and the system vibrations are minimized. During the selection of the optimal grasp posture matrix, constraints such as positions and forces at contact points should be investigated (G. Guo et al., 1992). Before continuing, we define some essential concepts:

Definition 1: The Grasp Postures Matrix G (or contact points distribution matrix) is:

$$G = [\vec{p}_1, \vec{p}_2, ..., \vec{p}_l] \in \Re^{3l} \qquad (17)$$

where l is the number of fingers and vector p is the contact points positions on the object relative to the center of mass, and $\vec{p}_i = [p_{ix}, p_{iy}, p_{iz}]^T \in \Re^3$.

Definition 2: The Grasping Stiffness Matrix K of a grasping system is:

$$K = -\frac{\partial Q}{\partial \delta x} \in \Re^{6 \times 6}, \tag{18}$$

in which $Q = J^T F$, $F = [f_1^T, f_2^T, ..., f_l^T]^T \in \Re^{3l}$ and $\delta \vec{x} = [\delta \vec{r}^T, \delta \vec{\varphi}^T]^T \in \Re^6$, and $\vec{f}_i$ is the force vector exerted on the object by the i-th fingertip. Small linear and angular displacements of the center of mass are denoted by $\delta \vec{r}$ and $\delta \vec{\varphi}$ respectively, and

$$J^T = \begin{bmatrix} I_3 & \cdots & I_3 \\ -P_1^T & \cdots & -P_l^T \end{bmatrix}, \quad P_i = \begin{bmatrix} 0 & -p_{iz} & p_{iy} \\ p_{iz} & 0 & -p_{ix} \\ -p_{iy} & p_{ix} & 0 \end{bmatrix} \tag{19}$$

where I_3 is the Identity matrix.

Definition 3: The Contact Stiffness Matrix K_C for a grasping system is:

$$K_C = -\frac{\partial F}{\partial \delta c} \in \Re^{3l \times 3l}, \tag{20}$$

in which $\delta c = [\delta \vec{c}_1^T, \delta \vec{c}_2^T, ..., \delta \vec{c}_l^T]^T \in \Re^{3l}$ and $\delta \vec{c}_i = \delta \vec{r} - \vec{p}_i \times \delta \vec{\varphi}$. Other symbols are defined as before.

Definition 4: A grasp is *stable* if and only if the Grasping Stiffness Matrix K is positive, which is guaranteed when K_C is positive and the matrix J is full rank.

For example, for the stability of a system of a convex object and the number of fingers $l \geq 3$ and rank of J is 6, the restrictive condition is that $K_C > 0$. Since the stiffness between the fingers is independent, it should be defined as $\min(K) > 0$.

If the object is defined as $S(p_{ix}, p_{iy}, p_{iz}) = 0$, then $S(p_{ix}, p_{iy}, p_{iz}) > 0$ indicates inside of the object. In the contact point p_i the normal unit vector will be

$$e_{Ni} = \frac{\nabla S(p_i)}{|\nabla S(p_i)|} \tag{21}$$

Since all contact points should lie on the object's surface, the surface equation of the object can be written as the constraint

$$S(p_i) = 0 \quad i = 1, 2, ..., l \tag{22}$$

In order for a contact to be stable, the following conditions should be satisfied:

1. The normal force at each contact point should be the surface inward. The uniaxial constraint (i.e., effective on only one axis) for a grasping system is $NF \geq 0$, where N is the unit normal matrix defined in (23) and F is the matrix of forces.

$$N = \begin{bmatrix} e_{N1}^T & 0 & \cdots & 0 \\ 0 & e_{N2}^T & & \\ \vdots & & \ddots & \\ 0 & & & e_{Nl}^T \end{bmatrix} \in \Re^{l \times 3l} \tag{23}$$

2. The forces of fingertips at the contact points should lie in the friction cone. The stable friction constraint is $NF \geq F_\mu$, in which μ is the coefficient of static friction, and

$$F_\mu = \left[\frac{|f_1|}{\sqrt{1+\mu_1^2}} \quad \frac{|f_2|}{\sqrt{1+\mu_2^2}} \quad \cdots \quad \frac{|f_i|}{\sqrt{1+\mu_i^2}} \right]^T \in \Re^l$$

(24)

The first condition is latent in the second condition.

Another issue during the grasping is the vibration of the system, which should attenuate as fast as possible for a stable grasp (Guo et al., 1992).

Calculating Stable Grasps using Friction Cones

Regarding the properties of friction cones described earlier, they can be used for calculating stable grasps. Consider a three-finger grasp with the contact points C_i, C_j, and C_k (Figure 22). Initially, the positions of the three friction cones must be determined (e.g. the friction cone of C_i is limited by the vectors n_{i1} and n_{i2}). In order to find the equilibrium of moments of contact forces f_i and f_j the following equation must be satisfied:

$$\overrightarrow{C_k C_i} \times f_i + \overrightarrow{C_k C_j} \times f_j = 0$$

(25)

Next, through a simple preprocessing method (called H disposition in (Li et al., 2003)), regions

Figure 22. Equilibrium of moments of contact forces f_i and f_j

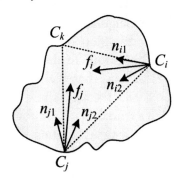

that are definitely outside the common subspace of the three friction cones are excluded to reduce future computations. A necessary and sufficient condition for three-finger equilibrium grasp is that the intersection of the three friction cones should not be empty, which is verified as follows:

It is easy to prove that the maximum number of points created by mutual intersections of boundary edges of three double-side friction cones is 12. The intersection of three double-side friction cones is not empty if and only if any intersection point of two boundary edges of two cones is not outside of the third cone. So, we need to calculate at most 12 intersection points and determine which one of them lies on or inside a certain double-side friction cone. If such a point exists, then the grasp is in equilibrium. If we name this intersecting point Bjk, then every point in its infinitesimal neighborhood will also lie in the three friction cones, which means that Equation (26) is satisfied. Such an equilibrium is non-marginal (Figure 19b) (Li et al., 2003). The previous fact is also true for 3D objects grasped at three contact points, provided that the three friction cones have intersection on the plane passing through the three contact points.

$$Sgn(\overrightarrow{C_i B_{jk}} \times n_{i1}) \cdot Sgn(\overrightarrow{C_i B_{jk}} \times n_{i2}) = 0$$

(26)

Note that the prior method only provides a procedure for verifying if a specific grasp posture is in equilibrium. In order to *plan* a stable grasp, the procedure must be incorporated in a higher-level method which suggests potentially feasible sets of contact points and forces. Thus a complete planner must be able to answer the three questions mentioned in Immobility Section.

Hitherto in this chapter is was explained that for performing a stable grasp by a multi-fingered robotic hand we need to specify and analyze concepts like immobility, form- and force-closure, Coulomb friction, etc., and determine or generate proper measures of grasp quality. Also, when the grasping procedure is complex, a pre-grasp analysis is also required. Afterwards, two grasp planning methods were discussed. In the Stiffness Matrix method, the optimal grasp posture was calculated based on positions of fingertips, contact constraints, Static friction constraints, and contact forces. Although this method can accommodate unlimited number of fingers, it needs considerable amount of calculations for grasping stiffness and contact stiffness matrices. In the Friction Cones method for grasp planning, an equilibrium grasp is verifiable by having only the positions of three contact points, the geometry of friction cones originated from them, and the intersection points of the three friction cones. Although this method applied only to three-finger instances, it requires simple calculations and is in fact a special and efficient case of the Stiffness Matrix method.

Regrasp Planning

During manipulation of an object, a robotic hand may encounter constraints arisen from joint angle limits, kinematic constraints, or limited maneuverability due to nearby obstacles. As a result, the grasping position must change in order to adapt to the conditions. This action is called *regrasping*, and planning for it is called *regrasp planning*. Through surveying the relevant literature, we have identified seven major approaches for regrasp planning, each of which encompasses a number of more specific methods, generally developed for parallel grippers or multi-fingered hands. Some methods are based on sensors that are embedded in grippers or fingers.

The main regrasping approaches are introduced, along with an example reference:

- **Finger Gaiting:** In this approach or more fingers lift off the object while other fingers stay on the object without unfixing it or loosing force-closure. The detached

fingers then rest on the object in new positions (Omata & Farooqi, 1997).

- **Finger Sliding:** In this approach one or more fingers slide on the object sequentially or simultaneously in a controlled manner. Fingers should have a trajectory on the object's surface and must hold a force-closure grasp during sliding (Cole, Hsu, & Sastry, 1989).

- **Finger Rolling:** In this approach fingertips roll on the object's surface to transfer it from one configuration to another. It is usually used along with finger gaiting in order to resolve situations when gaiting fails to generate a solution (Han & Trinkle, 1998).

- **Pick and Place:** In this approach the object is placed in an intermediate situation (e.g. on a table surface) and then picked up again. The emphasis is on planning a legal and feasible sequence of grasping-placing-grasping operations (Terasaki, Hasegawa, Takahashi, & Arakawa, 1991).

- **Throwing and Catching:** In this approach, for changing a grasp configuration the object is thrown upwards by the robot's hand and then caught at a new configuration. The final grasp configuration is exactly defined and must be achieved (Furukawa, Namiki, Taku, & Ishikawa, 2006).

- **Bimanual Regrasping:** In this approach the object which is initially grasped by a hand in its starting configuration is grasped by another hand and moved to a new configuration. This cycle is repeated until the final configuration is achieved (Saut, Gharbi, Cortés, Sidobre, & Siméon, 2010).

- **Human Imitation:** In this approach, robotic regrasping is based on imitation from human hand. Through some sensors attached to a human hand, information about relative movements of fingers and their

Figure 23. An example of human imitation (Vinayavekhin, et al., 2011)

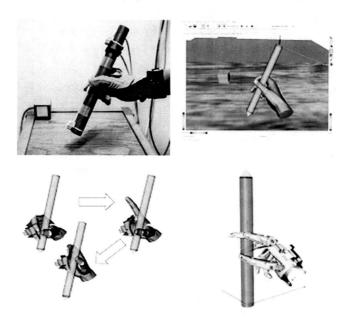

exerted forces are extracted, analyzed, and then replicated on a robotic hand (Figure 23) (Vinayavekhin, Kudohf, & Ikeuchi, 2011). Here the regrasping approach employed by the human (e.g. finger gaiting, rolling, sliding) is not important since no planning is done by the robot which acts as a slave agent and just imitates the human. After sufficient training of the robotic hand, it can then perform regrasping operations independently.

Among the prior approaches, Finger gaiting and Pick and place approaches have been researched more and therefore are more matured. Conversely, the Human Imitation is a relatively new and young approach, which is rooted in the biomimetic science, which focuses on the nature as an inspiration for creating more sophisticated mechanisms and methodologies and helps conscious abstraction of new principles and ideas and seeking new intelligent-based paradigms. Nowadays roboticists look for nature-inspired robots that exhibit more robustness and efficiency in unstructured and dynamic environments. Some robots are inspired from biological forms and mechanisms, while others imitate certain intelligent behaviors of living beings (Habib, Watanabe, & Izumi, 2007).

A possible trend of grasping and regrasping approaches could be to extract certain behavioral primitives and patterns of the human hand and implement this knowledge in developing new intelligent grasping and regrasping methods through irregular and unconventional sequence of postures.

CONCLUSION

Modern day robots can be carefully hand-programmed to carry out many complex manipulation tasks, ranging from using tools for assembling complex machinery to balancing a spinning top on the edge of a sword. Robotic grasping is an important element in automatic manipulation, which contributes greatly to improving the production quality and reducing cycle time and costs in manufacturing, assembly and other industrial fields. The significance of automatic grasping and its related issues is ensued from the prevalence and importance of assembly and fixturing in all industries, since in every assembly operation there is a need for grasping, manipulation, and placing of different parts. For example, when parts travel on a conveyor belt toward an assembly station, they may have different orientations which must be grasped by a robotic hand and manipulated to its correct final position. Also, in performing dangerous or sensitive tasks such as manipulating toxic materials or neutralizing bombs, stable grasps by robotic grippers must be ensured and guaranteed.

In this chapter three basic topics are presented: in the first section the major mechanisms and types of robotic grippers are introduced. In the next section, part immobility –which is a fundamental concept in grasping– is discussed in a mathematical and geometrical manner, and the first- and second-order immobility are introduced. After analyzing and identifying stable grasp positions for a part, planning of such grasps can be done using different approaches which try to optimize some measures of grasp quality, which are presented in Table 2. When moving an object from an initial configuration to a goal configuration, a robotic gripper or hand may not be able to perform the task directly because of surrounding obstacles, kinematic or geometric limitations, or incompetent grasp configurations. Therefore, the robot will need to change the grasp configuration in intermediate stages, before the final configuration is reached, which gives rise to the notion of Regrasping and Regrasp Planning, which have some major approaches and methods briefly introduced in the last section.

It is hoped that this chapter will provide the interested researchers with essentials of this evolving subfield of robotics.

REFERENCES

Brown, E., Rodenberg, Amend, Mozeika, Steltz, Zakin, Jaeger. (2010). Universal robotic gripper based on the jamming of granular material. *The National Academy of Sciences, 107*(44), 18809–18814. doi:10.1073/pnas.1003250107.

Brown, E. Rodenberg, Amend, Mozeika, Steltz, Zakin, Jaeger. (2013). Universal robotic gripper based on the jamming of granular material. *Cornell Creative Machines Lab*. Retrieved from http://creativemachines.cornell.edu /jamming_gripper?q=jamming_gripper

Chen, Y.-C., & Trinkle, J. C. (1993). On the form-closure of polygonal objects with frictional and frictionless contact models. In *Proceedings IEEE International Conference on Robotics and Automation* (pp. 963–970). IEEE. Doi:10.1109/Robot.1993.292268

Chen, & Walker. (1994). Visualization of form-closure and force distribution for grasping solid objects. In Proceedings of the IEEE International Conference on Systems, Man, and Cybernetics, 1994. Humans, Information and Technology. IEEE. DOI: 10.1109/ICSMC.1994.399828

Ciocarlie. (2010). *Low-dimensional robotic grasping: Eigengrasp subspaces and optimized underactuation.* New York: Columbia University.

Cole, Hsu, & Sastry. (1989). Dynamic regrasping by coordinated control of sliding for a multifingered hand. In *Proceedings of the 1989 IEEE International Conference* (pp. 781 – 786). IEEE. Doi:10.1109/Robot.1989.100079

Czyzowicz, Stojmenovic, & Urrutia. (1999). Immobilizing a shape. *International Journal of Computational Geometry & Applications*, 181–206. doi:10.1142/S0218195999000133.

Furukawa, N. Taku, & Ishikawa. (2006). Dynamic regrasping using a high-speed multifingered hand and a high-speed vision system. In *Proceedings of the 2006 IEEE International Conference* (pp. 181–187). IEEE. Doi:10.1109/Robot.2006.1641181

Guo, Gruver, & Jin. (1992). Grasp planning for multifingered robot hands. In *Proceedings of the IEEE International Conference on Robotics and Automation*. IEEE. Doi:10.1109/Robot.1992.219919

Habib, M. K. (2008). Interdisciplinary mechatronics engineering and science: Problem-solving, creative-thinking and concurrent design synergy. *International Journal of Mechatronics and Manufacturing Systems*, *1*(1), 4–22. doi:10.1504/IJMMS.2008.018272.

Habib, Watanabe, & Izumi. (2007). Biomimetics robots: From bio-inspiration to implementation. In *Proceedings of the 33rd Annual Conference of the IEEE Industrial Electronics Society*. IEEE.

Han, & Trinkle. (1998). Dextrous manipulation by rolling and finger gaiting. In *Proceedings of the 1998 IEEE International Conference* (pp. 730–735). IEEE. Doi:10.1109/Robot.1998.677060

Hasegawa & Morookalaboratory. (2012). *Multifingered robot hand*. Retrieved from http://fortune.ait.kyushu-u.ac.jp/research-e.html

Krug, D. Krzysztof, & Boyko. (2010). On the efficient computation of independent contact regions for force closure grasps. In *Proceedings of the IEEE/RSJ International Conference* (pp. 586–591). IEEE.

Krut, Bégoc, Dombre, & Pierrot. (2010). Extension of the form-closure property to underactuated hands. *IEEE Transactions on Robotics*, 853–866. Doi:10.1109/Tro.2010.2060830

Li, Liu, & Cai. (2003). On computing three-finger force-closure grasps of 2-D and 3-D objects. *IEEE Transactions on Robotics and Automation*, 155–161. Doi:10.1109/Tra.2002.806774

Museum, C. H. (2012). *Robots- Utah/M.I.T. dextrous hand closeup*. Retrieved from http://www.computerhistory.org/collections/accession/102693567

Nair. (2013). Types of artificial gripper mechanisms. *Society of Robotics & Automation*. Retrieved from http://sra.vjti.info/knowledgebase/posts-on-a-roll-1/gripper-mechanisms

Nguyen. (1986). Constructing force-closure grasps. In *Proceedings IEEE International Conference on Robotics and Automation*. IEEE. Doi:10.1109/Robot.1986.1087483

Omata, & Farooqi. (1997). Reorientation planning for a multifingered hand based on orientation states network using regrasp primitives. In *Proceedings of the Intelligent Robots and Systems, 1997*. IEEE. Doi:10.1109/Iros.1997.649067

Overmars, R. Schwarzkopf, & Wentink. (1995). Immobilizing polygons against a wall. In *Proceedings of the Symposium on Computational Geometry*. IEEE. Doi:10.1145/220279.220283

Phoka, Niparnan, & Sudsang. (2006). Planning optimal force-closure grasps for curved objects by genetic algorithm. In *Proceedings of the IEEE Conference on Robotics, Automation and Mechatronics*. IEEE. Doi:10.1109/Ramech.2006.252683

Prado, & Suarez. (2005). Heuristic grasp planning with three frictional contacts on two or three faces of a polyhedron. In *Proceedings of the IEEE International Symposium on Assembly and Task Planning*. IEEE. Doi:10.1109/Isatp.2005.1511459

Rimon, & Burdick. (1996). On force and form closure for multiple finger grasps. In *Proceedings of the IEEE International Conference on Robotics and Automation*. IEEE. Doi:10.1109/Robot.1996.506972

Saut, Gharbi, Cortés, Sidobre, & Siméon. (2010). Planning pick-and-place tasks with two-hand regrasping. In *Proceedings of Intelligent Robots And Systems (IROS)* (pp. 4528–4533). IEEE. Doi:10.1109/Iros.2010.5649021

Terasaki, H. Takahashi, & Arakawa. (1991). Automatic grasping and regrasping by space characterization for pick-and-place operations. In *Proceedings IROS '91 IEEE/RSJ International Workshop*. IEEE. Doi:10.1109/Iros.1991.174426

van der Stappen, Wentink, & Overmars. (2000). Computing immobilizing grasps of polygonal parts. *The International Journal of Robotics Research*, *19*(5), 467–479. doi:10.1177/02783640022066978.

van der Stappen, Wentink, & Overmars. (1999). Computing form-closure configurations. In *Proceedings of the IEEE International Conference on Robotics And Automation*. IEEE. Doi:10.1109/Robot.1999.770376

Vinayavekhin, Kudohf, & Ikeuchi. (2011). Towards an automatic robot regrasping movement based on human demonstration using tangle topology. In *Proceedings of the 2011 IEEE International Conference* (pp. 3332 – 3339). IEEE. Doi:10.1109/Icra.2011.5979648

Zhang, & Gu. (2012). Grasp planning of 3D objects using genetic algorithm. In *Proceedings of the IEEE International Conference on Automation and Logistics*. IEEE. Doi:10.1109/Ical.2012.6308157

Zuo, & Sun. (1991). Force/torque closure grasp in the plane. In *Proceedings of the IEEE International Workshop on Intelligent Robots and Systems*. IEEE.

Chapter 17
Cyberinfra Product Concept and its Prototyping Strategies

Balan Pillai
Stanford University, USA

Vesa Salminen
Lappeenranta University of Technology, Finland & HAMK University of Applied Sciences, Finland

ABSTRACT

The Knowledge-Intensive Sustainable Evolution Dynamics (KISBED) (patent pending), a platform the authors use in their "use-cases," shows that it works. Cyber, infrastructure, and product are integrated in the Cyberinfra Product "function." The perception properties are not long tagged or have no carriers, and the signal travels a short distance before it collides. The authors prove the KISBED through some examples.

INTRODUCTION

One branch of the philosophy of science, *methodology*, is closely related to the theory of knowledge. It discovers the methods by which science get there at its posited truths concerning the world and critically explores alleged foundations for these methods. In industry, they view that some crucial concepts that arise from the scientific knowledge, are of any use while sensing the credibility of its outcome. Truth is that the industry is hectic in their day-to-day arena. The performances are targeting to get a product out of the process within 90 days. We have seen this problem today with cell-phones, Internet-games and home-theaters etc. Therefore, the research and development team and their strategies must have to be focused on to a queue of potential versions of products and services. In fragile demand entangled, based on market segments. In any circumstances a product would be a mini robot, camera or cell phone. They could be large scale mechatronic or process equipments such as fighter-aircrafts, paper machines, and turbines, etc.

DOI: 10.4018/978-1-4666-4225-6.ch017

Scientific methodologists over and over again state that science is characterized by *convergence*. This is the claim that scientific theories in their historical path are converging to an ultimate, final, and ideal theory. However, sometimes this final theory is said to be true since it corresponds to the "real world," as in pragmatic accounts of convergence. It would like that the exact nature of light is a profound question, which seems to be not yet fully answered. Luckily, one does not need to know exactly what light is in order to understand how it behaves and utilize it (Pillai, 2011). This is a bridging effect between "superfast" and "cyber-infraproduct concept," which we are presenting here. Further on to light the topic. There are two convenient ways to describe the propagation of light and its interactions with materials. Neither system is sufficient alone nor are they completely ample together. At least the two systems are not contradictory. In the age of ubiquitous Internet; intelligent devices, and mobile displays that are good for more than just ruining your eyes, but also companies who can no longer afford to sidestep the expanding mobile market (Pillai, 2012).

Engineers have to design their products ingeniously to avoid any "technical dead-locks," and assure the security of its use. In those products, it is necessary that it should perform well. At any level, the wave train of radiation can be completely described by two vectors that are perpendicular to the direction of travel of the ray, which is again perpendicular to each other. In this context, it is perhaps in order to say that in decades we learned to produce fighter aircrafts, to use them strategically and counter attacking probes produced widely around the globe. In reality, they are mechatronic-products, and to some extent, they express themselves as to be the romantic yield - as robotic engineering. Here we made the bridge. Nevertheless, let us now focus on the cyberinfra product concept at this chapter. Let us begin in putting the ingredients to essential, and or potential theory-mixes. We identify and build the product-platform type. We plan to map out the prototyping strategies, and test-bed opportunities.

"'If I have seen farther, it is by standing on the shoulders of giants,' wrote Isaac Newton in a letter to Robert Hooke in 1676" (Hawking, 2002). While Newton was referring to his discoveries in optics rather than his more important work on gravity and the laws of motion, it is an appropriate comment on how science, and unquestionably the whole of civilization, is a series of incremental advances, each structure on what went before. This is a fascinating theme. This statement consistently discloses the interoperability between the gravity of science and civilization. To trace the evolution of today's portrait of the businesses from the gravity of science; is the avant-garde state of Internet liberation. Due to this and several other covering law models[1], the view of scientific explanation as a deductive argument which contains non-vacuously at least one universal law among one its premises. There by to scrutinize the business laws, this is looking for scalable, adaptive, cost-effective, collective, and pinpointed solutions (CE-NET, 2010). Clearly, there is a need to develop a systematic and holistic approach. Virtual Enterprise is the answer to it. It is triggered with ubiquitous (it is anywhere, anytime) technology. This technology is within your reach (meaning easy and cost-effective), produced it and at the collaborative environments. The system is able to maintain the overall security. The interoperability is as well tagged to the system. Ultimate goal is to cement the vision of the business world, in turning them "inside-out," as a plug-and-play Internet business community (Pallot, Salminen, Pillai, & Kulvant, 2004). In the practical terms, so-called the "Extended or Virtual Enterprise" approaches would create a life-size dilemma in the future. This is because of each time, when a new partner is entering in to the system, those results increasing the job of management potentials including its integration costs. This is mostly due to value clogging visions and or miss-interpretations, which dragons to ideal collaboration among trading partners.

Recently there has been a great interest in the Semantic Web and issues related to specification and exploration of semantics on the World Wide

Web. Berners-Lee, the initiator of the *World Wide Web* (www), lays stress on the importance of the *"Semantic Web"* for machine-understandable Web contents and emphasizes the need for ontology (Berners-Lee, Hendler, & Lassila, 2001). Though a self-describing protocol *extended Mark-Up Language* (XML) is in charge of the syntax of the Web contents, clear definitions of the semantics of the domain knowledge required to implement machine-understandable contents. Ontology and ontological analysis needed fundamentally to represent knowledge about the domain and be able to share the information (Chandrasekharan, Josephson, & Benjamins, 1999). In specific, the shared ontologies are being proposed for representing the core knowledge that forms the foundation for semantic information on the Web. We identify broad thrust related to ontologies:

- Approaches to standardize the formal semantics of information to enable machine processing. Work is to be done as a part of the W3C working group.
- Approaches to define real world semantics linking machine processable content with meaning for humans based on engineering terminology.

Creating standards, especially standards that generate information industry infrastructure, is difficult, time-consuming and at constant risk for irrelevance and failure. One way to mitigate this risk, and secure the participation of the diverse interest groups required making such standards a success is to focus on process—as in the process that produces and maintains a good standard. This is in contrast to an approach that says some existing artifact selected from a list will be the standard, and all the others will not be the standard. An observation that we attribute is that it does not matter where you start, that is, it does not much matter which terminology or terminologies one selects as a starting point, instead what does matter is the process by which the proposed standard evolves to

achieve and sustain the desired degree of quality, comprehensiveness, and functionality. The process is what determines where the standard ends up.

We have seen in this light, change, even a large amount of change, will be a feature of successful formal terminology, or ontology. We hope to demonstrate the feasibility and utility of this approach. The challenge in the context of the Semantic Web is to choose a representation for change that makes it explicit. We assume that the engineering tool kit viewed in this way the Semantic Web would be part of the semantics. The challenge with this approach is the formulation of the units of change and the creation of ontology of these change units. Making change part of the Semantic Web would preserve that consistency. One way to focus the development of the desired units, interrelationships, and uses is to solve real problems and gain experience from deployments of these solutions; we propose to do this by formulating, deploying, and evaluating what we now call "The New Engineering Transaction." This transaction needs to supply reverse engineering; operating schemes and control information systems with the requisite formal definition of a new "insert," given a reference model or application, and do so at Web scale. The main challenge is how to do this in a way that first avoids breaking working applications that use the engineering tool kit terminology and second preserves the longitudinal value of existing and future engineering innovations of implementation.

At a deeper level, we believe that Semantic Web is an opportunity to shrink the "formalization gap" between engineering disciplines. We argue that overcoming this gap is the fundamental change in engineering society. This discontinuity in formalization between human connected information process and the machine code necessary to accomplish comparable ends begins at a very high descriptive level and it is not itself a concern of computer science. If this concern is to be given a name at all, it must be regarded as concerning engineering applications, and it is increasingly being referred to as "engineering information science" in Finland and "Engineering Informatics" in Europe and the US. It will be the task of this new discipline to better understand and define the engineering information processes we have considered here, in order that appropriate activities will be chosen for computerization, and to improve the man-machine system for better interoperability.

DEFINITION OF CYBERINFRA PRODUCT CONCEPT

It is essential to define the cyberinfrastructure as a set and product concept. The definition here for methodical purposes and also for the chapter in general is at first, what are cyber, infrastructure; and product concepts in sequence.

- **Cyber:** Is a virtual imitation of things that are anywhere and everywhere. It would that who exercises the surveillance. It would produces, stimulate, propagate and polarize the signals; through all the possible media. It is simply a carier, an agent.
- **Infrastructure:** Is as networks' like traffic roads or nodes; employing the communications that is transported by the Cyber. We combine cyber and infrastructure, we get the Cyberinfrastrure. In short, let us call it as *"Cyberinfra."*
- **Product Concept:** Is proof of some idea or knowledge which turned out to be a physical product[2]. This product has a thought or concept, based on its function, behavior, utilization, or segment of a community. While we integrate all those three words into "function" as our topic demands – it becomes as Cyberinfraproduct concept. We shall further define it. It has a meaning of "being everywhere, anywhere." We call it, for this purpose, as cyberinfraproduct concept. This concept, as a technological niche, is rich with opportunities and challenges. This technology niche would be

able to access, in this case the entire signal trajectories that are reflected, communicated, and or fragmented. The frequency can be low or high in an environment. This environment can either be closed or opened loop, and or naturally or through a media or several spread signals and are capable to route the propensity, which is being processed, used collectively, individually, single-in, and or single-out mode. The perception properties are not long tagged or carriers. It travels a short distance before it collides. It is not necessarily, so to say, a physical product anymore neither it has any more a physical outlook nor it could simply be bits-bytes.

TECHNOLOGY ROADMAP FOR CYBERINFRA INTEGRATION

We summarize the vision of the change of Technology Roadmap. It visions a timeframe for 5 years. The mission accomplishes to fit a suitable level at 2010. This is due to NIST[3]. NIST was proposed early 2000 that the Semantic Web would become a reality in industry and a new "Knowledge Web", that is new Internet, would revolutionize once again the world. Figure 1 describes an onion model of the roadmap towards enterprise integration and full use of semantic infrastructure in open system architecture with interoperability and plug and play capability.

CYBERINFRA ARCHITECTURE

The underlying difficulty is not with speed or quantity alone, but with relevance. How does a system, given all that it knows about aardvarks, Alabama, and ax handles, "home in on" the pertinent fact that bananas don't get hungry, in the fraction of a second it can afford to spend on the pronoun "it"? The answer proposed is both simple and powerful: common sense is not just randomly

Figure 1. Technology roadmap- Cyberinfra integration

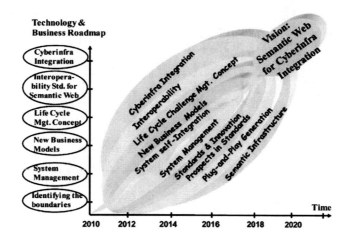

stored information, but is instead highly organized by topics, with lots of indexes, cross-references, tables, hierarchies, and so on. Therefore, it is necessary to have Cybrainfra architecture. The architectural notion or vocabulary in engineering at this point, is the information, which is stored either in a database or bases, and is connected to server or several servers. In this context let us use or reframe the terminology to get an answer the meaning of servers, which are ontology servers. Like words in the sentence itself trigger the "articles" on monkeys, bananas, hunger, and so on, and these quickly reveal that monkeys are mammals, hence animals, that bananas are fruit, hence from plants, that hunger is what animals or we human-beings feel when need to eat – and that settles it. The issue of relevance is solved instead by the antecedent structure in the stored knowledge itself. These types of servers "normalize" terminology functions for enterprises, some at Web scale. We believe that such servers will be, essential to support the Semantic Web, and as usual at the Web. The challenge will be how to maintain them in loose synchrony as appropriate.

The fact is that the formal terminologies would always construct and maintained by geographically divided domain experts. This means that we need additional things for "configuration management," which supports the conflicting resolutions, and so

on. One short-term reality is the need for what we call the "local enhancement". In another words, the ontology builders and their server, must specify clearly the "business requirements" that are to be addressed for common use in electronic commerce and *Business-to-Business, B2B*. Let us turn to an Architecture that can be drawn for simplifying the concept (Figure 2).

Here the builder uses an object-oriented knowledge representation model based on and compatible with knowledge model and is designed to use the best practices from other frame-based systems. This is assumed to supports operations on classes, slots, facets, and individuals. Interoperability, knowledge sharing, and reuse are important goals and works as a fully compliant server. This should support a meta-class architecture that is in an Open Environment to allow introduction of flexible and customizable behaviors into an engineering ontology. It is able to predefine certain system constants, classes, and primitives in a default upper engineering ontology that can extend or refine to change the knowledge model and behaviors within the system.

By threading the vision, the study may in future concentrate on cutting-edge at industry cases. A flash search will be conducted on the Industry needs and their product development capability and Systems. While surveying the requirements

Figure 2. The architecture of engineering ontology

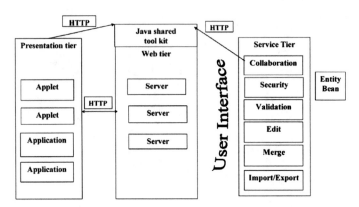

of an industry; we penetrate to verify the capabilities of Finnish Industries. We mirror them with US-industries and their peers, such as government and universities.

ROBUST DESIGN KNOWLEDGE AND CHALLENGES

We have now bridged and are able to process on further phases that are necessary to this chapter. Engineering Design Creation at this juncture of the chapter, one would define is the application of scientific and technical knowledge to solve a number of problems in and around a visible or imaginary product. Engineers incredibly use imagination, judgment and reasoning to apply; science, technology, mathematics, and practical experience in satisfying the required demand that exist, or being created in a market environment; at the time of a product invention. The outcome; is the design, production, and operation of useful objects or processes.

Business world is characterized by an increased demand for innovation, shorter product life cycles, and amazing pressures to launch new products fast. At the same time, product development teams face cost-cutting challenges. Even in the face of these pressures, no business can afford to sacrifice robust design. To fulfill the ultimate customer

promise, products must perform as expected in the true world, every day and in every circumstance. Salminen and Pillai (2004) defined and tested in their one of the works and supported this with a case study called "the Life-cycle Challenge Management (L_{cMgt})," at this phase. According to this work, it is pretty evident that the hectic industry is in that mode at this era of the century.

In mechanical engineering designs, it is known for their assembly and working that mechanical system contains complex assemblies of interconnected parts to function. Consider a large overall motion, such as at the ground vehicle suspension assemblies, robotic manipulators in their manufacturing processes, and aircraft landing gear systems. For a faster, more efficient solution to this problem class, ANSYS (n.d.)[4] provides a rigid multi-body dynamics module. *Figure 3*, shows a Bladed disc and its assembly in a passenger aircraft used by Lufthansa Airlines of Germany.

Figure 3 identifies that the rotor dynamics applications in the sense of design, it serves to spot the behavior and diagnosis of rotating structures. The capability is commonly used to analyze the behavior of structures that is ranging from jet engines and steam turbines, to auto engines and computer disk storage. Rotor dynamics can effectively compute critical speeds and the effect of unbalanced loads on a structure. They allow creation of *Campbell plots* to identify critical

Figure 3. A bladed disc and the passenger air craft

Passenger Aircraft Design by Ansys tool

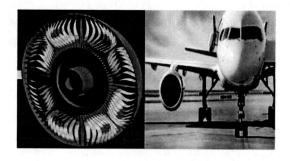

speeds of single or multiple spool systems, for beams, shells and solid elements. The knowledge-intensive scavenge is, therefore, the most difficult scopes in smart product design. Today the three dimensional graphics (3D), are a critical asset to engineers in all disciplines. This would play countless roles in product creation, from idea to prototyping and to test data visualization.

The first law of thermodynamics, it says, "when work is expanded in producing heat, the quantity of heat generated is proportional to the work done, and conversely, when heat is employed to do work, a quantity of heat precisely equivalent to the work done disappears" (Goodenough, n.d.). In the case of an engine driving a tiny break, every millimeter of mechanical work done raises the temperature of the air surrounding the brake, or of the water used to cool it. In another word a smart product that is to survive in a cyberinfra, that would be opened and or closed. It needs a lot of potential knowledge that is too intensive enough to attract, say, signals surrounded and is able to process in feeding back-and-forth. Bandwidth is not bound to a critical fallacy. Such as the second law of thermodynamics, which is based on the fact that heat will not flow from a body of lower temperature to one of higher temperature. This fundamental law would be found responsible for the fact that no process of converting heat into work can ever be complete! Nevertheless, it is never possible or meaningful to make all of the given heat quantity

enter the conversion process. And at the end of it, a certain amount of unconverted heat would always be left over. It is the same phenomena of whatever the new cyberinfra product designed have a certain amount of unused or unexpected clogging to the end of it. A design pattern is against the next "best" out of scratch (Pillai, Pyykkonen, & Salminen, 2009).

DESIGN KNOWLEDGE-INTENSITY AND CYBER-CHALLENGES

Product Design knowledge has reasonably been studied. They are based on which an immense number of modeling techniques have been developed. A large amount of them are tailored to precise products or specific uniqueness of the design activities. To fit an ideal semantic logic, the geometric modeling, is mainly used for supporting detailed designs. While knowledge modeling is working for supporting the conceptual designs, as we discuss this bit in detail later at this chapter. The National Institute of Standards and Technology *(NIST)* had set up a project that is based on the previous said techniques "a design repository." *The NIST* team has attempted to model, three fundamental facets of an artifact representation, such as the physical layout of the artifact (*form*), an indication of the overall effect that the artifact creates (*function*), and a causal account of the operation of the artifact (*behavior*) (Szykman, Sriram, & Regli, 2001). The *NIST* (Jia, 2007) has recently made an effort via this study. It offers that the development of the basic foundations of the next generation of Computer-Aided Designing (CAD) systems, which would able to gain a core representation for design information, called the *NIST* Core Product Model (CPM) (Fenves, 2001). It has a set of derived models defined as extensions of the CPM (e.g. Zha & Sriram, 2004; Economist, 2012). *The NIST* core product model has been developed to unify and integrate product or assembly information. The CPM provides a base-level product model that is not tied to any vendor software. It is meant to perform in an open, non-proprietary, expandable and independent of any one product development process. It is capable of capturing the engineering context, which is most commonly shared in product development activities. The mechanism is supposed to function on artifact representation that includes function, form, behavior, material, physical, and functional decompositions. It consistently interlinks the relationships among these concepts that are addressed previously. The entity-relationship data model influences the model a great deal. Accordingly, it consists of two sets of classes, called object and relationship, this is equivalent to the Unified Modeling Language (UML) class and association class, respectively. However, the Entity–Relationship-model (ER model) was developed by Chen (1976) in a software engineering pack, is an abstract way to describe a database. Chen's diagram is seen here as Figure 4.

An entity may be defined as a thing, which is recognized as being capable of an independent existence and which can be uniquely identified. An entity is an abstraction from the complexities of a domain. When one speaks of an entity, we normally speak of some aspect of the authentic world, which can be distinguished from other aspects of the real world (Beynon-Davies, 2004). An entity may be a physical object such as a house or a car, an event such as a house sale or a car service, or a concept such as a customer transaction or order. Although the term entity is the one most commonly used, following Chen we should really distinguish between an entity and an entity-type. An entity-type is a category. An entity, strictly speaking, is an instance of a given entity-type. There are usually many instances of an entity-type. Because the term entity-type is somewhat cumbersome, most people tend to use the term entity as a synonym for this term. Entities can be thought of as nouns. This can be ruled out in examples such as a computer, an employee, a song, a mathematical theorem. They are matured to use from the sky, such as ontology connected configuration. While would extract the knowledge. The other main orientation toward artificial intelligence is the pattern-based approach—often called "connectionism" or "parallel distributed processing"—reemerged from the shadow of symbol processing only in the 1980s. In many cases, better knowledge can be more important

Figure 4. Simple model that connects relationship entities

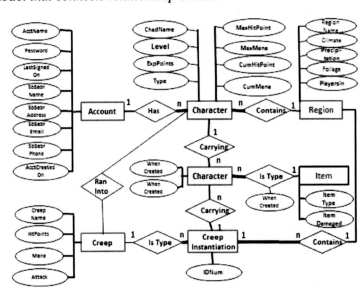

for solving a task than better algorithms. To have truly intelligent systems, knowledge needs to be captured, processed, reused, and communicated. Ontology-based system would support all these tasks, says Salminen and Pillai (2007). They researched out the ontology phenomenon, which was just fossilized out at the Silicon Valley and Maryland, USA. While in *NIST,* one of the authors, an immense number of standardization agenda were hanging out in ontology. It is time to dip out here a bit deeper the application trends in the case of Internet and electronic manufacturing sector.

The term "ontology" would be defined. It is simply an explicit specification of conceptualization. Ontologies are able to capture the structure of the domain that is the conceptualization. This includes the model of the domain with possible restrictions. The conceptualization describes the knowledge about the domain, not about the particular state of affairs in the domain. In other words, the conceptualization is not changing, or is changing very rarely. Ontology then is the specification of this conceptualization. That is the conceptualization, which is specified by using a particular modeling language and particular terms. Formal specification is required, in order to be able to process ontologies and operate on ontologies automatically. Ontology describes a domain, whiles a knowledge-base (based on ontology), and describes particular state of affairs. Each knowledge based *system* or agent has its own knowledge base, and only what can be expressed using ontology can be stored and used in the knowledge-base. When an agent wants to communicate to another agent, it uses the constructs from some ontology. In order to understand the communication, ontologies must be shared between agents. Although it is required from ontology, to be formally defined, there is no common definition of the term "ontology" itself (Pillai, 2002). The definitions can be categorized into roughly three groups:

1. Ontology is a term in philosophy and its meaning is "theory of existence".
2. Ontology is an explicit specification of conceptualization.
3. Ontology is a body of knowledge describing some domain, typically common sense knowledge domain.

CYBERINFRA METHODOLOGY DEVELOPMENT

There is a great need for most of the businesses to develop new product and service in Open System Architecture. A sustainable growth of the business lies on services that Corporations are offering (Salminen & Pillai, 2005). The Smart Products, or say cyberinfra products, of today are increasingly embedding with intelligent. Therefore, the role of product is very important. The best possible product architecture for the optimized product platform thus needs a method. It is done by organizing and recognizing the product modeling and managing the dependency matrix of the product. In some cases, it is to be re-engineered to achieve an optimum cash return.

The methodology phase contains here a freshly created new knowledge. It is attributing the redundancy or not. This phase draw ups the possibilities of self-integrating schemes. This includes the manufacturing system integrations and netting them into Business loop and thus creating B2B platform to serve B2C (Salminen & Pillai, 2007). Figure 5 introduces now the agent mechanisms for Internet based intelligent and electronic manufacturing.

Figure 6 describes the self-integration patterns by using semantic infrastructure.

Figure 7 explains the connectivity of Semantic Web into intelligent and electronic manufacturing concept. The methodology of this is outlined in Figures 7 and 8. Intelligent and Electronic Manufacturing System requires in building the

Figure 5. Agent-based mechanisms over internet browser for –intelligent and electronic manufacturing

Adapted from NIST

Figure 6. Self-integration environment

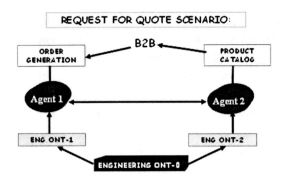

foundation that route the Semantic Web by "Plug-and-Play" format (Salminen & Pillai, 2007). The repository system feeds, the strong linguistic representation via on-line modeling and plugging them between the structures and classes. Internal system has sustained with a search engine that would simulate and fractionate via re-framing the engineering entities into an understandable semantic structure. The semantic infrastructure for Plug and Play Collaboration was created by the National Institute of Standards and Technology, NIST on their Test-bed facility and or so called information platform. Process Semantic Language, PSL, was also implemented on the test bed (Figure 8).

The time this methodology was introduced for applying at the Test-bed facility, it was too early. It is tempting to say that sheer speed will no longer suffice, and that more knowledge of chess, or something else, is needed. What this sketch shows is that a state transition must be represented as a pair forward >< backward. The forward is the information about how to get to the state from the

Figure 7. Intelligent and electronic manufacturing system integration for semantic web

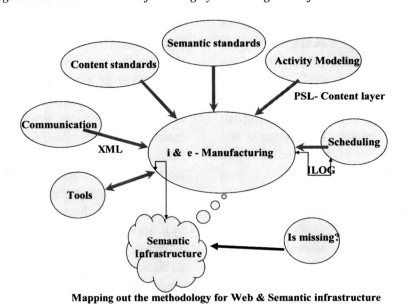

Mapping out the methodology for Web & Semantic infrastructure

Figure 8. Methodology on intelligent and electronic manufacturing system integration for semantic web

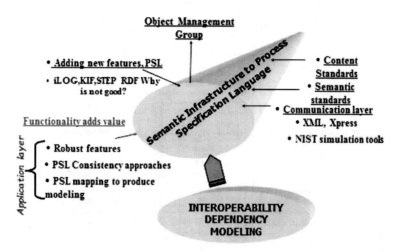

immediately preceding state. The backward is the information about how to undo forward, that is how to go back to the immediately preceding state. A procedure was initiated through this sketch. Process specification language and the semantic impacts were tried to coin by adding a modeling tool. This tool models the context into machine understandable language. Thus is pinning a representation that adds value to this new standard. In practice, it has proven that it does not rule the goof. The purpose of this exercise was due to an evaluation of this methodology - while building the semantic infrastructure, where one leads to more functionality to PSL. However, it is learned that the representation needs to be evaluated before one could bring more functionality. In this case we were proposing a requirement analyzing and mapping tool called Optiwise®[5]. We have also experimented to see the impact on, while adding the RDF *(Resource Description Format)* with ontology, for bring representation task at the application layer. The result was not encouraging at all.

PERFORMANCE MODEL: SEMANTIC DEFINITION

Michael Faraday (1791-1867) said long time ago very cleverly that "Nothing is too wonderful to be true if it be consistent with the laws of nature, and in such things as these, experiment is the best test of such consistency" (wikiversity.org). When a request involved in any format should answer the end-to-end performance. The Semantic definition is to be clear from the request structure. There will be three structural classes, such as mechanical, process and controls. A database is linking automatically to a system when specified in the process. A process model is based on this statement spread-out in Figure 9 for easy representation.

Here the tag associated, is to mean that it belongs to structural identification. This should belong to either Mechanical, Process, or Control properties and are associated to service, product and or both. This is a theoretical model that is tested in practice. Its content would represent a portion of undefined but used at paper engineering as a tool kit. It is formularized and used for

Figure 9. Identification of the tags (mechanical, process, control)

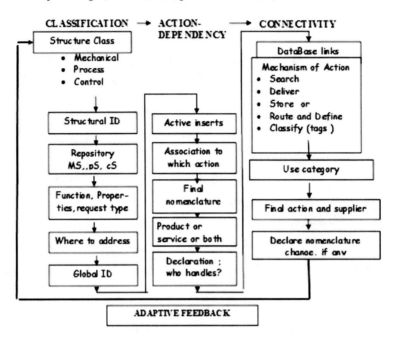

common process automation. It is primarily based on public and private vocabularies of the pre-defined instances or industry.

CYBERINFRA PROCESS

The model presented in Figure 9 is further said to be little more than an academic exercise. As a result, there is no warehouse of engineering *tool-kit* descriptions that can be reached over time. The changes across in the terminologies used there to formalize a solution, to the common understanding through the Web. The engineering tool-kit describes the repositories that support such *"time-travel,"* without queuing is so in the same manner. And none uses the existing or proposed standards. An explicit goal of this project is to begin to overcome this shortfall at least in the context of engineering. The first step is in making the formal terminology that change in itself into a new terminology/ontology, the "thing," or a "unit." This new one is further on to create a unit of change that has the same general properties as

any other "thingness" unit. For example, when an appropriate reference of given taxonomies, which is used to (in the Structural Logic sense) "classify" an engineering notion. One can create a desired reference terminology—by adding the definition of each application, this would allow again a new application at a time. Frequently, new application comes with new mechanisms of action. Furthermore, new indications (implementation) would set at and thus the corresponding "new thing" may need to update the reference taxonomies, before adding the new definition of the new thing and so on. To make this simple, there is one potential, which is closer to a term that is being updated. This new term now created, is not requiring update for to the reference. This is due to:

- New thing "publish," as XML, a newly "structured" version of the package insert or, "label," designed to "explain" that the everything to both human and computers.
- Do further processing and enhance the parts of the label that can be processed usefully by computers and then "publish" it,

once again in XML. The "enhancements" may include connections to the mechanical, process and control engineering literature, related to terminology and foreign language names.

- Applications or servers electing to process the new thing transaction will see that the XML indicates that is an "add," the simplest kind of transaction to process. That is the transaction, which would add a new concept—the new thing, the appropriate relationships to other concepts influence on the various reference taxonomies, and attributes.

It is not hard to imagine that most applications would be relevant of such insertion and subsequently "do the right action." However, the problem with this simple form of the new thing transaction is that as described by domain experts. Most new things represent "changes in understanding," and it is not at all clear how existing applications would deal with such changes in understanding automatically, or know when they need help from human-interactions. In the context, "changes in understanding," is represented by changes in the reference taxonomies.

CYBERINFRA: ONTOLOGY

We view a formal terminology or ontology as a corpus of "facts" or assertions. It is collected over time, and then one can contemplate ontology of such facts, or changes. The goal of such operation, is to evaluate and adapt semantic Web infrastructure and implement the ontologies for B2B (Business-to-Business) processes for engineering systems (Pillai & Salminen, 2004), and that is sighted at this reference. B2B of engineered products is very complex and has no foundation for easy sharing of product, process, or production information. The opportunity is a new semantic language by encoding meaning, which is being developed, or

already now existing for the Web. This is the basis for B2B of engineered products and services. The difficulty is defining and implementing the semantics to be attached to each type of change unit. One step toward such semantics is the simple expedient of tagging each terminological unit—concept, term, relationship, and attribute—with a "start entity" and "end entity." More disciplined and complete forms of such semantics are what needed to preserve the longitudinal functionality of systems. This uses the ontology, and what will be needed to transfer knowledge gained from a successful test of the new thing transaction to the Semantic Web (Salminen & Pillai, 2007). Therefore, even when the user interface returns an exact equivalent for the casual term, users may choose a "better" formal term from the displayed semantic neighborhood. The simple explanation of this phenomenon is that humans are better at recognition than recall. Those who are developing ontologies are familiar with the phenomenon, once domain experts can "see" a domain model, they can almost always make it better. Doumeingts, G, et al, of the European Union (EU), for instance, proposed a draft to the CEN (European Committee for Standardization) framework, on Interoperability Schemes (Doumeingts, Li, Piddington, & Ruggaber, 2005), though not widely explored. Figure 10 shows the same approach as we explored it elsewhere in this chapter.

KNOWLEDGE-INTENSIVE SOLUTION-BASED CYBERINFRA DESIGN

We are now on the next step to reach out at intelligence as a whole, particularly the artificial side of it. It is precious to know as to how that works in this concept. In control theory (Jia, 2007), we do dwell with the system performance. As well as its functions and this would equally affect when create an engineering product either imaginary or physical. While designing an embedded system

Figure 10. CEN framework for enterprise interoperability6

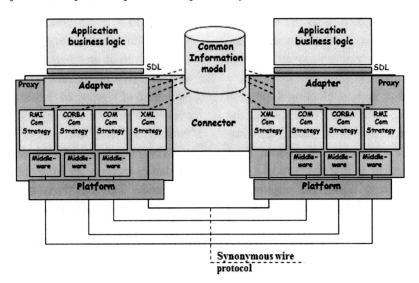

often reside in machines. Those machines are expected to run continuously for years without errors, and in some cases, it should recover by itself, if an error occurs. Therefore, the software is developed and tested more cautiously. This is more vigorous than for personal computers. It is entirely obsolete when there are unreliable mechanical moving parts resting at, such as disk drives, switches or buttons. The old way of making things involved, taking lots of parts and screwing or welding them together. Now a product can be designed on a computer and "printed" at a 3D printer; which creates a solid object by building up successive layers of material. The digital design can be tweaked with a few mouse-clicks. A 3D printer can run unattended, and can make many things, which are too complex for a traditional factory to handle. The applications of 3D printing are especially mind-boggling as an exercise of the present day technology of this decade. Already, hearing aids and high-tech parts of military jets are being printed in customized shapes. The geography of supply chains will change or nor more exist. An engineer working in the middle of a wilderness, who finds that he lacks a certain tool, no longer has to have it delivered from the nearest city. He can simply download the design and print it. The

days when projects ground to a halt for want of a piece of kit; or when customers complained that they could no longer find spare parts for things they had bought, will one day seem quaint.

At this point, it is worth to note that on the design point of view, which the material used in manufacturing are lighter, stronger and more durable than the old ones. Let us look quickly at the Carbon fiber, which is replacing steel and aluminum in products ranging from aircrafts to mountain bikes. Right now new techniques, let engineers shape objects at a nano-scale. Nanotechnology is giving products better features, such as bandages that help to cure cuts. Engines that run pretty efficiently and crockery that cleans more easily. In the Internet, it is allowing more and more designers to work in partnership on new products, the barriers to entry are falling.

CYBERINFRA DEFINITION FOR INVENTION

The current techniques are for totally custom design type of droplet-based "digital" biochips do not scale well for side-by-side entity or next-generation System-On-Chip (SOC) patterns.

These are expected to include in microfluidic components. Micro fluids and biochips are the new inventions. Cyberinfra is defined earlier where lonely visionaries would encounter with the comparable associations which are tent to be an invention in its nature. This is like the same principles as parties think of creating something new or being modified to fit the existing, or both. This is our intention. While each of these formulas says how much information is generated by the selection of a specific message, communication theory is seldom primarily interested in these measures.

Case: Cyber-Keksimö

Let us begin to think of Keksimö, a typical Finnish-language word, has an anew meaning of its use.

In this context, came to thinking about what the term "Keksimö"[7] exactly means. We invented[8] it! Recent past history taught us and inspired by Gabriel Tarden and Joseph Schumpeter, in their thought on the neoclassic economic theory. According to them, the "innovative economy" would require pragmatic grounding efforts or put them into practice. "Innovative" expresses the term "innovation" or "innovare," which means "to make a new." Novelty value and importance that is always determined in relation to a specific user

base or format. The term "Inventio" comes from the word "invenire" on the other hand, the "come in, to find." Therefore, inventions may not be recognized at first the "new." This was also the case "Keksimö." The Keksimö is close to the familiar words, such as "invent" and "invention" but in this sense that is as a place where discoveries are made and processed, the term is fresh and new. Keksimö, therefore, is an "invention," because the term is "found" familiar to the existing natural logic and "introduced" *or* "passed" in to a new logical sense of application. This is so-called *novel-era of use.* It does not necessarily want to understand or barely stay unnoticed. The Keksimö has become a secondary question in relation to how much it affects the way we perceive, act, think, and discuss.

INVENTION PROCEDURES

The phenomenon of *normal* inventors[9] are very dare some - How to, and where to begin; due to heavy bureaucratic set up in Europe. Figure 11 wraps up the true story of a Finn in Finland. He begins to think new ideas to realize as a product or service. In materializing, he is in tight track for something. He likes perhaps to solve some existing problems. He goes the following steps, as in Figure 11.

Figure 11. Inventor-investor conspiracy and state of confusion

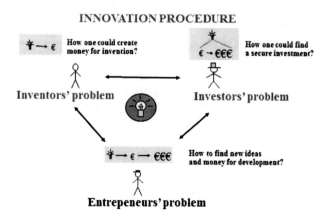

Let us look at this in a generic term the performance, substantives, or activities of inventions in ones' day-to-day life spectrum. The transition logic to this aspect could be drafted as in Figure 12.

CYBER-PLATFORM DESIGN: KISBED

A sporadic violence would be right at this point such as to bring in to share the fundamental concept of prototyping the KISBED platform for indulge to spurn (See Figure 13). The pique in a nascent form is diluted at a three step to trap. At the orchestrating front is every single grossularite, should have to divide their roles before in the defragmenting phase.

Let us now add-up this step to the orchestrating, which is the initial front at sight; the grossularite moves to a comforting segment where other activities of philharmonic would be set to timely perform. This Figure 13 faces the synchronizing phase in Figure 14. Notwithstanding, the KISBED method describes in several steps that are inside the camera of functions and the initial steps of "invention" meets the most sturdy bricks before passing the level to an entrepreneurship.

USE CASE: KNOWLEDGE INTENSIVE CYBERINFRA PLATFORM

We known and as well reported (Economist, 2012) that the first industrial revolution began in Great Britain at the late 18th century with the mechanization of the textile industry. Tasks up to that time done laboriously by hand in hundreds of weavers' cottages were brought together in a single cotton mill, and the factory was born. The second industrial revolution bumped up in the early 20th century, when Henry Ford mastered the moving assembly line at manufacturing and boosted to the age of mass production. The first two industrial revolutions made people richer and more urban. Now a third revolution is under way. The mechatronics represents a unifying engineering science paradigm. This science sector is bound to an interdisciplinary knowledge area that addresses interactions, in terms of the ways of it is working and thinking, and is able to spin together the practical experiences that are being extracted from the theoretical knowledge (Habib, 2008). The physical, solid and the entire manufacturing scheme are going digital. An amazing number of high technologies are converging. The system is in contriving with intelligent software,

Figure 12. Invention phases: generic setup

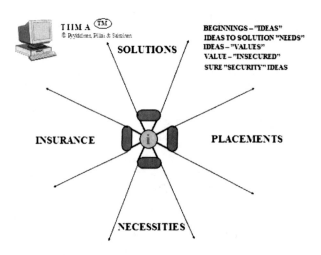

316

Figure 13. Paradigm of KISBED method

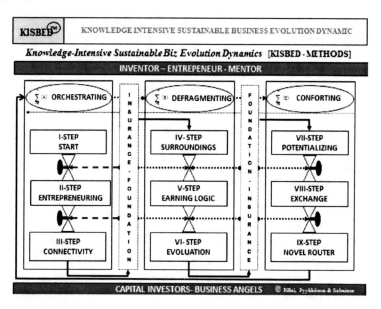

Figure 14. Synchronized phases few elements at the KISBED

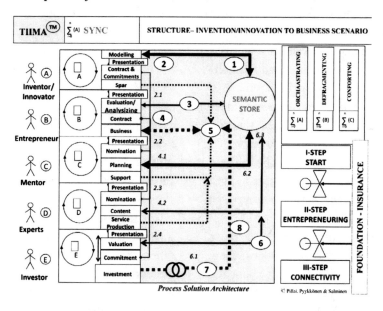

novel materials, more dexterous robots with new processes (including the three-dimensional printing) and a whole range of Web-based or enabled services. The factory of the past was based on cranking out zillions of identical products. Ford famously said that car-buyers could have any color they liked, as long as it was black. But the cost of producing much smaller batches of a wider variety with each product tailored precisely to each customer's whims is falling. The factory of the future will focus on mass customization—and may look more like those weavers' cottages than Ford's assembly line.

Haves (2013) explained the computation problems so well in his one of the articles at the American Scientists (Haves, 2013) those roughly 30 years ago, Harold J. Morowitz, who was then at the Yale, set forth a bold plan for molecular biology (Barile & Razin, 1989). Morowitz was outlined a campaign to study one of the smallest single-celled organisms, a bacterium of the genus *Mycoplasma*. The first step was to be to decipher its complete genetic sequence, which in turn would reveal the amino acid sequences of all the proteins in the cell. In the 1980s reading an entire genome was not the routine task as it is today, though Morowitz argued that the analysis should be possible if the genome was small enough. He calculated the information content of mycoplasma DNA to be about 160,000 bits, and then added on the other hand, this much DNA will code for about 600 proteins—which suggested that the logic of life can be written in 600 steps. Completely understanding the operations of a *prokaryotic cell* is a possible concept that can be visualized and one that is within the range of the possibility. There was one more fascinating element to Morowitz's plan at 600 steps, a computer model become feasible. And every experiment that can be carried out in the laboratory can also be carried out on the computer. The extent to which, these match measures the completeness of the paradigm of molecular biology in the past of our modern scientific history.

Today, when one is looking back on these proposals, the genomics and proteomics, there is no doubt that Morowitz was right about the viability of collecting sequence data (Bonarius, Schmid, & Tramper, 1997). On the other hand, the challenges of writing down "the logic of life" in 600 steps and "completely understanding" a living cell still look fairly daunting. And what about the computer program that would simulate a living cell well enough to match the experiments that was carried out on real organisms? A computer program with exactly that goal was published last summer by Covert of Stanford University with his coworkers (Brenner, 2010). The program, called the WholeCell[10] simulation that describes the full life cycle of *Mycoplasma genitalium*, which is a bacterium from the genus that Morowitz was suggested. The model included, are all the major processes of life. The transcription is of DNA into RNA. A translation is of RNA into protein, metabolism of nutrients to produce energy. Their structural constituents have replication of the genome, and ultimately its reproduction by cell fission. The outputs of the simulation do seem to match experimental results (Brenner, 2010).

While considering the immense work of bright people out there, this niche approach seems to be genuine in three ways in adapting at a scenario of natural life spectrum. Initially we know already that there are billions of "ones-and-zeros" at our cyberspace. Now we have the problem to store them, process or re-use. This is the first way. Second way, the zeros-and-ones have some meaningful kernel when they are on transition. Third way, is the ontology adaption to understand and translate the meaning to the next phase of orderly use that we might process in Clouds. The Clouds technology we refer here is part of our foundation where we have the Knowledge-Intensive cyber-platform – an improved one. We fit this next generation technology here lively (meaning the improved and advanced user capability). The fitting would be on the form of an Application Programming Interface (API). This API would affect in cost reduction and it is also device or location independent with multi-tenancy. In another words, it would be sharing the resources. The centralization infrastructure would be an additional feature in which the system would be steady in controlling the peak-load capacity, and would add efficiency. Count on other features of the system such as that is scalable, would have enforced or remarkable security. It would be easy to maintain and measure the system in anytime, anywhere. It is also assured that the system available whenever needs. Figure 15 is shown here as a solution, which is apparently very close to the patent-pending[11] concept.

Figure 15. KISBED: a global approach where it plays

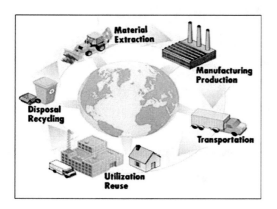

Figure 15 is a generic approach of the Knowledge-intensive platform (Pillai, Pyykkonen, & Salminen, 2009). The peer's active roles are plotted better in Figure 16.

This platform is based on an embedded system with a strong algorithm where the communication would simply be implemented in ZigBee, Wi-Fi, UWB. It is possible also to run any latest protocols that are standard. However, it is now assumed that it would be specific to the normal frequency level between 486 MHz and 2.6 GHz (Covert, Xiao, Chen, & Karr, 2008). On the other hand, this is not the final specification to the platform. When application-specific systems are to be implemented then said points above, are to be

Figure 16. Communication at the grid-cyber-semantic clouds

treated only as guidelines. Further investigation and or implementation strategies are drawn as in Figure 17.

The knowledge-intensive methodology is a technical invention, which was plotted in the year 2000. This came out as a series of researches that exceed the ambition and expectation of the team partially in the Silicon Valley, USA, and the rest in Finland. The motivation resulted at the end of the day with a final touch that drawn the work as an international patent concept. This patent is now pending in many countries. However, the Finnish Patent authority has approved this as patent in December 2009. This patent was instrumental in digging deeper to find evidence with several academic researches for implementing into real world. Those studies were led to light in diversified application protocols. Methodology of KISBED is very complex, though mathematically it seems to be almost perfect when one may see the trend in Clouds and Semantic-cyber-grids. Pipe-line structure has its root where most of the works with and without database; virtually anywhere and everywhere. One of the applications is tagged and that is shown in Figure 18.

CASE FORMULA 1: KISBED APPLICATION SCENARIO

One of the test-bed implementations shows the competence of KISBED-cyberinfra method here in real time motion feature at the Formula 1 car-race competition. Personalized experience to each race while you are seeing the ground actions through your Personal Digital Assistant (PDA). One could reach and feel the performances of the Formula 1 driver on the fly. We had experimented with joy this with Nokia 800 Tablet. The Viewer would have a number of options, while on motion – the driver, Car speed, built-in-cameras or simply associated advertisements that are randomly provided at the net. Alongside the video stream, one would be able to browse through their favorite driver's statistics

Figure 17. A similar approach as this one for test-bed in manufacturing for industrial applications

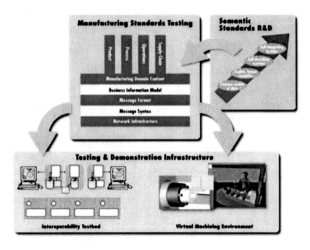

Figure 18. The experience system architecture for formula 1

Figure 19. Personalized experience while formula 1 driver is in action at his race track

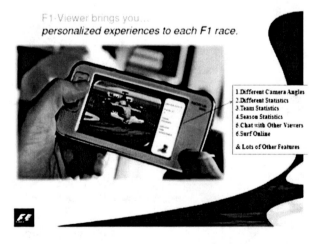

and earlier performances at the same time, while watching the race. Only the limit is the imagination of the viewer with his or her blended lifetime experiences at the race-area in-time. All what one need to a simple device that is operated at the wireless or cyberinfrastructure stadium and one is absorbing the entire "happening" in his or her chair where located—near and far—though this is true to believe. Figure 19 shows the real-time performance with KISBED tool in Nokia's Tablet.

SUMMARY OF CYBERINFRA PRODUCT STRATEGY

In this chapter, we are successful in offering a realistic groundwork. We have designed and implemented the cyberinfra product, and we presented here a concept. We were able to suggest that the interoperability schemes functions when applied at the real industrial environment. Today, the concept is almost ready and behaves right for the future. We have also specified the requirements and its challenges. We have introduced here also a very modern concept with various views of it. We have integrated all but possible elements to have an affordable pack. The entire system would be needed to run the concept creation and prototyping, perhaps in an open semantic infrastructure. Figure 1, as a system model in the beginning, is showing the technology roadmap with the steps to reach the vision on the top. To create a cyperinfra product with it, new concepts while engaging the interoperability scheme and plug and play capability is sometime very bold. Manufacturing industry has many numbers of stumbling blocks which are not only in developing systems and software, but also for a smart environment that is a daunting task. There is sensor hardware and software perceiving the environment. The application software that interprets and reasons about perception data and the effecter control the software acting on the environment. As well as many support systems that make the challenges to standards in posting them to Semantic Infrastructure or *per se* Semantic Web.

This software is commonly called "middleware." It lies between the software applications it assists and the platform it is based on. Middleware classically resides in a layer which is built directly on other layers of middleware. This characteristically is forming the higher abstractions with each other layer. Middleware must be designed by the API (Application Programming Interface). And it is providing the applications with the protocol(s) it supports (Bernstein, 1996). At the end of the chapter, we have introduced case studies of how this concept creation and prototyping functionality is working. One of the case studies is also verified the scenario of KISBED applications.

In concluding, we do say and see that there are a number of approaches and paradigm combinations to explore the Semantic Web Infrastructure. We have used here a basic perceive-reason-act artificial intelligence approach (Russell & Norvig, 1995), and this was also meant as a building-block to categorize a distinctive "set of needs." Additionally, it is also linking up "some wants," and it tries to address the desirable uniqueness. One might regard this in future studies.

REFERENCES

Ansys. (n.d.). *Introduction*. Retrieved from www.ansys.com

Barile, M. F., & Razin, S. (Eds.). (1989). *The mycoplasmas*. New York: Academic Press.

Bernstein, P. A. (1996). Middleware: A model for distributed system services. *Communications of the ACM, 39*(2), 86–98. doi:10.1145/230798.230809.

Beynon-Davies, P. (2004). *Database systems*. Basingstoke, UK: Palgrave.

Bonarius, H. P. J., Schmid, G., & Tramper, J. (1997). Flux analysis of underdetermined metabolic networks: The quest for the missing constraints. *Trends in Biotechnology, 15*, 308–314. doi:10.1016/S0167-7799(97)01067-6.

Brenner, S. (2010). Sequences and consequences. *Philosophical Transactions of the Royal Society of London*, *365*, 207–212. doi:10.1098/rstb.2009.0221 PMID:20008397.

CE-NET. (2010). Concurrent enterprising network of excellence. *Concurrent Engineering Roadmap Vision 2010*. Retrieved from http://www.ce-net.org

Chandrasekharan, B., Josephson, J. R., & Benjamins, V. R. (1999). What are ontologies, and why do we need them? *IEEE Intelligent Systems*, *14*(1), 20–26. doi:10.1109/5254.747902.

Chen, P.-S. (1976). *A sample entity – relationship diagram*. Retrieved from www.isu.edu

Covert, M. W., Xiao, N., Chen, T. J., & Karr, J. R. (2008). Integrating metabolic, transcriptional regulatory and signal transduction models. *Escherichia coli. Bioinformatics (Oxford, England)*, *24*, 2044–2050. doi:10.1093/bioinformatics/btn352 PMID:18621757.

Doumeingts, G., Li, M.-S., Piddington, C., & Ruggaber, R. (2005). *Enterprise interoperability – Research roadmap document*.

Economist. (2012, April 21), Third industrial revolution. *The Economist*.

Fenves, S. J. (2001). *A core product model for representing design information*. Gaithersburg, MD: NIST.

Goodenough. (n.d.). Principles of thermodynamics. In *Diesel Engineering Handbook* (5th ed.). New York: Diesel Publication, Inc..

Habib, M. K. (2008). Interdisciplinary mechatronics: Problem solving, creative thinking and concurrent design synergy. *International Journal of Mechatronics and Manufacturing Systems*, *1*(1), 264–269. doi:10.1504/IJMMS.2008.018272.

Haves, B. (2013). Imitation of life. *American Scientist*, 10–15.

Hawking, S. (2002). *On the shoulders of giants – The great works of physics and astronomy*. London: Penguin Books.

Jia, Y. M. (2007). *Robust h control*. Retrieved from www.sciencep.com

Pallot, M., Salminen, V., Pillai, B., & Kulvant, P. (2004). Business semantics: The magic instrument enabling plug & play collaboration? In *Proceedings of ICE 2004*. ICE. Berners-Lee, T., Hendler, J., & Lassila, O. (2001). The semantic web. *Scientific American*.

Pillai, B. (2002). *Process control and non-linear system using ontology*. Gaithersburg, MD: NIST.

Pillai, B. (2011). *Energy saving solutions at the wireless environment*. Paper presented at the Automation Systems 2011. Helsinki, Finland.

Pillai, B. (2012). *Bio-sensor implant for human body – Technology solutions*. Paper presented at the 8th International Conference on Humanized Systems. Daejeon, South Korea.

Pillai, B., Pyykkönen, M., & Salminen, V. (2009). *Knowledge-intensive arrangement of scattered data*. Patent No. 120639 dated December 31, 2009. Helsinki, Finland: Finnish Patent Office.

Pillai, B., & Salminen, V. (2004). *Trends in design and connectivity to broadband*. Paper presented at the 14th CIRP Design 2004. Cairo, Egypt.

Russell, S. J., & Norvig, P. (1995). *Artificial intelligence: A modern approach*. Upper Saddle River, NJ: Prentice Hall.

Salminen, V., & Pillai, B. (2004). Methodology on product life cycle challenge management for virtual enterprises. In Camarinho-Matos, L. M., & Afsarmanesh, H. (Eds.), *Processes and Foundations for Virtual Organizations*. Dordrecht, The Netherlands: Kluver Academic Publishers.

Salminen, V., & Pillai, B. (2005). *Integration of products and services – Towards system and performance.* Paper presented at the International Conference on Engineering Design. Melbourne, Australia.

Salminen, V., & Pillai, B. (2007). *Interoperability requirement challenges - Future trends.* Paper presented at the International Symposium on Collaborative Technologies and Systems. Orlando, FL.

Szykman, S., Sriram, R. D., & Regli, W. (2001). The role of knowledge in next generation product development systems. *ASME Journal of Computing and Information Science in Engineering, 1*(1), 3–11. doi:10.1115/1.1344238.

Zha, X. F., & Sriram, S. D. (2004). Feature-based component model for design of embedded system. In Gopalakrishnan, B. (Ed.), *Intelligent Systems in Design and Manufacturing, Proceedings of SPIE* (*Vol. 5605*, pp. 226–237). Bellingham, WA: SPIE. doi:10.1117/12.571612.

ENDNOTES

[1] The names of this view include "Hempel's model", "Hempel-Oppenheim (HO) Model," "Popper-Hempel Model," "Deductive-Nomological (D-N) Model," and the "Subsumption Theory."

[2] Physical products vacuum cleaners, fridges, cars, aircrafts, turbine engines, etc. There are things that are distributed via Internet also called as products. In other words, one may say that Microsoft's or Nokia's Internet products other than cell phones are not always physical.

[3] NIST: National Institute of Science and Technology, Department of Commerce, USA.

[4] ANSYS is a system provider and designing team in the US.

[5] Optiwise is a system owned by Real Time Systems Inc., Sunnyvale, CA, USA.

[6] Adopted courtesy of the Network Enabled Abilities (CEN) CWA draft 9–N 27 Research Roadmap Working Group.

[7] Keksimö is Finnish, and it means "the Inventors' factory," where new "things" are invented and/or created from ideas. This name or term "Keksimö" was first used in practice by Balan Pillai, Matti Pyykkönen, and Vesa Salminen in their patent, which is approved by the Finish Government Patent Authority with the patent number: 120639 and the World patent No: PCT/FI2006/050479.

[8] The Inventors are Balan Pillai, Matti Pyykkönen, and Vesa Salminen. All are from Finland.

[9] There is no one called a normal inventor, or explicitly, there is one who is normal. We are all extraordinary individuals, and we exist only once on this planet.

[10] www.wholecell.ed.

[11] An international patent is pending for Pillai, Pyykkönen, and Salminen. In Finland it is approved, and the patent number is 120639.

Compilation of References

Abbot, J. J., & Okamura, A. M. (2003). Virtual fixture architecture for telemanipulation. In *Proceedings of the IEEE International Conference on Robotics and Automation*, (pp. 2798-2805). IEEE.

Abdollahi, S. E., & Vahedi, A. (2004). *Dynamic modelling of micro-turbine generation systems using MATLAB/SIMULIN*. Tehran, Iran: Department of Electrical Engineering, Iran University of Science and Technology.

Aguero, J. L., Beroqui, M. C., & Di Pasquo, H. (2002). *Gas turbine control modifications for: Availability and limitation of spinning reserve and limitation of non-desired unloading*. Pluspetrol Energy, SA: Facultad de Ingeniería Universidad Nacional de La Plata, and Central Térmica Tucumán.

Aguiar, A. B. M., Pinto, J. O. P., & Nogueira, L. A. H. (2007). Modelling and simulation of natural gas microturbine for residential complexes. In *Proceedings of the World Congress on Engineering and Computer Science*. San Francisco, CA: WCECS.

Ahn, C. K., & Lee, M. C. (2000). An off-line automatic teaching by vision information for robotic assembly task. In *Proceedings of IEEE International Conference on Industrial Electronics, Control and Instrumentation* (pp. 2171–2176). IEEE.

Ailer, P., Santa, I., Szederkenyi, G., & Hangos, K. M. (2001). Nonlinear model-building of a low-power gas turbine. *Periodica Polytechnica Ser. Transp.*, *29*(1-2), 117–135.

Albath, J., Leopold, J., Sabharwal, C., & Maglia, A. (2010). *RCC-3D: Qualitative spatial reasoning in 3D*. Paper presented at the 23rd International Conference on Computer Applications in Industry and Engineering (CAINE). Las Vegas, NV.

Alexandrov, Frolov, & Massion. (1998). Axial synergies during human upper trunk bending. *Experimental Brain Research*, *118*, 210–220. doi:10.1007/s002210050274 PMID:9547090.

Alexandrov, Frolov, & Massion. (2001). Biomechanical analysis of movement strategies in human forward trunk bending. *Biological Cybernetics*, *84*, 425–434. doi:10.1007/PL00007986 PMID:11417054.

Al-Hamdan, Q. Z., & Ebaid, M. S. Y. (2006). Modelling and simulation of a gas turbine engine for power generation. *Journal of Engineering for Gas Turbines and Power*, *128*, 302–311. doi:10.1115/1.2061287.

Alise, M., Roberts, et al. (2009). On extending the wave variable method to multiple-DOF teleoperation systems. *IEEE/ASME Transactions on Mechatronics*, *14*(1), 55–63. doi:10.1109/TMECH.2008.2006181.

Alvarado, V. Y. (2007). *Design of biomimetic compliant devices for locomotion in liquid environments* (Vol. 68).

Álvares, A. J., & Ferreira. (2006). Webturning: Teleoperation of a CNC turning center through the internet. *Journal of Materials Processing Technology*, *179*, 251–259. doi:10.1016/j.jmatprotec.2006.03.096.

Amishev, D., & Evanson. (2011). Walking machines in forest operations. *FFR Technical Note*, *3*(9).

Amishev, D., Evanson, et al. (2009). Felling and bunching on steep terrain - A review of the literature. *FFR Technical Note*, *1*(7).

Anderson, J. M., & Chhabra, N. K. (2002). Maneuvering and stability performance of a robotic tuna. *Integrative and Comparative Biology*, *42*(1), 118–126. doi:10.1093/icb/42.1.118 PMID:21708700.

Anderson, N. H., & Butzin, C. A. (1974). Performance = motivation x ability: An integration-theoretical analysis. *Journal of Personality and Social Psychology, 30*(5), 598–604. doi:10.1037/h0037447.

Anderson, R. J., & Spong, M. W. (1989). Bilateral control of teleoperators with time delay. *IEEE Transactions on Automatic Control, 34*, 494–501. doi:10.1109/9.24201.

Ang, W. T., & Riviere, G. N. (2001). Neural network methods for error canceling in human-machine manipulation. In *Proceedings of the 23rd Annual International Conference of Engineering in Medicine and Biology*, (pp. 3462-3465). IEEE.

Ansys. (n.d.). *Introduction*. Retrieved from www.ansys.com

Aoi & Tsuchiya. (2005). Locomotion control of a biped robot using nonlinear oscillators. *Autonomous Robots, 19*(3), 219–232. doi:10.1007/s10514-005-4051-1.

Aoi & Tsuchiya. (2007). Adaptive behavior in turning of an oscillator-driven biped robot. *Autonomous Robots, 23*(1), 37–57. doi:10.1007/s10514-007-9029-8.

Aoi, Egi, & Sugimoto, Yamashita, Fujiki, & Tsuchiya. (2012). Functional roles of phase resetting in the gait transition of a biped robot from quadrupedal to bipedal locomotion. *IEEE Transactions on Robotics, 28*(6), 1244–1259. doi:10.1109/TRO.2012.2205489.

Aoi, Ogihara, & Funato, Sugimoto, & Tsuchiya. (2010). Evaluating functional roles of phase resetting in generation of adaptive human bipedal walking with a physiologically based model of the spinal pattern generator. *Biological Cybernetics, 102*(5), 373–387. doi:10.1007/s00422-010-0373-y PMID:20217427.

Arcara, P., & Melchiorri, C. (2002). Control schemes for teleoperation with time delay: A comparative study. *Robotics and Autonomous Systems, 28*, 49–64. doi:10.1016/S0921-8890(01)00164-6.

Argyros, A., Georgiadis, P., Trahanias, P., & Tsakiris, D. (2002). Semi-autonomous navigation of a robotic wheelchair. *Journal of Intelligent & Robotic Systems, 34*, 315–329. doi:10.1023/A:1016371922451.

Arkov, V., Kulikov, G., & Breikin, T. (2002). Life cycle support for dynamic modelling of gas turbines. In *Proceedings of the 15th Triennial World Congress*. Barcelona, Spain: World Congress.

Arkov, V., Evans, C., Fleming, P. J., Hill, D. C., Norton, J. P., & Pratt, I. et al. (2000). System identification strategies applied to aircraft gas turbine engine. *Annual Reviews in Control, 24*, 67–81.

Army, U. A. S. CoE Staff. (2010). U. S. army roadmap for unmanned aircraft systems 2010-2035. Fort Rucker, AL: U. S. Army UAS Center of Excellence (ATZQ-CDI-C).

Arrigada, J., Genrup, M., Loberg, A., & Assadi, M. (2003). Fault diagnosis system for an industrial gas turbine by means of neural networks. In *Proceedings of the International Gas Turbine Congress 2003*. Tokyo, Japan: IGTC.

Artigas, J., Preusche, G., & Hirzinger, G. (2006). Time domain passivity-based telepresence with time delay. In *Proceedings of IEEE/RSJ International Conference Intelligent Robotics Systems (IROS)*, (pp. 4205-4210). IEEE.

Asada, N., Matsuki, H., Minami, K., & Esashi, M. (1994). Silicone micromachined two-dimensional galvano optical scanner. *IEEE Transactions on Magnetics, 30*, 4647–4649. doi:10.1109/20.334177.

Ashikaga, M., Kohno, Y., Higashi, M., Nagai, K., & Ryu, M. (2003). A study on applying nonlinear control to gas turbine systems. In *Proceedings of the International Gas Turbine Congress*. Tokyo, Japan: IGTC.

Astrom, K., Block, D. J., & Spong, M. W. (2007). *The reaction wheel pendulum*. New York: Morgan and Claypool.

Ayers, J., Wilbur, C., & Olcott, C. (2000). *Lamprey robots*.

Aziminejad, A., Tavakoli, et al. (2008). Transparent time-delayed bilateral teleoperation using wave variables. *IEEE Transactions on Control Systems Technology, 16*(3), 548–555. doi:10.1109/TCST.2007.908222.

Baisch, A. T., Sreetharan, P. S., & Wood, R. J. (2010). Biologically-inspired locomotion of a 2g hexapod robot. In *Proceedings of EEE IROS 2010*, (pp. 5360-5365). IROS. doi:10.1109/IROS.2010.5651789

Bakkum, D. J., Gamblen, P. M., Ben-Ary, G., Chao, Z. C., & Potter, S. M. (2007). MEART: The semi-living artist. *Frontiers in Neurorobotics*, *1*(5), 1–10. doi: doi:10.3389/neuro.12.005.2007 PMID:18958272.

Bakkum, D. J., Shkolnik, A. C., Ben-Ary, G., Gamblen, P., DeMarse, T. B., & Potter, S. M. (2004). Removing some 'A' from AI: Embodied cultured networks. In Iida, F., Pfeifer, R., Steels, L., & Kuniyoshi, Y. (Eds.), *Embodied Artificial Intelligence* (pp. 130–145). New York: Springer. doi:10.1007/978-3-540-27833-7_10.

Balabanovic, M., & Shoham, Y. (1997). Fab: Content-based, collaborative recommendation. *Communications of the ACM*, *40*(3), 66–72. doi:10.1145/245108.245124.

Balakirsky, S., Kootbally, Z., Schlenoff, C., Kramer, T., & Gupta, S. (2012). *An industrial robotic knowledge representation for kit building applications*. Paper presented at the International Robots and Systems (IROS) Conference Vilamoura. Algarve, Portugal.

Balamurugan, S., Xavier, R. J., & Jeyakumar, A. E. (2008). ANN controller for heavy-duty gas turbine plant. *International Journal of Applied Engineering Research*, *3*(12), 1765–1771.

Balchen, J. G. (1996). Model based teleoperation of untethered underwater vehicles with manipulators. *Modeling. Identification and Control*, *17*, 37–45. doi:10.4173/mic.1996.1.4.

Banker, G. A., & Cowan, W. M. (1977). Rat hippocampal neurons in dispersed cell culture. *Brain Research*, *126*, 397–342. doi:10.1016/0006-8993(77)90594-7 PMID:861729.

Bard, J. F. (1986). An assessment of industrial robots: Capabilities, economics, and impacts. *Journal of Operations Management*, *6*, 99–124. doi:10.1016/0272-6963(86)90020-3.

Barile, M. F., & Razin, S. (Eds.). (1989). *The mycoplasmas*. New York: Academic Press.

Bartolini, C. M., Caresana, F., Comodi, G., Pelagalli, L., Renzi, M., & Vagni, S. (2011). Application of artificial neural networks to micro gas turbines. *Energy Conversion and Management*, *52*, 781–788. doi:10.1016/j.enconman.2010.08.003.

Basanez, L. (2009). *Surrez, et al.* Teleoperation.

Basso, M., Giarre, L., Groppi, S., & Zappa, G. (2004). NARX models of an industrial power plant gas turbine. *IEEE Transactions on Control Systems Technology*.

Bateman, J., & Farrar, S. (2006). *Spatial ontology baseline version 2.0. OntoSpace Project Report - Spatial Cognition SFB/TR 8: 11* [OntoSpace]. University of Bremen.

Batsomboon, P., & Tosunoglou, S. (1996). A review of teleoperation and telesensation system. In *Proceedings of the 1996 Florida Conference on Recent Advances in Robotics*. Florida Atlantic University.

Beal, D. N. (2003). *Propulsion through wake synchronization using a flapping foil*. Cambridge, MA: Massachusetts Institute of Technology.

Beaudoin, V. Collard, & Escazu. (2006). Deceptiveness and neutrality: The ND family of fitness landscapes. In *Proceedings of the 2006 Conference on Genetic and Evolutionary Computation*. GECCO.

Belhai, S., & Tagina, M. (2009). Modeling and prediction of the Internet delay using recurrent neural networks. *Journal of Networks*, *4*, 528–535.

Berestesky, P., Chopra, N., & Spong, M. W. (2004). Discrete-time passivity in bilateral teleoperation over the internet. In *Proceedings IEEE International Conference in Robotics and Automation*, (pp. 4557-4564). IEEE.

Bergkvist, I., Nordén, et al. (2006). Innovative unmanned harvester system. *Skogforsk Results*.

Bernard, E., Francois, C., & Patrick, R. (1992). A new approach to visual servoing in robotics. *IEEE Transactions on Robotics and Automation*, *8*(3), 313–326. doi:10.1109/70.143350.

Bernstein, P. A. (1996). Middleware: A model for distributed system services. *Communications of the ACM*, *39*(2), 86–98. doi:10.1145/230798.230809.

Bettocchi, R., Pinelli, M., Spina, P. R., & Venturini, M. (2005). Artificial intelligent for the diagnostics of gas turbine, part 1: Neural network approach. In *Proceedings of the ASME Turbo Expo 2005*. ASME.

Bettocchi, R., Pinelli, M., Spina, P. R., Venturini, M., & Burgio, M. (2004). Set up of a robust neural network for gas turbine simulation. In *Proceedings of the ASME Turbo Expo 2004*. Vienna, Austria: ASME.

Beynon-Davies, P. (2004). *Database systems*. Basingstoke, UK: Palgrave.

Bhardwaj, J. K., & Ashraf, H. (1995). Advanced silicon etching using high-density plasmas. In *Proceedings of SPIE Micromachining and Micro Fabrication Process Technology*, (Vol. 2639, pp. 224-233). SPIE. doi:10.1117/12.221279

Bianchi, Angelini, Orani, & Lacquaniti. (1998). Kinematic coordination in human gait: Relation to mechanical energy cost. *Journal of Neurophysiology*, *79*, 2155–2170. PMID:9535975.

Billingsley, J., Arto, et al. (2007). *Robotics in agriculture and forestry*.

Billingsley, J., Oetomo, et al. (2009). Agricultural robotics. *IEEE Robotics & Automation Magazine*, 19.

Bird & Layzell. (2002). The evolved radio and its implications for modelling the evolution of novel sensors. In *Proceedings of the Evolutionary Computation on 2002. CEC '02*, (pp. 1836–1841). Washington, DC: IEEE Computer Society.

Bleuler, B. Thiele & Zitzler. (2001). Multiobjective genetic programming: Reducing bloat using spea2. In *Proceedings of the 2001 Congress on Evolutionary Computation*, (vol. 1, pp. 536 –543). IEEE.

Bodor, R., Jackson, B., & Papanikolopoulos, N. (2003). *Vision-based human tracking and activity recognition*. Paper presented at the 11th Mediterranean Conference on Control and Automation. Athens, Greece.

Bonarius, H. P. J., Schmid, G., & Tramper, J. (1997). Flux analysis of underdetermined metabolic networks: The quest for the missing constraints. *Trends in Biotechnology*, *15*, 308–314. doi:10.1016/S0167-7799(97)01067-6.

Boukhnifer, M., & Ferreira. (2006). Stability and transparency for scaled teleoperation system. In *Proceedings of the 2006 IEEE/RSJ International Conference on Intelligent Robots and Systems*, (pp. 4217-4222). Bejing, China: IEEE.

Boyce, M. P. (2002). *Gas turbine engineering handbook* (2nd ed.). New York: Butterworth-Heinemann.

Boyer, F., Chablat, D., Lemoine, P., & Wenger, P. (2009). The eel-like robot. *Arxiv preprint arXiv:0908.4464*.

Boyle, B. G., McMaster, et al. (1995). Concept evaluation trials of teleoperation system for control of an underwater robotic arm by graphical simulation techniques. *Transactions of the Institute of Measurement and Control*, *17*(5), 242–250. doi:10.1177/014233129501700504.

Brace, M. (2009). Automating the mine. *Earthmatters*, (19).

Brander, M., Eriksson, et al. (2004). Automation of knuckleboom work can increase productivity. *Skogforsk Results*.

Brenner, S. (2010). Sequences and consequences. *Philosophical Transactions of the Royal Society of London*, *365*, 207–212. doi:10.1098/rstb.2009.0221 PMID:20008397.

Brooks, R. (1986). A robust layered control system for a mobile robot. *IEEE Journal on Robotics and Automation*, *2*, 14–23. doi:10.1109/JRA.1986.1087032.

Brown, E., Rodenberg, Amend, Mozeika, Steltz, Zakin, Jaeger. (2010). Universal robotic gripper based on the jamming of granular material. *The National Academy of Sciences*, *107*(44), 18809–18814. doi:10.1073/pnas.1003250107.

Brown, H. J., & Xu, Y. (1997). A single wheel, gyroscopically stabilized robot. *IEEE Robotics & Automation Magazine*, *4*(3), 39–44. doi:10.1109/100.618022.

Burke, Degtyarenko, & Simon. (2001). Patterns of locomotor drive to motoneurons and last-order interneurons: Clues to the structure of the CPG. *Journal of Neurophysiology*, *86*, 447–462. PMID:11431524.

Burns, R. S. (2001). *Advanced control engineering*. New York: Butterworth-Heinemann Publications.

Buss, M., & Schmidt, G. (1999). Control problems in multi-modal telepresence systems. In *Proceedings of the European Control Conference (ECC'99)*, (pp. 65-101). ECC.

Cai, W. C., Yu, Q., & Wang, H. (2004). A fast contour-based approach to circle and ellipse detection. In *Proceedings of IEEE 5th World Conference on Intelligence Control and Automation*, (vol. 5, pp. 4686-4690). IEEE. doi: 10.1109/WCICA.2004.1342408

Camporeale, S. M., Fortunato, B., & Mastrovito, M. (2006). A modular code for real time dynamic simulation of gas turbines in SIMULINK. In *Proceedings of ASME,* (Vol. 128). ASME.

CCC. (2009). *A roadmap for US robotics: From internet to robotics.* Retrieved from http://www.us-robotics.us/reports/CCC Report.pdf

CE-NET. (2010). Concurrent enterprising network of excellence. *Concurrent Engineering Roadmap Vision 2010.* Retrieved from http://www.ce-net.org

Centeno, P., Egido, I., Domingo, C., Fernandez, F., Rouco, L., & Gonzalez, M. (2002). *Review of gas turbine models for power system stability studies.* Madrid, Spain: Universidad Pontificia Comillas and Endesa Generación.

Chabrol, J. (1987). Industrial robot standardization at ISO. *Robotics, 3*(2). doi:10.1016/0167-8493(87)90012-X.

Cha, D. H., Cho, H. S., & Kim, S. (1996). Design of a force reflection controller for telerobot systems using neural network and fuzzy logic. *Journal of Intelligent & Robotic Systems, 16,* 1–24. doi:10.1007/BF00309653.

Cha, D.-H., & Cho, H. S. (2002). A neurofuzzy algorithm-based advanced bilateral controller for telerobot systems. *ICASE: The Institute of Control Automation and Systems Engineers, 4,* 100–107.

Chan, Man, Tang, & Kwong. (2008). A jumping gene paradigm for evolutionary multiobjective optimization. *IEEE Transactions on Evolutionary Computation, 12,* 143–159.

Chandrasekharan, B., Josephson, J. R., & Benjamins, V. R. (1999). What are ontologies, and why do we need them? *IEEE Intelligent Systems, 14*(1), 20–26. doi:10.1109/5254.747902.

Chao, Z. C., Bakkum, D. J., & Potter, S. M. (2007). Region-specific network plasticity in simulated and living cortical networks: Comparison of the center of activity trajectory (CAT) with other statistics. *Journal of Neural Engineering, 4,* 1–15. doi:10.1088/1741-2560/4/3/015 PMID:17409475.

Chao, Z. C., Bakkum, D. J., & Potter, S. M. (2008). Shaping embodied neural networks for adaptive goal-directed behavior. *PLoS Computational Biology, 4*(3). doi:10.1371/journal.pcbi.1000042 PMID:18369432.

Charlie I. (n.d.). Retrieved 2012 http://Web.mit.edu/towtank/www-new/Tuna/Tuna1/pictures.html

Chen, P.-S. (1976). *A sample entity–relationship diagram.* Retrieved from www.isu.edu

Chen, Q., Quan, J., & Xia, J. (2007). Neural network based multiple model adaptive predictive control for teleoperation. In *Proceedings 4th International Symposium on Neural Networks: Advances in Neural Networks,* (pp. 64-69). Berlin: Springer.

Chen, X. (2012). *Mobile robotics in agriculture and horticulture.*

Chen, Y.-C., & Trinkle, J. C. (1993). On the form-closure of polygonal objects with frictional and frictionless contact models. In *Proceedings IEEE International Conference on Robotics and Automation* (pp. 963–970). IEEE. Doi:10.1109/Robot.1993.292268

Chen, J. Y. C., & Thropp. (2007). Review of low frame rate effects on human performance. *IEEE Transactions on Systems, Man, and Cybernetics. Part A, Systems and Humans, 37*(6), 1063–1076. doi:10.1109/TSMCA.2007.904779.

Chen, J., Sun, J., Liu, G., & Rees, D. (2010). new delay dependent stability criteria for neural networks with time varying interval delay. *Physics Letters. [Part A], 374,* 4397–4405. doi:10.1016/j.physleta.2010.08.070.

Chiras, N., Evans, C., & Rees, D. (2001). Nonlinear gas turbine modelling using NARMAX structures. *IEEE Transactions on Instrumentation and Measurement, 50*(4), 893–898. doi:10.1109/19.948295.

Chiras, N., Evans, C., & Rees, D. (2001). *Nonlinear modelling and validation of an aircraft gas turbine engine.* Wales, UK: School of Electronics, University of Glamorgan.

Chiras, N., Evans, C., & Rees, D. (2002). *Nonlinear gas turbine modelling using feedforward neural networks.* Wales, UK: School of Electronics, University of Glamorgan. doi:10.1115/GT2002-30035.

Chiras, N., Evans, C., & Rees, D. (2002). Global nonlinear modelling of gas turbine dynamics using NARMAX structures. *Journal of Engineering for Gas Turbines and Power, 124,* 817. doi:10.1115/1.1470483.

Chong, N. Y., Kotoku, et al. (2003). A collaborative multi-site teleoperation over an ISDN. *Mechatronics*, (13): 957–979. doi:10.1016/S0957-4158(03)00010-2.

Chopra, N., Berestesky, et al. (2008). Bilateral teleoperation over unreliable communication networks. *IEEE Transactions on Control Systems Technology, 16*, 304–313. doi:10.1109/TCST.2007.903397.

Choudhury, T., & Borriello, G. (2008). *The mobile sensing platform: An embedded system for activity recognition.* IEEE Pervasive Magazine. doi:10.1109/MPRV.2008.39.

Cho, Y. C., & Park, J. H. (2003). Stable bilateral teleoperation under a time delay using a robust impedance control. *Journal of Mechatronics, 15*, 1–10.

Ciocarlie. (2010). *Low-dimensional robotic grasping: Eigengrasp subspaces and optimized underactuation.* New York: Columbia University.

Clifton, D. (2006). *Condition monitoring of gas turbine engines.* St. Cross College.

Cole, Hsu, & Sastry. (1989). Dynamic regrasping by co-ordinated control of sliding for a multifingered hand. In *Proceedings of the 1989 IEEE International Conference* (pp. 781 – 786). IEEE. Doi:10.1109/Robot.1989.100079

Colgate, J. E., & Lynch, K. M. (2004). Mechanics and control of swimming: A review. *IEEE Journal of Oceanic Engineering, 29*(3), 660–673. doi:10.1109/JOE.2004.833208.

Conti, F., & Khatib, O. (2005). Spanning large work spaces using small haptic devices. In *Proceedings 1st Joint Eurohaptics Conference*, (pp. 183-188). Pisa, Italy: Eurohaptics.

Conway, Hultborn, & Kiehn. (1987). Proprioceptive input resets central locomotor rhythm in the spinal cat. *Experimental Brain Research, 68*, 643–656. doi:10.1007/BF00249807 PMID:3691733.

Covert, M. W., Xiao, N., Chen, T. J., & Karr, J. R. (2008). Integrating metabolic, transcriptional regulatory and signal transduction models. *Escherichia coli. Bioinformatics (Oxford, England), 24*, 2044–2050. doi:10.1093/bioinformatics/btn352 PMID:18621757.

Cozzi, L., D'Angelo, P., Chiappalone, M., Ide, A. N., Novellino, A., Martinoia, S., & Sanguineti, V. (2005). Coding and decoding of information in a bi-directional neural interface. *Neurocomputing, 65*, 783–792. doi:10.1016/j.neucom.2004.10.075.

Cross, R. (2006). The fall and bounce of pencils and other elongated objects. *American Journal of Physics, 74*(1), 26–30. doi:10.1119/1.2121752.

Czyzowicz, Stojmenovic, & Urrutia. (1999). Immobilizing a shape. *International Journal of Computational Geometry & Applications*, 181–206. doi:10.1142/S0218195999000133.

Daenotes-Electronics Notes, Lectures, Theory and Projects for Engineering Students. (2012). Retrieved from http://www.daenotes.com

Dag, G. (2009). *Fish robots search for pollution in the waters.* Retrieved from http://www.robaid.com/bionics/fish-robots-search-for-pollution-in-the-waters.htm

Danica, K., & Henrik, I. C. (2002). Survey on visual servoing for manipulation. *Technical Report ISRN KTH/NA/P-02/01-SE, CVAP259.* Retrieved from http://citeseerx.ist.psu.edu/

Danna-dos-Santos, Slomka, Zatsiorsky, & Latash. (2007). Muscle modes and synergies during voluntary body sway. *Experimental Brain Research, 179*, 533–550. doi:10.1007/s00221-006-0812-0 PMID:17221222.

DARPA. (2012). *Information innovation office: Mind's eye program.* Retrieved from http://www.darpa.mil/Our_Work/I2O/Programs/Minds_Eye.aspx

d'Avella & Bizzi. (2005). Shared and specific muscle synergies in natural motor behaviors. *Proceedings of the National Academy of Science USA, 102*(8), 3076-3081.

d'Avella, Saltiel, & Bizzi. (2003). Combinations of muscle synergies in the construction of a natural motor behavior. *Nature Neuroscience, 6*, 300–308. doi:10.1038/nn1010 PMID:12563264.

Deb, Pratap, Agarwal, & Meyarivan. (2002). A fast and elitist multiobjective genetic algorithm: Nsga-ii. *IEEE Transactions on Evolutionary Computation, 6*(2), 182–197. doi:10.1109/4235.996017.

Dede, M., & Tosunoglu. (2006). Fault-tolerant teleoperation systems design. *Industrial Robot: An International Journal, 33*, 365–372. doi:10.1108/01439910610685034.

DeJong, B. P., Faulring, et al. (2006). Lessons learned from a novel teleoperation testbed. *Industrial Robot: An International Journal, 33*(3), 187–193. doi:10.1108/01439910610659097.

DeMarse, T. B., Wagenaar, D. A., Blau, W. P. A., & Potter, S. M. (2001). The neurally controlled animat: Biological brains acting with simulated bodies. *Autonomous Robots, 11*, 305–310. doi:10.1023/A:1012407611130 PMID:18584059.

Demolombe, R., Mara, A., & Fern, O. (2006). *Intention recognition in the situation calculus and probability theory frameworks.* Paper presented at the Computational Logic in Multi-Agent Systems (CLIMA) Conference. New York, NY.

Deng, L. F. (2004). *Comparison of image-based and position-based robot visual servoing methods and improvements.* Waterloo, Canada: Waterloo University.

Dodd, N., & Martin, J. (1997, June). Using neural networks to optimize gas turbine aero engines. *Computing & Control Engineering Journal*, 129-135.

Doerr, G. Hebbinghaus, & Neumann. (2007). A rigorous view on neutrality. In Proceedings of Evolutionary Computation, (pp. 2591–2597). IEEE.

Dogangil, G., Ozcicek, E., & Kuzucu, A. (2005). *Design, construction, and control of a robotic dolphin.* Paper presented at the IEEE International Conference on Robotics and Biomimetics (ROBIO). New York, NY.

Dominici, Ivanenko, Cappellini, d'Avella, Mondì, Cicchese, Lacquaniti. (2011). Locomotor primitives in newborn babies and their development. *Science, 334*, 997–999. doi:10.1126/science.1210617 PMID:22096202.

Donald, B. R., Levey, C. G., McGray, C. D., Paprotny, I., & Rus, D. (2006). An untethered, electrostatic, globally controllable MEMS micro-robot. *Journal of Microelectromechanical Systems, 15*, 1–15. doi:10.1109/JMEMS.2005.863697.

Doumeingts, G., Li, M.-S., Piddington, C., & Ruggaber, R. (2005). *Enterprise interoperability – Research roadmap document.*

Dowling. (1999). Limbless locomotion: Learning to crawl. In *Proceedings of Robotics and Automation,* (vol. 4, pp. 3001–3006). IEEE.

Drew, Kalaska, & Krouchev. (2008). Muscle synergies during locomotion in the cat: A model for motor cortex control. *The Journal of Physiology, 586*(5), 1239–1245. doi:10.1113/jphysiol.2007.146605 PMID:18202098.

Duff, E., Caris, et al. (2009). *The development of a telerobotic rock breaker.*

Duff, E., Usher, et al. (2007). *Web-based tele-robotics revisited.*

Duysens. (1977). Fluctuations in sensitivity to rhythm resetting effects during the cat's step cycle. *Brain Research, 133*(1), 190-195.

Economist. (2012, April 21), Third industrial revolution. *The Economist.*

Edqvist, E., Snis, N., Mohr, R. C., Scholz, O., Corradi, P., & Gao, J. et al. (2009). Evaluation of building technology for mass producible millimeter-sized robots using flexible printed circuit boards. *Journal of Micromechanics and Microengineering*, 1–11. doi: doi:10.1088/0960-1317/19/7/075011.

Emde, M., Krahwinkler, et al. (2012). Sensor fusion in forestry. *GPS World, 21*, 48.

Epstein, M., Colgate, J. E., & MacIver, M. A. (2006). *Generating thrust with a biologically-inspired robotic ribbon fin.* Paper presented at the IEEE/RSJ International Conference on Intelligent Robots and Systems. New York, NY.

Evans, C., Chiras, N., Guillaume, P., & Rees, D. (2001). *Multivariable modelling of gas turbine dynamics.* Wales, UK: School of Electronics, University of Glamorgan.

Evans, C., Rees, D., & Hill, D. (1998). Frequency domain identification of gas turbine dynamics. *IEEE Transactions on Control Systems Technology, 6*(5), 651–662. doi:10.1109/87.709500.

Evanson, T., & Amishev. (2010). A steep slope excavator feller buncher. *FFR Technical Note, 3*, 1-8.

Fast, M. (2010). *Artificial neural networks for gas turbine monitoring.* (Doctoral Thesis). Lund University, Lund, Sweden.

Fast, M., Assadi, M., & De, S. (2008). Condition based maintenance of gas turbines using simulation data and artificial neural network: A demonstration of feasibility. In *Proceedings of the ASME Turbo Expo 2008*. Berlin, Germany: ASME.

Fast, M., Palme, T., & Genrup, M. (2009). A novel approach for gas turbines monitoring combining CUSUM technique and artificial neural network. In *Proceedings of ASME Turbo Expo 2009*. Orlando, FL: ASME.

Fast, M., Palme, T., & Karlsson, A. (2009). Gas turbines sensor validation through classification with artificial neural networks. In *Proceedings of ECOS 2009*. ECOS.

Fast, M., Assadi, M., & De, S. (2009). Development and multi-utility of an ANN model for an industrial gas turbine. *Journal of Applied Energy*, 86(1), 9–17. doi:10.1016/j.apenergy.2008.03.018.

Fast, M., & Palme, T. (2010). Application of artificial neural network to the condition monitoring and diagnosis of a combined heat and power plant. *Journal of Energy*, 35(2), 114–1120.

Fenves, S. J. (2001). *A core product model for representing design information*. Gaithersburg, MD: NIST.

Ferguson, K. R. (1978). *Past and future challenges in developing remote systems technology*.

Fischler, M. A., & Bolles, R. C. (1981). Random sample consensus: a paradigm for model fitting with applications to image analysis and automated cartography. *Communications of the ACM*, 24, 381–395. doi:10.1145/358669.358692.

Flemmer, H. (2004). *Aspects of using passivity in bilateral telemanipulation* (Tech. Report No. TRITA-MMK 2004:16). Stockholm, Sweden: Royal Institute of Technology.

Fong, T., & Thorpe. (2001). Vehicle teleoperation interfaces. *Autonomous Robots*, 9–18. doi:10.1023/A:1011295826834.

Freitas, Duarte, & Latash. (2006). Two kinematic synergies in voluntary whole-body movements during standing. *Journal of Neurophysiology*, 95, 636–645. doi:10.1152/jn.00482.2005 PMID:16267118.

Freksa, C. (1992). Using orientation information for qualitative spatial reasoning. In Frank, A. U., Campari, I., & Formentini, U. (Eds.), *Theories and methods of spatio-temporal reasoning in geographic space* (pp. 162–178). Heidelberg, Germany: Springer. doi:10.1007/3-540-55966-3_10.

Funato, Aoi, Oshima, & Tsuchiya. (2010). Variant and invariant patterns embedded in human locomotion through whole body kinematic coordination. *Experimental Brain Research*, 205(4), 497–511. doi:10.1007/s00221-010-2385-1 PMID:20700732.

Furukawa, N. Taku, & Ishikawa. (2006). Dynamic regrasping using a high-speed multifingered hand and a high-speed vision system. In *Proceedings of the 2006 IEEE International Conference* (pp. 181 – 187). IEEE. Doi:10.1109/Robot.2006.1641181

Galv, Poli, Kattan, O'Neill, & Brabazon. (2011). Neutrality in evolutionary algorithms: What do we know? *Evolving Systems*, 2, 145–163. doi:10.1007/s12530-011-9030-5.

Ganjefar, S., Momeni, H., Sharifi, F. J., & Behesthi Hamidi, M. T. (2003). Behaviour of Smith predictor in teleoperation systems with modeling and delay time error. In *Proceedings IEEE Conference on Control Applications*, (pp. 1176-1180). IEEE.

Gao, J., Bi, S., Li, J., & Liu, C. (2009). *Design and experiments of robot fish propelled by pectoral fins*. Paper presented at the IEEE International Conference on Robotics and Biomimetics (ROBIO). New York, NY.

Gao, J., Bi, S., Xu, Y., & Liu, C. (2007). *Development and design of a robotic manta ray featuring flexible pectoral fins*. Paper presented at the IEEE International Conference on Robotics and Biomimetics (ROBIO). New York, NY.

Garcia-Valdovinos, L. G., Parra-Vega, V., & Arteaga, M. A. (2007). Observer-based sliding-mode impedance control of bilateral teleoperation under constant unknown time-delay. *International Journal of Robotics and Autonomous Systems*, 55, 609–617. doi:10.1016/j.robot.2007.05.011.

Ge, D. F., Takeuchi, Y., & Asakawa, N. (1993). Automation of polishing work by an industrial robot – 2nd report, automatic generation of collision- free polishing path. *Transactions of the Japan Society of Mechanical Engineers*, 59(561), 1574–1580. doi:10.1299/kikaic.59.1574.

Ghanbari, A. Abdi, et al. (2010). Haptic guidance for microrobotic intracellular injection. In *Proceedings of the 2010 3rd IEEE RAS & EMBS*. IEEE.

Ghorbani, H., Ghaffari, A., & Rahnama, M. (2008). Constrained model predictive control implementation for a heavy-duty gas turbine power plant. *WSEAS Transactions on Systems and Control, 6*(3), 507–516.

Giampaolo, T. (2009). *Gas turbine handbook – Principles and practice* (4th ed.). The Fairmont Press, Inc..

Goodenough. (n.d.). Principles of thermodynamics. In *Diesel Engineering Handbook* (5th ed.). New York: Diesel Publication, Inc..

Grasser, F., D'Arrigo, A., Colombi, S., & Rufer, A. (2002). Joe: A mobile, inverted pendulum. *IEEE Transactions on Industrial Electronics, 49*(1), 107–114. doi:10.1109/41.982254.

Griffiths, G., & Edwards, I. (2003). AUVs: Designing and operating next generation vehicles. *Elsevier Oceanography Series, 69*, 229–236. doi:10.1016/S0422-9894(03)80038-7.

Grillner. (1975). Locomotion in vertebrates: Central mechanisms and reflex interaction. *Physiology Review, 55*(2), 247-304.

Gross, G. W., Rieske, E., Kreutzberg, G. W., & Meyer, A. (1977). A new fixed-array multimicroelectrode system designed for long term monitoring of extracellular single unit neuronal activity in vitro. *Neuroscience Letters, 6*, 101–105. doi:10.1016/0304-3940(77)90003-9 PMID:19605037.

Gross, G. W., Williams, A. N., & Lucas, J. H. (1982). Recording of spontaneous activity with photo etched microelectrode surfaces from mouse spinal neurons in culture. *Journal of Neuroscience Methods, 5*(1–2), 13–22. doi:10.1016/0165-0270(82)90046-2 PMID:7057675.

Guarnieri, M., Takao, I., Fukushima, E., & Hirose, S. (2007). Helios VIII search and rescue robot: Design of an adaptive gripper and system improvements. In *Proceedings of Intelligent Robots and Systems, 2007* (pp. 1775–1780). IEEE. doi:10.1109/IROS.2007.4399372.

Guez, A., & Selinsky, J. (1989). Neuro-controller design via supervised and unsupervised learning. *Journal of Intelligent & Robotic Systems, 2*, 307–335. doi:10.1007/BF00238695.

Guil, N., & Zapata, E. L. (1997). Lower order circle and ellipse hough transform. *Pattern Recognition, 30*(10), 1729–1744. doi:10.1016/S0031-3203(96)00191-4.

Gunn, C., & Zhu. (2010). *Haptic tele-operation of industrial equipment.*

Guo, Gruver, & Jin. (1992). Grasp planning for multifingered robot hands. In *Proceedings of the IEEE International Conference on Robotics and Automation*. IEEE. Doi:10.1109/Robot.1992.219919

Habib, M. K. (2006). Mechatronics engineering the evolution, the needs and the challenges. In *Proceedings of the 32nd Annual Conference of IEEE Industrial Electronics Society (IECON 2006)*, (pp. 4510-4515). IEEE.

Habib, M. K., Watanabe, K., & Izumi, K. (2007). Biomimetcs robots: From bio-inspiration to implementation. In *Proceedings of the 33rd Annual Conference of the IEEE Industrial Electronics Society*, (pp. 143-148). IEEE. doi:10.1109/IECON.2007.4460382

Habib, M. K. (2007). Mechatronics: A unifying interdisciplinary and intelligent engineering paradigm. *IEEE Industrial Electronics Magazine, 1*(2), 12–24. doi:10.1109/MIE.2007.901480.

Habib, M. K. (2008). Interdisciplinary mechatronics: Problem solving, creative thinking and concurrent design synergy. *International Journal of Mechatronics and Manufacturing Systems, 1*(1), 264–269. doi:10.1504/IJMMS.2008.018272.

Habib, M. K. (2011). Biomimetcs: Innovations and Robotics. *International Journal of Mechatronics and Manufacturing Systems, 4*(2), 113–134. doi:10.1504/IJMMS.2011.039263.

Hainsworth, D. W. (2001). Teleoperation user interfaces for mining robotics. *Autonomous Robots, 11*, 19–28. doi:10.1023/A:1011299910904.

Han, & Trinkle. (1998). Dextrous manipulation by rolling and finger gaiting. In *Proceedings of the 1998 IEEE International Conference* (pp. 730–735). IEEE. Doi:10.1109/Robot.1998.677060

Hannaford, B. (1989). A design framework for teleoperators with kinesthetic feedback. *IEEE Transactions on Robotics and Automation, 5*(4), 426–434. doi:10.1109/70.88057.

Hannett, L. N., Jee, G., & Fardanesh, B. (1995). A governer/turbine model for a twin-shaft combustion turbine. *IEEE Transactions on Power Systems, 10*(1). doi:10.1109/59.373935.

Hansson, A., & Servin. (2010). Semi-autonomous shared control of large-scale manipulator arms. *Control Engineering Practice, 18*(9), 1069–1076. doi:10.1016/j.conengprac.2010.05.015.

Hao, M., Sun, Z. Q., & Fujii, M. (2007). Ellipse detection based long range robust visual servoing. *Journal of Central South University, 38*, 432–439.

Harmelen, F., & McGuiness, D. (2004). *OWL web ontology language overview*. Retrieved from http://www.w3.org/TR/2004/REC-owl-features-20040210/

Hasegawa & Morookalaboratory. (2012). *Multi-fingered robot hand*. Retrieved from http://fortune.ait.kyushu-u.ac.jp/research-e.html

Hashemzadeh, F., Hassanzadeh, I., Tavakoli, M., & Alizadeh, G. (2000). A new method for bilateral teleoperation passivity under varying stability. In *Mathematical Problems in Engineering*. Hindawi Publ. Corp..

Hashtrudi-Zaad, K., & Salcudean. (2002). Transparency in time-delayed systems and the effect of local force feedback for transparent teleoperation. *IEEE Transactions on Robotics and Automation, 18*(1), 108–114. doi:10.1109/70.988981.

Haves, B. (2013). Imitation of life. *American Scientist*, 10–15.

Hawking, S. (2002). *On the shoulders of giants – The great works of physics and astronomy*. London: Penguin Books.

Hayati, S., Volpe, R., Backes, P., Balaram, J., Welch, R., Ivlev, R., & Laubach, S. (1997). The rocky 7 rover: A mars sciencecraft prototype. In Proceedings of Robotics and Automation, 1997 (vol. 3, pp. 2458 2464). IEEE. doi: doi:10.1109/ROBOT.1997.619330.

Hayn, H., & Schwarzmann. (2010). A haptically enhanced operational concept for a hydraulic excavator. *Robert Bosch GmbH*, 199-220.

Hebb, D. O. (1949). *The organization of behaviour: A neuropsychological theory*. New York: John Wiley & Sons.

Hellström, T. (2005). *Intelligent vehicles in forestry.*

Hellström, T., Lärkeryd, et al. (2008). *Autonomous forest machines - Past, present and future.*

Hellström, T., Lärkeryd, et al. (2009). Autonomous forest vehicles: Historic, envisioned, and state-of-the-art. *International Journal of Forest Engineering, 20*(1).

Hirabayashi, T., Akizono, et al. (2006). Teleoperation of construction machines with haptic information for underwater applications. *Automation in Construction*, (15): 563–570. doi:10.1016/j.autcon.2005.07.008.

Hirata, K. (2001). *Up-down motion for a fish robot*. Retrieved from http://www.nmri.go.jp/eng/khirata/fish/general/updown/index_e.html

Hirose. (1993). *Biologically inspired robots: Snake-like locomotors and manipulators*. Oxford, UK: Oxford University Press.

Ho, C. T., & Chen, L. H. (1995). A fast ellipse/circle detector using geometric symmetry. *Pattern Recognition, 28*(1), 117–124. doi:10.1016/0031-3203(94)00077-Y.

Hokayem, P. F., & Spong. (2006). Bilateral teleoperation: An historical survey. *Automatica*, 2035–2057. doi:10.1016/j.automatica.2006.06.027.

Homma, D. (2003). Metal artificial muscle bio metal fiber. *RSJ, 21*, 22–24.

Hoogs, A., & Perera, A. G. A. (2008). *Video activity recognition in the real world*. Paper presented at the American Association of Artificial Intelligence (AAAI) Conference. New York, NY.

Hu, H., Liu, J., Dukes, I., & Francis, G. (2006). *Design of 3D swim patterns for autonomous robotic fish.*

Hu, J., Ren, J., Thompson, J., & Sheridan, T. (1996). Fuzzy sliding control of a force reflecting teleoperator system. In *Proceedings 5th IEEE International Conference on Fuzzy Sytems*, (pp. 2162-2167). New Orleans, LA: IEEE.

Huang, J. (2002). Neurocontrol of telerobotic systems with time delays. In Lewis, F. L., Campos, J., & Selmic, R. (Eds.), *Neuro-Fuzzy Control of Industrial Systems with Actuator Nonlinearities*. Philadelphia: SIAM Publishing. doi:10.1137/1.9780898717563.ch8.

Hu, H., Yu, et al. (2001). Internet-based robotic systems for teleoperation. *Assembly Automation, 21,* 143–151. doi:10.1108/01445150110388513.

Hung, N., Narikiyo, T., & Tuan, H. (2003). Nonlinear adaptive control of master slave system in teleoperation. *Control Engineering Practice, 11,* 1–10. doi:10.1016/S0967-0661(02)00068-0.

Hu, T., Shen, L., Lin, L., & Xu, H. (2009). Biological inspirations, kinematics modeling, mechanism design and experiments on an undulating robotic fin inspired by gymnarchus niloticus. *Mechanism and Machine Theory, 44*(3), 633–645. doi:10.1016/j.mechmachtheory.2008.08.013.

Huynen, Stadler, & Fontana. (1996). Smoothness within ruggedness: The role of neutrality in adaptation. *Proceedings of the National Academy of Sciences of the United States of America, 93,* 397–401. doi:10.1073/pnas.93.1.397 PMID:8552647.

Ijspeert. (2008). Central pattern generators for locomotion control in animals and robots: a review. *Neural Networking, 21*(4), 642-653.

Ikemoto, T., Nagashino, H., Kinouchi, Y., & Yoshinaga, T. (1997). oscillatory mode transitions in a four coupled neural oscillator model. In *Proceedings of the International Symposium on Nonlinear Theory and its Applications,* (pp. 561-564). Retrieved from http://ci.nii.ac.jp/naid/110003291511/en/

Iqbal, A., Roth, H., & Abu-Zaitoon, M. (2005). Stabilization of delayed teleoperation using predictive time-domain passivity control. In *Proceedings IASTED International Conference on Robotics and Applications,* (pp. 20-25). Cambridge, MA: IASTED.

Isermann, R. (1997). Mechatronics systems: A challenge for control engineering. In *Proceedings American Control Conference,* (pp. 2617-2632). ACC.

Ivanenko, Cappellini, & Dominici, Poppele, & Lacquaniti. (2005). Coordination of locomotion with voluntary movements in humans. *The Journal of Neuroscience, 25*(31), 7238–7253. doi:10.1523/JNEUROSCI.1327-05.2005 PMID:16079406.

Ivanenko, Poppele, & Lacquaniti. (2004). Five basic muscle activation patterns account for muscle activity during human locomotion. *The Journal of Physiology, 556,* 267–282. doi:10.1113/jphysiol.2003.057174 PMID:14724214.

Ivanenko, Poppele, & Lacquaniti. (2006). Motor control programs and walking. *The Neuroscientist, 12*(4), 339–348. doi:10.1177/1073858406287987 PMID:16840710.

Jakuba, M. V. (2000). *Design and fabrication of a flexible hull for a bio-mimetic swimming apparatus.*

Jarvis, P. A., Lunt, T. F., & Myers, K. L. (2005). Identifying terrorist activity with AI plan-recognition technology. *AI Magazine, 26*(3), 9.

Jelali, M., & Kroll, A. (2004). *Hydraulic servo-systems: Modelling, identification, and control.* Berlin: Springer Publications.

Jenelle, A. P., & Harvey, L. (2003). Uncalibrated eye-in-hand visual servoing. *The International Journal of Robotics Research, 22,* 805–819. doi:10.1177/0278364 90302210002.

Jeon, H., Kim, T., & Choi, J. (2008). *Ontology-based user intention recognition for proactive planning of intelligent robot behavior.* Paper presented at the International Conference on Multimedia and Ubiquitous Engineering. Busan, Korea.

Jia, Y. M. (2007). *Robust h control.* Retrieved from www.sciencep.com

Jimbo, Y., Robinson, H. P., & Kawana, A. (1993). Simultaneous measurement of intracellular calcium and electrical activity from patterned neural networks in culture. *IEEE Transactions on Bio-Medical Engineering, 40,* 804–810. doi:10.1109/10.238465 PMID:8258447.

Jimbo, Y., Tateno, T., & Robinson, H. P. C. (1999). Simultaneous induction of pathway-specific potentiation and depression in networks of cortical neurons. *Biophysical Journal, 76,* 670–678. doi:10.1016/S0006-3495(99)77234-6 PMID:9929472.

Junzhi, Y., Min, T., Shuo, W., & Erkui, C. (2004). Development of a biomimetic robotic fish and its control algorithm. *IEEE Transactions on Systems, Man, and Cybernetics. Part B, Cybernetics*, *34*(4), 1798–1810. doi:10.1109/TSMCB.2004.831151.

Jurado, F. (2005). Nonlinear modelling of micro-turbines using NARX structures on the distribution feeder. *Energy Conversion and Management*, *46*, 385–401. doi:10.1016/j.enconman.2004.03.012.

Kaikko, J., Talonpoika, T., & Sarkomma, P. (2002). Gas turbine model for an on-line condition monitoring and diagnostic system. In *Proceeding of the Australasian Universities Power Engineering Conference (AUPEC2002)*. Melbourne, Australia: AUPEC.

Kaiser, U., & Steinhagen, W. (1995). A low-power transponder IC for high- performance identification systems. *IEEE Journal of Solid-State Circuits*, *30*, 306–310. doi:10.1109/4.364446.

Kaplan, F., Oudeyer, P.-Y., Kubinyi, E., & Miklosi, A. (2002). Robotic clicker training. *Robotics and Autonomous Systems*, *38*(3-4), 197–206. doi:10.1016/S0921-8890(02)00168-9.

Karibe, H. (2008). *Easy book for non-contact IC card.* Tokyo, Japan: Nikkan Kogyo Shinbun Publishing.

Kato, N. (n.d.). *Fish fin motion.* Retrieved from http://www.naoe.eng.osaka-u.ac.jp/~kato/fin9.html

Kato, N., Wicaksono, B. W., & Suzuki, Y. (2000). *Development of biology-inspired autonomous underwater vehicle BASS III with high maneuverability.* Paper presented at the International Symposium on Underwater Technology. New York, NY.

Kato, N. (2000). Control performance in the horizontal plane of a fish robot with mechanical pectoral fins. *IEEE Journal of Oceanic Engineering*, *25*(1), 121–129. doi:10.1109/48.820744.

Kawato, M. (2008). Brain controlled robots. *HFSP Journal*, *2*(3), 136–142. doi:10.2976/1.2931144 PMID:19404467.

Kawato, M., Uno, Y., Isobe, M., & Suzuki, R. (1988). Hierarchical neural network model for voluntary movement with application to robotics. *IEEE Control Systems Magazine*, 8–16. doi:10.1109/37.1867.

Kelley, R., Tavakkoli, A., King, C., Nicolescu, M., Nicolescu, M., & Bebis, G. (2008). *Understanding human intentions via hidden Markov models in autonomous mobile robots.* Paper presented at the 3rd ACM/IEEE International Conference on Human Robot Interaction. Amsterdam, The Netherlands.

Khosravi-el-Hossani, M., & Dorosti, Q. (2009). Improvement of gas turbine performance test in combine-cycle. World Academy of Science, Engineering and Technology, 58.

Kikuchi, K. Takeo, & Kosuge. (1998). Teleoperation system via computer network for dynamic environment. In *Proceedings of the IEEE International Conference on Robotics and Automation*, (pp. 3534-3539). Leuven, Belgium: IEEE.

Kikuchi, K., Sakaguchi, K., Sudo, T., Bushida, N., Chiba, Y., & Asai, Y. (2008). A study on a wheel-based stair-climbing robot with a hopping mechanism. *Mechanical Systems and Signal Processing*, *22*(6), 1316–1326. doi:10.1016/j.ymssp.2008.03.002.

Kim, H., Lee, B., & Kim, R. (2007). *A study on the motion mechanism of articulated fish robot.* Paper presented at the International Conference on Mechatronics and Automation (ICMA). New York, NY.

Kim, W. S. (1989). Developments of new force reflecting control schemes and an application to a teleoperator training simulator. In *Proceedings IEEE International Conference on Robotics and Automation*, (pp. 1764-1767). IEEE.

Kim, D., Kim, et al. (2009). Excavator tele-operation system using a human arm. *Automation in Construction*, *18*(2), 173–182. doi:10.1016/j.autcon.2008.07.002.

Kim, S. K., Pilidis, P., & Yin, J. (2000). *Gas turbine dynamic simulation using SIMULINK.* Society of Automation Engineers, Inc. doi:10.4271/2000-01-3647.

Kimura, Fukuoka, & Cohen. (2007). Adaptive dynamic walking of a quadruped robot on natural ground based on biological concepts. *The International Journal of Robotics Research*, *26*(5), 475–490. doi:10.1177/0278364907078089.

Kishigami, J. (2005). *Textbook on RFID – Whole RF IC tags towards ubiquitous society.* Tokyo, Japan: ASKII.

Kiyohara, A., Taguchi, T., & Kudoh, S. N. (2011). Effects of electrical stimulation on autonomous electrical activity in a cultured rat hippocampal neuronal network. *IEEJ Transactions on Electrical and Electronic Engineering*, 6(2), 163–167. doi:10.1002/tee.20639.

Klang, H., & Lindholm, A. (2005). *Modelling and simulation of a gas turbine*. (PhD Thesis). Linkopings University, Linkopings, Sweden.

Kodati, P., Hinkle, J., Winn, A., & Deng, X. (2008). Microautonomous robotic ostraciiform (MARCO), hydrodynamics, design, and fabrication. *IEEE Transactions on Robotics*, 24(1), 105–117. doi:10.1109/TRO.2008.915446.

Kohonen, T. (2001). *Self-organizing maps* (3rd ed.). New York: Springer. doi:10.1007/978-3-642-56927-2.

Komatsu. (2012). *KOMATSU: Autonomous haulage system - Komatsu's pioneering technology deployed at Rio Tinto Mine in Australia.*

Konda, T., Tensyo, S., & Yamaguchi, T. (2002). Learning: Phased method for average reward reinforcement learning - Preliminary results. In Ishizuka & Sattar (Eds.), PRICAI2002: Trends in Artificial Intelligence (LNAI), (vol. 2417, pp. 208-217). Berlin: Springer.

Konidaris, G., & Barto, A. (2006). Automonous shaping: Knowledge transfer in reinforcement learning. In *Proceedings of the 23rd International Conference on Machine Learning* (pp. 489-496). New York: ACM.

Koza. (1994). *Genetic programming II: Automatic discovery of reusable programs*. Cambridge, MA: MIT Press.

Koza. (2003). *Genetic programming IV: Routine human-competitive machine intelligence*. Boston: Kluver Academic Publishers.

Koza, Keane, & Yu, Bennett, & Mydlowec. (2000). Automatic creation of human-competitive programs and controllers by means of genetic programming. *Genetic Programming and Evolvable Machines*, 1, 121–164. doi:10.1023/A:1010076532029.

Krug, D. Krzysztof, & Boyko. (2010). On the efficient computation of independent contact regions for force closure grasps. In *Proceedings of the IEEE/RSJ International Conference* (pp. 586–591). IEEE.

Krut, Bégoc, Dombre, & Pierrot. (2010). Extension of the form-closure property to underactuated hands. *IEEE Transactions on Robotics*, 853 – 866. Doi:10.1109/Tro.2010.2060830

Kubo, R., Iiyama, N., Natori, K., Ohnishi, K., & Furukawa, H. (2007). Performance analysis of a three-channel control architecture for bilateral teleoperation with time delay. *IEEJ Transactions*, 1A(127), 1224–1230. doi:10.1541/ieejias.127.1224.

Kudoh, S. N., Hosokawa, C., Kiyohara, A., Taguchi, T., & Hayashi, I. (2007). Biomodeling system—Interaction between living neuronal networks and the outer world. *Journal of Robotics and Mechatronics*, 19(5), 592–600.

Kudoh, S. N., Nagai, R., Kiyosue, K., & Taguchi, T. (2001). PKC and CaMKII dependent synaptic potentiation in cultured cerebral neurons. *Brain Research*, 915(1), 79–87. doi:10.1016/S0006-8993(01)02835-9 PMID:11578622.

Kudoh, S. N., & Taguchi, T. (2003). Operation of spatio-temporal patterns in living neuronal networks cultured on a microelectrode array. *Journal of Advanced Computational Intelligence and Intelligent Informatics*, 8, 100–107.

Kudoh, S. N., Tokuda, M., Kiyohara, A., Hosokawa, C., Taguchi, T., & Hayashi, I. (2011). Vitroid—The robot system with an interface between a living neuronal network and outer world. *International Journal of Mechatronics and Manufacturing Systems*, 4(2), 135–149. doi:10.1504/IJMMS.2011.039264.

Kulikov, G. G., & Thompson, H. A. (2004). *Dynamic modelling of gas turbines*. London: Springer-Verlag.

Kumagai, M., & Ochiai, T. (2008). Development of a robot balancing on a ball. In Proceedings of Control, Automation and Systems, 2008 (pp. 433-438). ICCAS. doi: doi:10.1109/ICCAS.2008.4694680.

Kushida, D., Nakamura, M., Goto, S., & Kyura, N. (2001). Human direct teaching of industrial articulated robot arms based on force-free control. *Artificial Life and Robotics*, 5(1), 26–32. doi:10.1007/BF02481317.

Kuyucu, Trefzer, Greensted, Miller, & Tyrrell. (2008). Fitness functions for the unconstrained evolution of digital circuits. In *Proceedings of the 9th IEEE Congress on Evolutionary Computation (CEC08)*, (pp. 2589–2596). Hong Kong: IEEE.

Lachat, D., & Ijspeert, A. J. (2005). *BoxyBot, the fish robot: Project report.* Lusanne, France: Biologially Inspired Robotics Group, School of Computer and Communication Sciences at Ecole Polytechnique Federale de Lausanne.

Lafreniere-Roula & McCrea. (2005). Deletions of rhythmic motoneuron activity during fictive locomotion and scratch provide clues to the organization of the mammalian central pattern generator. *Journal of Neurophysiology, 94,* 1120–1132. doi:10.1152/jn.00216.2005 PMID:15872066.

Lamon, P., Krebs, A., Lauria, M., Siegwart, R., & Shooter, S. (2004). Wheel torque control for a rough terrain rover. In *Proceedings of Robotics and Automation, 2004 (Vol. 5,* pp. 4682–4687). IEEE. doi:10.1109/ROBOT.2004.1302456.

Langdon & Nordin. (2000). Seeding genetic programming populations. In *Proceedings of the European Conference on Genetic Programming,* (pp. 304–315). London, UK: Springer-Verlag.

Lankenau, A. (2001). Avoiding mode confusion in service robots – The Bremen autonomous wheelchair as an example. In *Proceedings of the 7th International Conference on Rehabilitation Robotics (ICORR 2001),* (pp. 162-167). Evry, France: ICORR.

Lauwers, T., Kantor, G. A., & Hollis, R. (2006). A dynamically stable single-wheeled mobile robot with inverse mouse-ball drive. In *Proceedings of the 2006 IEEE International Conference on Robotics and Automation (* (pp. 2884 - 2889). IEEE.

Lawrence, D. A. (1993). Stability and transparency in bilateral teleoperation. *IEEE Transactions on Robotics and Automation, 9*(5), 2649–2655. doi:10.1109/70.258054.

Lawton, D. T., Schoppers, et al. (1989). *Interactive model based vision system for telerobotic vehicles.*

Lazzaretto, A., & Toffolo, A. (2001). Analytical and neural network models for gas turbine design and off-design simulation. *International Journal of Applied Thermodynamics, 4*(4), 173–182.

Leblebici, T., Calli, B., Unel, M., Sabanovic, A., Bogosyan, S., & Gokasan, M. (2011). Delay compensation in bilateral control using a sliding mode observer. *Journal of Electrical Engineering and Computer Science, 19,* 851–859.

Lee, D., & Spong, M. W. (2006). Passive bilateral teleoperation with constant time delay. *IEEE Transactions on Robotics, 22,* 269–281. doi:10.1109/TRO.2005.862037.

Lee, H. K., & Chung, M. J. (1998). Adaptive controller of a master-slave system for transparent teleoperation. *Journal of Robotic Systems, 15,* 465–475. doi:10.1002/(SICI)1097-4563(199808)15:8<465::AID-ROB3>3.0.CO;2-J.

Lee, S., & Lee, H. S. (1993). Modeling, design and evaluation of advanced teleoperator control system with short time delay. *IEEE Transactions on Robotics and Automation, 9,* 607–623. doi:10.1109/70.258053.

Lee, S., Park, J., & Han, C. (2007). Optimal control of a mackerel-mimicking robot for energy efficient trajectory tracking. *Journal of Bionics Engineering, 4*(4), 209–215. doi:10.1016/S1672-6529(07)60034-1.

Lee, Y. K., Mavris, D. N., Volovoi, V. V., Yuan, M., & Fisher, T. (2010). A fault diagnosis method for industrial gas turbines using bayesian data analysis. *Journal of Engineering for Gas Turbines and Power, 132,* 041602–1. doi:10.1115/1.3204508.

Lehmann, E. A. (2006). Particle filtering approach to adaptive time-delay estimation. In *Proceedings 2006 International Conference IEEE Acoustic, Speech and Signal Processing,* (pp. 1129-1132). Toulouse, France: IEEE.

Lenz, J., & Edelstein, S. A. (2006). Magnetic sensors and their applications. *IEEE Sensors Journal, 6,* 631–649. doi:10.1109/JSEN.2006.874493.

Lewis, F. L., Liu, K., & Yesildirek, A. (1995). Neural net robot controller with guaranteed tracking performance. *IEEE Transactions on Neural Networks, 6,* 703–715. doi:10.1109/72.377975 PMID:18263355.

Li, Liu, & Cai. (2003). On computing three-finger force-closure grasps of 2-D and 3-D objects. *IEEE Transactions on Robotics and Automation,* 155 – 161. Doi:10.1109/Tra.2002.806774

Liang, J., Wang, T., & Wen, L. (2011). Development of a two-joint robotic fish for real-world exploration. *Journal of Field Robotics, 28*(1), 70–79. doi:10.1002/rob.20363.

Ligozat, G. (1993). Qualitative triangulation for spatial reasoning. In Campari, I., & Frank, A. U. (Eds.), *CPSIT 1993 (Vol. 716,* pp. 54–68). Heidelberg, Germany: Springer.

Lin, Q., & Kuo. (1999). Assisting the teleoperation of an unmanned underwater vehicle using a synthetic subsea scenario. *Presence (Cambridge, Mass.), 8*(5), 520–530. doi:10.1162/105474699566431.

Lin, W., Tsai, C., & Liu, J. (2001). Robust neuro-fuzzy control of multivariable system by tuning consequent membership function. *Fuzzy Sets and Systems, 124.*

Liu, J. (2006). *Welcome! Essex robotic fish.* Retrieved from http://dces.essex.ac.uk/staff/hhu/jliua/index.htm#Profile

Liu, J., Dukes, I., & Hu, H. (2005). *Novel mechatronics design for a robotic fish.* Paper presented at the IEEE/RSJ International Conference on Intelligent Robots and Systems (IROS). New York, NY.

Liu, J., Dukes, I., Knight, R., & Hu, H. (2004). Development of fish-like swimming behaviours for an autonomous robotic fish. *Proceedings of the Control, 4.*

Liu, Sheng, & Jiao. (2009). Gene transposon based clonal selection algorithm for clustering. In *Proceedings of the 11th Annual Conference on Genetic and Evolutionary Computation,* (pp. 1251–1258). IEEE.

Liu, Habib, Watanabe, & Izumi. (2008). Central pattern generators based on Matsuoka oscillators for the locomotion of biped robots. *Artificial Life and Robotics, 12*(1), 264–269. doi:10.1007/s10015-007-0479-z.

Liu, J., & Hu, H. (2004). A 3D simulator for autonomous robotic fish. *International Journal of Automation and Computing, 1*(1), 42–50. doi:10.1007/s11633-004-0042-5.

Ljung, L., & Glad, T. (1994). *Modelling of dynamic systems.* Englewood Cliffs, NJ: PTR Prentice Hall.

Lobo, Miller, & Fontana. (2004). Neutrality in technological landscapes. *Santa Fe Working Paper.*

Low, K. H. (2006). *Maneuvering and buoyancy control of robotic fish integrating with modular undulating fins.* Paper presented at the IEEE International Conference on Robotics and Biomimetics, ROBIO '06. New York, NY.

Low, K. H., & Willy, A. (2005). *Development and initial investigation of NTU robotic fish with modular flexible fins.* Paper presented at the IEEE International Conference on Mechatronics and Automation. New York, NY.

Low, K. H., & Yu, J. (2007). *Development of modular and reconfigurable biomimetic robotic fish with undulating fin.* Paper presented at the IEEE International Conference on Robotics and Biomimetics (ROBIO). New York, NY.

Low, K. H., Prabu, S., Yang, J., Zhang, S., & Zhang, Y. (2007). *Design and initial testing of a single-motor-driven spatial pectoral fin mechanism.* Paper presented at the International Conference on Mechatronics and Automation (ICMA). New York, NY.

Low, K. H. (2009). Modelling and parametric study of modular undulating fin rays for fish robots. *Mechanism and Machine Theory, 44*(3), 615–632. doi:10.1016/j.mechmachtheory.2008.11.009.

Lozano, R., Chopra, N., & Spong, M. W. (2002). Passivation of force reflecting bilateral teleoperators with time varying delay. In *Proceedings Mechatronics '02.* Enschede, The Netherlands: Mechatronics.

MacDonncadha, M. (1997). *Japanese forestry and forest harvesting techniques: With emphasis on the potential use of Japanese harvesting techniques on steep & sensitive sites in Ireland.*

Maeda, Y., Ishido, N., Kikuchi, H., & Arai, T. (2002). Teaching of grasp/graspless manipulation for industrial robots by human demonstration. In *Proceedings IEEE/RSJ International Conference on Intelligent Robots and Systems* (pp. 1523–1528). IEEE.

Mahadevan, S. (1996). Average reward reinforcement learning: Foundations, algorithms, and empirical results. *Machine Learning, 22*(1-3), 159–195. doi:10.1007/BF00114727.

Mamdani, E. H. (1974). Application of fuzzy algorithms for the control of a simple dynamic plant. *Proceedings of the IEEE, 121*(12), 1585–1588.

MAN Diesel & Turbo Company. (n.d.). *THM gas turbine basic training.* Retrieved from http://www.mandiesel-turbo.com

Manifar, S. (2010). Application of GA based neuro-fuzzy automatic generation for teleoperation systems. In *Proceedings 6th International Conference on Digital Context, Multimedia Technology and its Applications (IDC-2010),* (pp. 296-301). IDC.

Mantzaris, J., & Vournas, C. (2007). Modelling and stability of a single-shaft combined-cycle power plant. *International Journal of Thermodynamics, 10*(2), 71–78.

Mao, W., & Gratch, J. (2004). *A utility-based approach to intention recognition*. Paper presented at the AAMAS Workshop on Agent Tracking: Modeling Other Agents from Observations. New York, NY.

Marthi, B. (2007). Automatic shaping and decomposition of reward functions. In *Proceedings of the 24th International Conference on Machine Learning*. ACM.

Martin, D., Burstein, M., Hobbs, J., Lassila, O., McDermott, D., McIlrath, S., et al. (2004). *OWL-S: Semantic markup of web services*. Retrieved from http://www.w3.org/Submission/OWL-S/

Marvel, J., Hong, T.-H., & Messina, E. (2012). *2011 solutions in perception challenge performance metrics and results*. Paper presented at the Performance Metrics for Intelligent Systems (PerMIS) Conference. College Park, Maryland.

Mason, R., & Burdick, J. (2000). Construction and modelling of a carangiform robotic fish. *Experimental Robotics, 6*, 235–242.

Massimono, M. J., Sheridan, J. B., & Roseborough, J. B. (1989). One handed tracking in six degrees of freedom. In *Proceedings IEEE International Conference Systems, Man, and Cybernetics*, (vol. 2, pp. 498-503). IEEE.

Matsuoka, K. (1987). Mechanism of frequency and pattern control in the neural rhythm generators. *Biological Cybernetics, 56*, 345–353. doi:10.1007/BF00319514 PMID:3620533.

Mavroidis, C., Pfeiffer, C., & Mosley, M. (2000). 5.1 conventional actuators, shape memory alloys, and electrorheological fluids. *Automation, Miniature Robotics, and Sensors for Nondestructive Testing and Evaluation, 4*, 189.

McClintock. (1950). The origin and behaviour of mutable loci in maize. *Proceedings of the National Academy of Sciences of the United States of America, 36*, 344–355. doi:10.1073/pnas.36.6.344 PMID:15430309.

McConaghy, Vladislavleva, & Riolo. (2010). Genetic programming theory and practice. In *Genetic programming theory and practice 2010: An introduction*, (pp. vii–xviii). New York: Springer.

McGregor & Harvey. (2005). Embracing plagiarism: Theoretical, biological and empirical justification for copy operators in genetic optimisation. *Genetic Programming and Evolvable Machines, 6*, 407–420. doi:10.1007/s10710-005-4804-9.

McIsaac, K. A., & Ostrowski, J. P. (2003). Motion planning for anguilliform locomotion. *IEEE Transactions on Robotics and Automation, 19*(4), 637–652. doi:10.1109/TRA.2003.814495.

Minh, V. T., Nitin, A., & Mansor, W. (2007). *Fault detection and control of process systems* (p. 80321). Article, ID: Mathematical Problems in Engineering. doi:10.1155/2007/80321.

Montana. (1995). Strongly typed genetic programming. *Evolutionary Computation, 3*(2), 199–230.

Moratz, R., Dylla, F., & Frommberger, J. (2005). *A relative orientation algebra with adjustable granularity*. Paper presented at the Workshop on Agents in Real-Time and Dynamic Environments. Edinburgh, UK.

Morgansen, K. A. (2003). Geometric methods for modeling and control of a free-swimming carangiform fish robot. In *Proceedings of the 13th Unmanned Untethered Submersible Technology*. UUST.

Morgansen, K. A., Triplett, B. I., & Klein, D. J. (2007). Geometric methods for modeling and control of free-swimming fin-actuated underwater vehicles. *IEEE Transactions on Robotics, 23*(6), 1184–1199. doi:10.1109/LED.2007.911625.

Mori, Mori, & Nakajima. (2006). Higher nervous control of quadrupedal vs bipedal locomotion in non-human primates: Common and specific properties. In H. Kimura, K. Tsuchiya, A. Ishiguro, & H. Witte (Eds.), *Adaptive Motion of Animals and Machines*, (pp. 53-65). New York: Springer.

Mori, Tachibana, & Takasu, Nakajima, & Mori. (2001). Bipedal locomotion by the normally quadrupedal Japanese monkey. *Acta Physiologica et Pharmacologica Bulgarica, 26*(3), 147–150. PMID:11695527.

Morowitz. (2002). *The emergence of everything: How the world became complex.* Oxford, UK: Oxford University Press.

Mostafavi, M., Alaktiwi, A., & Agnew, B. (1998). Thermodynamic analysis of combined open-cycle twin-shaft gas turbine (Brayton cycle) and exhaust gas operated absorption refrigeration unit. *Applied Thermal Engineering, 18*, 847–856. doi:10.1016/S1359-4311(97)00105-1.

MT1 Gallery. (n.d.). Retrieved from http://dces.essex.ac.uk/staff/hhu/jliua/images/gallery/MT1/P1010049.JPG

Mu, J., & Rees, D. (2004). Approximate model predictive control for gas turbine engines. In *Proceeding of the 2004 American Control Conference,* (pp. 5704-5709). Boston, MA: ACC.

Mu, J., Rees, D., & Liu, G. P. (2004). Advanced controller design for aircraft gas turbine engines. *Control Engineering Practice, 13*, 1001–1015. doi:10.1016/j.conengprac.2004.11.001.

Mulder, F., & Voorbraak, F. (2003). A formal description of tactical plan recognition. *Information Fusion, 4*(1). doi:10.1016/S1566-2535(02)00102-1.

Munir, S., & Book, W. J. (2002). Internet-based teleoperation using wave variables with prediction. *IEEE/ASME Transactions on Mechatronics, 7*, 119–127. doi:10.1109/TMECH.2002.1011249.

Murakami, N., Ito, et al. (2008). Development of a teleoperation system for agricultural vehicles. *Computers and Electronics in Agriculture, 63*, 81–88. doi:10.1016/j.compag.2008.01.015.

Murata, M., Ito, H., Taenaka, T., & Kudoh, S. N. (2011). Modification of activity pattern induced by synaptic enhancements in a semi-artificial network of living neurons. In Proceedings of the International Symposium on Micro-NanoMechatronics and Human Science (MHS), (pp. 250–254). MHS.

Museum, C. H. (2012). *Robots- Utah/M.I.T. dextrous hand closeup.* Retrieved from http://www.computerhistory.org/collections/accession/102693567

Nabney, I. T., & Cressy, D. C. (1996). Neural network control of a gas turbine. *Neural Computing & Applications, 4*, 198–208. doi:10.1007/BF01413818.

Nagamachi, M. (1986). Human factors of industrial robots and robot safety management in Japan. *Applied Ergonomics, 17*, 9–18. doi:10.1016/0003-6870(86)90187-0 PMID:15676565.

Nagata, F., Watanabe, K., & Izumi, K. (2001). Furniture polishing robot using a trajectory generator based on cutter location data. In *Proceedings of 2001 IEEE International Conference on Robotics and Automation* (pp. 319–324). IEEE.

Nagata, F., Watanabe, K., & Kiguchi, K. (2006). Joystick teaching system for industrial robots using fuzzy compliance control. In Industrial Robotics: Theory, Modelling and Control (pp. 799–812). INTECH.

Nagata, F., Kusumoto, Y., & Watanabe, K. (2009). Intelligent machining system for the artistic design of wooden paint rollers. *Robotics and Computer-integrated Manufacturing, 25*(3), 680–688. doi:10.1016/j.rcim.2008.05.001.

Nagpal, M., Moshref, A., Morison, G. K., & Kundur, P. (2001). Experience with testing and modelling of gas turbines. In *Proceedings of the Power Engineering Society Winter Meeting.* IEEE.

Nair. (2013). Types of artificial gripper mechanisms. *Society of Robotics & Automation.* Retrieved from http://sra.vjti.info/knowledgebase/posts-on-a-roll-1/gripper-mechanisms

Nair, P. S., & Saunder, A. T. (1997). Hough transform based ellipse detection algorithm. *Pattern Recognition Letters, 17*, 777–784. doi:10.1016/0167-8655(96)00014-1.

Najjar, Y. S. H. (1994). Performance of single-cycle gas turbine engines in two modes of operation. *Energy Conversion and Management, 35*(5), 433–441. doi:10.1016/0196-8904(94)90101-5.

Nakada, K., Asai, T., & Amemiya, Y. (2003). An analog CMOS central pattern generator for interlimb coordination in quadruped locomotion. *IEEE Transactions on Neural Networks, 14*, 1356–1365. doi:10.1109/TNN.2003.816381 PMID:18244582.

Nakajima, Mori, & Takasu, Mori, Matsuyama, & Mori. (2004). Biomechanical constraints in hindlimb joints during the quadrupedal versus bipedal locomotion of M. fuscata. *Progress in Brain Research, 143*, 183–190. doi:10.1016/S0079-6123(03)43018-5 PMID:14653163.

Nakanishi, Morimoto, & Endo, Cheng, Schaal, & Kawato. (2004). Learning from demonstration and adaptation of biped locomotion. *Robotics and Autonomous Systems, 47*(2-3), 79–91. doi:10.1016/j.robot.2004.03.003.

Nakanishi, Nomura, & Sato. (2006). Stumbling with optimal phase reset during gait can prevent a humanoid from falling. *Biological Cybernetics, 95*, 503–515. doi:10.1007/s00422-006-0102-8 PMID:16969676.

Nakashima, M., & Ono, K. (2002). Development of a two-joint dolphin robot. *Neurotechnology for Biomimetic Robots*, 309.

Nakashima, M., Takahashi, Y., Tsubaki, T., & Ono, K. (2004). Three-dimensional maneuverability of the dolphin robot (roll control and loop-the-loop motion). *Bio-Mechanisms of Swimming and Flying*, 79-92.

Natori, K., Tsuji, T., & Ohnisi, K. (2004). Robust bilateral control with internet communication. In *Proceedings of the 30th IEEE Annual Conference on Industrial Electronics*, (pp. 2321-2326). Busan, Korea: IEEE.

Nau, D., Ghallab, M., & Traverso, P. (2004). *Automated planning: Theory and practice*. San Francisco, CA: Morgan Kaufmann Publishers Inc..

Neto, P., Pires, J. N., & Moreira, A. P. (2010). CAD based off-line robot programming. In *Proceedings of IEEE International Conference on Robotics Automation and Mechatronics* (pp. 516–521). IEEE.

New Zealand Forestry Owners Association. (2011). *New Zealand Plantation forest industry - Facts & figures 2010/2011*.

New Zealand Ministry of Primary Industries. (2012). *New Zealand ministry of primary industries - International trade statistics*. Retrieved from http://www.mpi.govt.nz/news-resources/statistics-forecasting/international-trade.aspx

Newman, M., & Balakirsky, S. (2011). Contests in China put next-generation robot technology to the test. *IEEE Robotics & Automation Magazine*. doi:10.1109/MRA.2011.942540 PMID:23028210.

Ng, A. Y., Harada, D., & Russell, S. J. (1999). Policy invariance under reward transformations: Theory and application to reward shaping. In *Proceedings of the 16th International Conference on Machine Learning* (pp. 278-287). New York, NY: ACM.

Nguyen. (1986). Constructing force-closure grasps. In *Proceedings IEEE International Conference on Robotics and Automation*. IEEE. Doi:10.1109/Robot.1986.1087483

Nie, J., & Linkens, D. (1995). *Fuzzy-neural control: Principles, algorithms and applications*. Englewood Cliffs, NJ: Prentice Hall.

Niemeyer, G., & Slotine, J.-E. (1997). Using wave variables for system analysis and robot control. In *Proceedings IEEE International Conference on Robotics and Automation*. Albuquerque, NM: IEEE.

Niemeyer, G., Preusche, C., & Hirzinger, G. (2008). Telerobotics. In Siciliano & Habib (Eds.), Springer Handbook of Robotics, (pp. 741-757). Berlin: Springer.

Niemeyer, G., & Slotine, J. J. (1991). Stable adaptive teleoperation. *IEEE Journal of Oceanic Engineering, 16*, 152–162. doi:10.1109/48.64895.

Niemeyer, G., & Slotine, J.-J. E. (2004). Telemanipulation with time delays. *The International Journal of Robotics Research, 23*, 873–890. doi:10.1177/0278364904045563.

Nikkei Computer. (2003). *IC tag (RFID)*. Tokyo, Japan: Nikkei BP.

Nishino, T. (2000). Forestry principles in Japan. Encyclopedia of Life Support Systems.

Noetzel, R., Meisenberg, A., & Bartos, A. (2006). Customized GMR-spin valve sensors for low field applications. [Daegu, Korea: IEEE.]. *Proceedings of IEEE Sensors, 2006*, 1020–1023. doi:10.1109/ICSENS.2007.355798.

Nolfi, F. Miglino, & Mondada. (1994). How to evolve autonomous robots: Different approaches in evolutionary robotics. In *Proceedings of the 4th International Workshop on Artificial Life*. Boston: MIT Press.

Norgaard, M., Ravn, O., Poulsen, N. K., & Hansen, L. K. (2000). *Neural networks for modelling and control of dynamic systems*. Berlin: Springer Publications. doi:10.1007/978-1-4471-0453-7.

Northeastern Marine Science Center - Biomimetic Robots. (n.d.). Retrieved from http://www.expo21xx.com/automation21xx/17600_st2_university/default.htm

Novellino, A., D'Angelo, P., Cozzi, L., Chiappalone, M., Sanguineti, V., & Martinoia, S. (2007). Connecting neurons to a mobile robot: An in vitro bidirectional neural interface. *Journal of Computational Intelligence and Neuroscience*. doi: 10.1155/2007/12725

Nowacki, Higgins, & Maquilan, Swart, Doak, & Land-Weber. (2009). A functional role for transposases in a large eukaryotic genome. *Science, 324*(5929), 935–938. doi:10.1126/science.1170023 PMID:19372392.

Occupational Safety and Health Service. (1998). *Safety And health in forest operations - Approved code of practice*. Occupational Safety and Health Service.

Ogaji, S. O. T., Singh, R., & Probert, S. D. (2002). Multiple-sensor fault-diagnosis for a 2-shaft stationary gas turbine. *Applied Energy, 71*, 321–339. doi:10.1016/S0306-2619(02)00015-6.

Ogihara, Aoi, & Sugimoto, Tsuchiya, & Nakatsukasa. (2011). Forward dynamic simulation of bipedal walking in the Japanese macaque: Investigation of causal relationships among limb kinematics, speed, and energetics of bipedal locomotion in a non-human primate. *American Journal of Physical Anthropology, 145*(4), 568–580. doi:10.1002/ajpa.21537 PMID:21590751.

Ogihara, Makishima, & Aoi, Sugimoto, Tsuchiya, & Nakatsukasa. (2009). Development of an anatomically based whole-body musculoskeletal model of the Japanese macaque. *American Journal of Physical Anthropology, 139*(3), 323–338. doi:10.1002/ajpa.20986 PMID:19115360.

Okabe, H., & Wakaumi, H. (1990). Grooved bar-code pattern recognition system with magnetoresistive sensor. *IEEE Transactions on Magnetics, 26*, 1575–1577. doi:10.1109/20.104451.

Oka, H., Shimono, K., Ogawa, R., Sugihara, H., & Taketani, M. (1999). A new planar multielectrode array for extracellular recording: Application to hippocampal acute slice. *Journal of Neuroscience Methods, 93*, 61–67. doi:10.1016/S0165-0270(99)00113-2 PMID:10598865.

Okazaki, K., Ogiwara, T., Yang, D., Sakata, K., Saito, K., Sekine, Y., & Uchikoba, F. (2011). Development of pulse control type MEMS micro robot with hardware neural network. *Artificial Life and Robotics, 16*(2), 229–233. doi:10.1007/s10015-011-0925-9.

Omata, & Farooqi. (1997). Reorientation planning for a multifingered hand based on orientation states network using regrasp primitives. In *Proceedings of the Intelligent Robots and Systems, 1997*. IEEE. Doi:10.1109/Iros.1997.649067

Orlovsky, Deliagina, & Grillner. (1999). *Neuronal control of locomotion: From mollusk to man*. Oxford, UK: Oxford University Press.

Osafo-Yeboah, B., & Jiang. (2009). Usability evaluation of a haptically controlled backhoe excavator simulation. In *Proceedings of IIE Annual Conference*, (pp. 961-966). IIE.

Ota, I. (2001). The economic situation of small scale forestry in Japan. *EFI Proceedings, 36*, 29-41.

Overmars, R. Schwarzkopf, & Wentink. (1995). Immobilizing polygons against a wall. In *Proceedings of the Symposium on Computational Geometry*. IEEE. Doi:10.1145/220279.220283

Pallot, M., Salminen, V., Pillai, B., & Kulvant, P. (2004). Business semantics: The magic instrument enabling plug & play collaboration? In *Proceedings of ICE 2004*. ICE. Berners-Lee, T., Hendler, J., & Lassila, O. (2001). The semantic web. *Scientific American*.

Park, J. H., & Cho, H. C. (1999). Sliding-mode controller for bilateral teleoperation with varying time delay. In *Proceedings 1999 IEEE/ASME International Conference on Advanced Intelligent Mechatronics*, (pp. 311-316). Atlanta, GA: IEEE.

Parker, R. (2009). Robotics for Steep country tree felling. *FFR Technical Note, 2*(1).

Park, J. H., & Kwon, O. M. (2009). Delay-dependent stability criterion for bidirectional associative memory neural networks with interval time-varying delays. *Modern Physics Letters B, 23*(1), 35–46. doi:10.1142/S0217984909017807.

Parlos, A. G. (2002). Identification of the internet-end-to-end delay dynamics using multi step neuro predictors. In *Proceedings 2002 International Joint Conference on Neural Networks*, (pp. 2460-2465). Honolulu, HI: IEEE.

Parsons, M. H. (1986). Human factors in industrial robot safety. *Journal of Occupational Accidents, 8*, 25–47. doi:10.1016/0376-6349(86)90028-3.

Paulson, L. D. (2004). Biomimetic robots. *Computer, 37*(9), 48–53. doi:10.1109/MC.2004.121.

Pereira, L. M., & Ahn, H. T. (2009). *Elder care via intention recognition and evolution prospection*. Paper presented at the 18th International Conference on Applications of Declarative Programming and Knowledge Management (INAP'09). Evora, Portugal.

Perry. (1994). The effect of population enrichment in genetic programming. In *Proceedings of Evolutionary Computation*, (pp. 456–461). IEEE.

Pfeifer, R., & Scheier, C. (1999). *Understanding intelligence*. Cambridge, MA: The MIT Press.

Philipose, M., Fishkin, K., Perkowitz, M., Patterson, D., Hahnel, D., Fox, D., & Kautz, H. (2005). Inferring ADLs from interactions with objects. *IEEE Pervasive Computing / IEEE Computer Society [and] IEEE Communications Society*.

Phoka, Niparnan, & Sudsang. (2006). Planning optimal force-closure grasps for curved objects by genetic algorithm. In *Proceedings of the IEEE Conference on Robotics, Automation and Mechatronics*. IEEE. Doi:10.1109/Ramech.2006.252683

Pillai, B. (2011). *Energy saving solutions at the wireless environment*. Paper presented at the Automation Systems 2011. Helsinki, Finland.

Pillai, B. (2012). *Bio-sensor implant for human body – Technology solutions*. Paper presented at the 8th International Conference on Humanized Systems. Daejeon, South Korea.

Pillai, B., & Salminen, V. (2004). *Trends in design and connectivity to broadband*. Paper presented at the 14th CIRP Design 2004. Cairo, Egypt.

Pillai, B., Pyykkönen, M., & Salminen, V. (2009). *Knowledge-intensive arrangement of scattered data*. Patent No. 120639 dated December 31, 2009. Helsinki, Finland: Finnish Patent Office.

Pillai, B. (2002). *Process control and non-linear system using ontology*. Gaithersburg, MD: NIST.

Pine, J. (1980). Recording action potentials from cultured neurons with extracellular microcircuit electrodes. *Journal of Neuroscience Methods, 2*, 19–31. doi:10.1016/0165-0270(80)90042-4 PMID:7329089.

Polushin, I. G., Tayebi, A., & Marquez, J. (2005). Adaptive schemes for stable teleoperation with communication delay based on IOS small gain theorem. In *Proceedings 2005 American Control Conference*, (pp. 4143-4148). Portland, OR: ACC.

Pongaen, W. (2008). Using neuro-fuzzy control to enhance maneuverability of master-slave system in position feedback frameworks. In *Proceedings IEEE International Conference on Robotics and Biomimetics (ROBIO)*. IEEE.

Pongaen, W., Bicker, R., Hu, Z., & Burn, K. (2004). Approach to telerobotic control using neuro-fuzzy techniques. In *Proceedings 11th World Congress in Mechanism and Machine Science*, (pp. 1761-1766). IEEE.

Pongracz, B., Ailer, P., Hangos, K. M., & Szederkenyi, G. (2000). *Nonlinear reference tracking control of a gas turbine with load torque estimation*. Budapest, Hungary: Computer and Automation Research Institute, Hungarian Academy of Sciences.

Poppele & Bosco. (2003). Sophisticated spinal contributions to motor control. *Trends in Neurosciences, 26*, 269–276. doi:10.1016/S0166-2236(03)00073-0 PMID:12744844.

Potter, S. M. (2008). How should we think about bursts? In *Proceedings of the 6th International Meeting on Substrate-Integrated Microelectrodes*. ISBN 3-938345-05-5

Pounds, P., Mahony, et al. (2010). Modelling and control of a large quadrotor robot. *Control Engineering Practice, 18*(7), 691–699. doi:10.1016/j.conengprac.2010.02.008.

Prado, & Suarez. (2005). Heuristic grasp planning with three frictional contacts on two or three faces of a polyhedron. In *Proceedings of the IEEE International Symposium on Assembly and Task Planning*. IEEE. Doi:10.1109/Isatp.2005.1511459

Puterman, M. L. (1994). *Markov decision processes: Discrete stochastic dynamic programming*. New York, NY: John Wiley & Sons, Inc. doi:10.1002/9780470316887.

Randell, D., & Cui, Z. (1992). *A spatial logic based on regions and connection*. Paper presented at the 3rd International Conference on Representation and Reasoning. San Mateo, CA.

Ravi, N., Dandekar, N., Mysore, P., & Littman, M. (2005). *Activity recognition from accelerometer data*. Paper presented at the Seventeenth Conference on Innovative Applications of Artificial Intelligence (IAAI/AAAI). New York, NY.

Raymond, K. (2012). Innovation to increase profitability of steep terrain harvesting in New Zealand. *New Zealand Journal of Forestry, 57*(2), 19–23.

Reed, F., Feintuch, P., & Bershold, N. (1981). Time delay estimation using the LMS adaptive filter-static behavior. *IEEE Transactions on ASSP, 29*, 561–571. doi:10.1109/TASSP.1981.1163614.

RFID Technology Editorial Department. (2004). *All radio-frequency IC tags*. Tokyo, Japan: Nikkei BP.

Ricketts, B. E. (1997). Modelling of a gas turbine: A precursor to adaptive control. In *Proceedings of the IEE Colloquium on Adaptive Controllers in Practice '97 (Digest No: 1997/176)*. IEE.

Riecken, D. (2000). Introduction: Personalized views of personalization. *Communications of the ACM, 43*(8), 26–28. doi:10.1145/345124.345133.

Rimon, & Burdick. (1996). On force and form closure for multiple finger grasps. In *Proceedings of the IEEE International Conference on Robotics and Automation*. IEEE. Doi:10.1109/Robot.1996.506972

Ringdahl, O. (2011). *Automation in forestry - Development of unmanned forwarders*. Umeå University.

Ringdahl, O., Lindroos, et al. (2011). Path tracking in forest terrain by an autonomous forwarder. *Scandinavian Journal of Forest Research, 26*(4), 350–359. doi:10.1080/02827581.2011.566889.

Robert, A. M. (1998). Randomized hough transform: Improved ellipse detection with comparison. *Pattern Recognition Letters, 19*, 299–305. doi:10.1016/S0167-8655(98)00010-5.

Robinson, H. P., Kawahara, M., Jimbo, Y., Torimitsu, K., Kuroda, Y., & Kawana, A. (1993). Periodic synchronized bursting and intracellular calcium transients elicited by low magnesium in cultured cortical neurons. *Journal of Neurophysiology, 70*(4), 1606–1616. PMID:8283217.

Robotic Fish SPC-03, BUA - CASIA, China. (n.d.). Retrieved from http://www.robotic-fish.net/index.php?lang=en&id=robots#top

RoboTuna II. (n.d.). Retrieved from http://Web.mit.edu/towtank/www-new/Tuna/Tuna2/tuna2.html

Rodríguez, C. A., & Guerrero. (2010). Wireless robot teleoperation via internet using IPv6 over a bluetooth personal area network. *Rev. Fac. Ing. Univ. Antioquia*, 172-184.

Rofer, T., & Lankenau, A. (2000). Architecture and applications of the Bremen autonomous wheelchair. *Science Direct.com–. Information Sciences, 126*, 1–20. doi:10.1016/S0020-0255(00)00020-7.

Roland-Villasana, E. J., Vazquez, A., & Jimenez-Sanchez, V. M. (2010). Modelling of the simplified systems for a power plant simulator. In *Proceedings of the Fourth UKSim European Symposium on Computer Modelling And Simulation (EMS)*. EMS.

Rolston, J. D., Wagenaar, D. A., & Potter, S. M. (2007). Precisely-timed spatiotemporal patterns of neural activity in dissociated cortical cultures. *Neuroscience, 148*, 294–303. doi:10.1016/j.neuroscience.2007.05.025 PMID:17614210.

Rowen, W. I. (1992, July/August). Simplified mathematical representations of single-shaft gas turbines in mechanical derive service. *Turbomachinery International.*

Rowen, W. I. (1983). Simplified mathematical representations of heavy-duty gas turbines. *Journal of Engineering for Power, 105*(4). doi:10.1115/1.3227494.

Roy, P., Bouchard, B., Bouzouane, A., & Giroux, S. (2007). *A hybrid plan recognition model for Alzheimer's patients: Interleaved-erroneous dilemma.* Paper presented at the IEEE/WIC/ACM International Conference on Intelligent Agent Technology. New York, NY.

Ruanoa, A. E., Fleming, P. J., Teixeiraa, C., Vazquezc, K. R., & Fonsecaa, C. M. (2003). Nonlinear identification of aircraft gas turbine dynamics. *Neurocomputing, 55*, 551–579. doi:10.1016/S0925-2312(03)00393-X.

Russell, S. J., & Norvig, P. (1995). *Artificial intelligence: A modern approach.* Upper Saddle River, NJ: Prentice Hall.

Rybak, Shevtsova, Lafreniere-Roula, & McCrea. (2006). Modelling spinal circuitry involved in locomotor pattern generation: Insights from deletions during fictive locomotion. *The Journal of Physiology, 577*(2), 617–639. doi:10.1113/jphysiol.2006.118703 PMID:17008376.

Ryu, J.-H., Preusche, B., Hannaford, B., & Hirzinger, G. (2005). Time-domain passivity control with reference energy following. *IEEE Transactions on Control Systems Technology, 13*, 737–742. doi:10.1109/TCST.2005.847336.

Sadri, F. (2011). Logic-based approaches to intention recognition. In Chong, N.-Y., & Mastrogiovanni, F. (Eds.), *Handbook of Research on Ambient Intelligence and Smart Environments: Trends and Perspectives* (pp. 346–375). Academic Press. doi:10.4018/978-1-61692-857-5.ch018.

Saimek, S., & Li, P. Y. (2004). Motion planning and control of a swimming machine. *The International Journal of Robotics Research, 23*(1), 27–53. doi:10.1177/0278364904038366.

Saito, K., Matsuda, A., Saeki, K., Uchikoba, F., & Sekine, Y. (2011). Synchronization of coupled pulse-type hardware neuron models for CPG model. In *The relevance of the time domain to neural network models* (pp. 117–133). New York: Springer.

Saito, K., Takato, M., Sekine, Y., & Uchikoba, F. (2012). Biomimetics micro robot with active hardware neural networks locomotion control and insect-like switching behaviour. *International Journal of Advanced Robotic Systems*, 1–6. doi:10.5772/54129.

Sakai, Y., & Kitazawa, M. (1995). Human-centered wheelchair supervisory system utilizing human ideas for operation. *Biomedical Fuzzy Human Science, 1*, 57–70.

Salminen, V., & Pillai, B. (2005). *Integration of products and services – Towards system and performance.* Paper presented at the International Conference on Engineering Design. Melbourne, Australia.

Salminen, V., & Pillai, B. (2007). *Interoperability requirement challenges - Future trends.* Paper presented at the International Symposium on Collaborative Technologies and Systems. Orlando, FL.

Salminen, V., & Pillai, B. (2004). Methodology on product life cycle challenge management for virtual enterprises. In Camarinho-Matos, L. M., & Afsarmanesh, H. (Eds.), *Processes and Foundations for Virtual Organizations.* Dordrecht, The Netherlands: Kluver Academic Publishers.

Salumäe, T. (2010). *Design of a compliant underwater propulsion mechanism by investigating and mimicking the body a rainbow trout (oncorhynchus mykiss).* Tallinn University of Technology.

Saranli, U., Buehler, M., & Koditschek, D. E. (2001). Rhex: A simple and highly mobile hexapod robot. *The International Journal of Robotics Research, 20*(1), 616–631. doi:10.1177/02783640122067570.

Satoh, K., & Yamaguchi, T. (2006). Preparing various policies for interactive reinforcement learning. In *Proceedings of the SICE-ICASE International Joint Conference 2006 (SICE-ICASE 2006)* (pp. 2440-2444). New York: Institute of Electrical and Electronics Engineers (IEEE).

Saut, Gharbi, Cortés, Sidobre, & Siméon. (2010). Planning pick-and-place tasks with two-hand regrasping. In *Proceedings of Intelligent Robots And Systems (IROS)* (pp. 4528–4533). IEEE. Doi:10.1109/Iros.2010.5649021

Schafer, J. B., Konstan, J. A., & Riedl, J. (2001). E-commerce recommendation applications. *Journal of Data Mining and Knowledge Discovery*, 5, 115–153. doi:10.1023/A:1009804230409.

Schlenoff, C. (2012). *An approach to ontology-based intention recognition using state representations*. Paper presented at the Fourth International Conference on Knowledge Engineering and Ontology Development. Barcelona, Spain.

Schlenoff, C. (2012). *An IEEE standard ontology for robotics and automation*. Paper presented at the International Conference on Intelligent Robots and Systems (IROS). Algarve, Portugal.

Schlieder, C. (1995). Reasoning about ordering. In Kuhn, W., & Frank, A. U. (Eds.), *COSIT* (Vol. 988, pp. 341–349). Heidelberg, Germany: Springer.

Schmidt-Wetekam, C., & Bewley, T. (2011). An arm suspension mechanism for an underactuated single legged hopping robot. In *Proceedings of Robotics and Automation* (pp. 5529–5534). IEEE. doi:10.1109/ICRA.2011.5980339.

Schmidt-Wetekam, C., Zhang, D., Hughes, R., & Bewley, T. (2007). Design, optimization, and control of a new class of reconfigurable hopping rovers. In *Proceedings of Decision and Control* (pp. 5150–5155). IEEE. doi:10.1109/CDC.2007.4434975.

Schomburg, Petersen, Barajon, & Hultborn. (1998). Flexor reflex afferents reset the step cycle during fictive locomotion in the cat. *Experimental Brain Research*, 122(3), 339–350. doi:10.1007/s002210050522 PMID:9808307.

Schrempf, O., & Hanebeck, U. (2005). *A generic model for estimating user-intentions in human-robot cooperation*. Paper presented at the 2nd International Conference on Informatics in Control, Automation, and Robotics ICINCO 05. Barcelona, Spain.

Schweitzer, G. (1996). Mechatronics for the design of human-oriented machines. *IEEE/ASME Transactions on Mechatronics*, 1(2), 120–126. doi:10.1109/3516.506148.

Seth, H., Gregory, D. H., & Peter, I. C. (1996). A tutorial on visual servo control. *IEEE Transactions on Robotics and Automation*, 12(5), 651–670. doi:10.1109/70.538972.

Sfakiotakis, M., Lane, D. M., & Davies, J. B. C. (1999). Review of fish swimming modes for aquatic locomotion. *IEEE Journal of Oceanic Engineering*, 24(2), 237–252. doi:10.1109/48.757275.

Sgouros, N. M. (2002). Qualitative navigation for autonomous wheelchair robots in indoor environments. *Autonomous Robots*, 12, 257–266. doi:10.1023/A:1015265514820.

Shalan, H. E. (2011). Parameter estimation and dynamic simulation of gas turbine model in combined-cycle power plants based on actual operational data. *Journal of American Science*, 7(5).

Sheridan, T. B. (1992). *Telerobotics, automation and human supervisory control*. Cambridge, MA: MIT Press.

Sheridan, T. B. (1995). Teleoperation, telerobotics and telepresence: A progress report. *Control Engineering Practice*, 3, 205–214. doi:10.1016/0967-0661(94)00078-U.

Shibata, T., Aoki, Y., Otsuka, M., Idogaki, T., & Hattori, T. (1997). Microwave energy transmission system for microrobot. *IEICE Transactions on Electronics*, 80(2), 303–308. Retrieved from http://search.ieice.org/bin/summary.php?id=e80-c_2_303.

Shichel & Sipper. (2011). Gp-rars: Evolving controllers for the robot auto racing simulator. *Memetic Computing*, 3, 89–99. doi:10.1007/s12293-011-0056-9.

Shik & Orlovsky. (1976). Neurophysiology of locomotor automatism. *Physiological Reviews*, 56(3), 465–501. PMID:778867.

Siahmansouri, M., Ghanbari, A., & Fakhrabadi, M. M. S. (2011). Design, implementation and control of a fish robot with undulating fins. *International Journal of Advanced Robotic Systems*, 8(5), 61–69.

Sim & Costa. (2000). Using genetic algorithms with asexual transposition. In *Proceedings of the Genetic and Evolutionary Computation Conference* (pp. 323–330). San Francisco, CA: Morgan Kaufmann.

Simani, S., & Patton, R. J. (2008). Fault diagnosis of an industrial gas turbine prototype using a system identification approach. *Control Engineering Practice, 16,* 769–786. doi:10.1016/j.conengprac.2007.08.009.

Simoes, C. Sim, & Costa. (1999). Transposition: A biologically inspired mechanism to use with genetic algorithms. In *Proceedings of the Fourth International Conference on Neural Networks and Genetic Algorithms (ICANNGA'99),* (pp. 612–619). Berlin: Springer-Verlag.

Slawinski, E., & Mut, W. (2008). Transparency in time for teleoperation systems. In *Proceedings in IEEE International Conference on Robotics and Automation.* Pasadena, CA: IEEE.

Slawiñski, E., Postigo, et al. (2007). Bilateral teleoperation through the internet. *Robotics and Autonomous Systems, 55,* 205–215. doi:10.1016/j.robot.2006.09.002.

Slotine, J. J., & Li, W. (1991). *Applied nonlinear control.* Englewood-Cliffs, NJ: Prentice Hall.

Smith, A., & Mohashtrudi-Zaad, K. (2005). Neural network-based teleoperation using Smith predictors. In *Proceedings IEEE International Conference on Mechatronics and Automation (ICMA 05).* IEEE.

Smith, A.C., & Hashtrudi-Zaad. (2005). Adaptive teleoperation using newral network-based predictive control. In *Proceedings of the 2005 IEEE Conference on Control Applications,* (pp. 1269-1274). Toronto, Canada: IEEE.

Smith. (2004). *Open dynamics engine.*

Smith, A. C., & Hashtrudi-Zaad, K. (2006). Smith predictor type control architectures for time delayed teleoperation. *The International Journal of Robotics Research, 25,* 797–818. doi:10.1177/0278364906068393.

Sniegowski, J. J., & Garcia, E. J. (1996). Surface-micromachined gear trains driven by an on-chip electrostatic microengine. *IEEE Electron Device Letters, 17,* 366–368. doi:10.1109/55.506369.

Song, G., & Wang, H. (2007). A fast and robust ellipse detection algorithm based on pseudo-random sample consensus. *Lecture Notes in Computer Science, 4673,* 669–676. doi:10.1007/978-3-540-74272-2_83.

Spina, P. R., & Venturini, M. (2007). Gas turbine modelling by using neural networks trained on field operating data. In *Proceedings of ECOS 2007.* Padova, Italy: ECOS.

Spirov, Kazansky, Zamdborg, Merelo, & Levchenko. (2009). *Forced evolution in silico by artificial transposons and their genetic operators: The John Muir ant problem.* Technical Report arXiv:0910.5542, Oct 2009. Comments: 33 pages.

Spong, M. W., Corke, P., & Lozano, R. (2001). Nonlinear control of the reaction wheel pendulum. *Automatica, 37*(11), 1845–1851. doi:10.1016/S0005-1098(01)00145-5.

Spong, M. W., & Vidyasagar, M. (1989). *Robot dynamics and control.* New York: Wiley.

Steingrube, Timme, Wörgötter, & Manoonpong. (2010). Self-organized adaptation of a simple neural circuit enables complex robot behaviour. *Nature Physics, 6,* 224–230. doi:10.1038/nphys1508.

Stilman, M., Olson, J., & Gloss, W. (2010). Golem Krang: Dynamically stable humanoid robot for mobile manipulation. In Proceedings of Robotics and Automation (pp. 3304-3309). IEEE. doi: doi:10.1109/ROBOT.2010.5509593.

Stramiglioli, S., van der Schaft, A., Maschke, B., & Melchiori, C. (2002). Geometric scattering in robotic telemanipulation. *IEEE Transactions on Robotics and Automation, 18,* 588–596. doi:10.1109/TRA.2002.802200.

Strand & McDonald. (1985). Copia is transcriptionally responsive to environmental stress. *Nucleic Acids Research, 13*(12), 4401–4410. doi:10.1093/nar/13.12.4401 PMID:2409535.

Suebromran, A., & Parnichkun. (2005). Disturbance observer-based hybrid control of displacement and force in a medical tele-analyzer. *International Journal of Control, Automation, and Systems, 3*(1), 70–78.

Suematsu, H., Kobayashi, K., Ishii, R., Matsuda, A., Sekine, Y., & Uchikoba, F. (2009). MEMS type micro robot with artificial intelligence system. In *Proceedings of International Conference on Electronics Packaging*, (pp. 975-978). ICEP.

Sugitani, Y., Kanjo, Y., & Murayama, M. (1996). Systemization with CAD/CAM welding robots for bridge fabrication. In *Proceedings of 4th International Workshop on Advanced Motion Control* (pp. 80–85). IEEE.

Sugita, S., Itaya, T., & Takeuchi, Y. (2003). Development of robot teaching support devices to automate deburring and finishing works in casting. *International Journal of Advanced Manufacturing Technology, 23*(3/4), 183–189.

Sukthanker, G., & Sycara, K. (2001). *Team-aware robotic demining agents for military simulation*. Paper presented at the Innovative Applications of Artificial Intelligence (IAAI). New York, NY.

Sun, Z. Q. (1995). *Robot intelligence control*. Beijing, China: Beijing Education Press.

Surbled, P., Clerc, C., Pioufle, B. L., Ataka, M., & Fujita, H. (2001). Effect of the composition and thermal annealing on the transformation temperature sputtered TiNi shape memory alloy thin films. *Thin Solid Films, 401*, 52–59. doi:10.1016/S0040-6090(01)01634-0.

Sutton, R., & Barto, A. (1998). *Reinforcement learning: An introduction*. Cambridge, MA: MIT Press.

Suzuki, Y., Tani, K., & Sakuhara, T. (1999). Development of a new type piezo electric micromotor. In *Proceedings of Transduceres '99* (pp. 1748–1751). Transduceres.

Szabo, S., Norcross, R., & Shackleford, W. (2011). *Safety of human-robot collaboration systems project*. Retrieved from http://www.nist.gov/el/isd/ps/safhumrobcollsys.cfm

Szykman, S., Sriram, R. D., & Regli, W. (2001). The role of knowledge in next generation product development systems. *ASME Journal of Computing and Information Science in Engineering, 1*(1), 3–11. doi:10.1115/1.1344238.

Takadama, K., Sato, F., Otani, M., Hattori, K., Sato, H., & Yamaguchi, T. (2012). Preference clarification recommender system by searching items beyond category. In *Proceedings of IADIS (International Association for Development of the Information Society) International Conference Interfaces and Human Computer Interaction 2012* (pp. 3-10). Lisbon, Portugal: IADIS Press.

Takeda, M. (2001). Applications of MEMS to industrial inspection. In *Proceedings of IEEE MEMS 2001*, (pp. 182-191). IEEE. doi:10.1109/MEMSYS.2001.906510

Tanev & Shimohara. (2008). Co-evolution of active sensing and locomotion gaits of simulated snake-like robot. In *Proceedings of the 10th Annual Conference on Genetic and Evolutionary Computation*, (pp. 257–264). New York, NY: ACM.

Tanev. (2004). Dom/xml-based portable genetic representation of the morphology, behavior and communication abilities of evolvable agents. *Artificial Life and Robotics, 8*, 52–56.

Tanev, Ray, & Buller. (2005). Automated evolutionary design, robustness and adaptation of sidewinding locomotion of simulated snake-like robot. *IEEE Transactions on Robotics, 21*, 632–645. doi:10.1109/TRO.2005.851028.

Tang, W. C., Nguyen, T. H., & Howe, R. T. (1989). Laterally driven poly silicon resonant microstructure. In *Proceedings of IEEE Micro Electro Mechanical Systems: An Investigation of Micro Structures, Sensors, Actuators, Machines and Robots*, (pp. 53-59). IEEE. doi:10.1016/0250-6874(89)87098-2

Tanner, N. A., & Niemeyer, G. (2006). High-frequency acceleration feedback in wave variables telerobotics. *IEEE/ASME Transactions on Mechatronics, 11*, 119–127. doi:10.1109/TMECH.2006.871086.

Tavakoli, M. R. B., Vahidi, B., & Gawlik, W. (2009). An educational guide to extract the parameters of heavy-duty gas turbines model in dynamic studies based on operational data. *IEEE Transactions on Power Systems, 24*(3). doi:10.1109/TPWRS.2009.2021231.

Teeyapan, K., Wang, J., Kunz, T., & Stilman, M. (2010). Robot limbo: Optimized planning and control for dynamically stable robots under vertical obstacles. In *Proceedings of Robotics and Automation* (pp. 4519–4524). IEEE. doi:10.1109/ROBOT.2010.5509334.

Teranno, T., Asai, K., & Sugeno, M. (1992). *Fuzzy systems theory and its applications*. Boston, MA: Academic Press.

Terasaki, H. Takahashi, & Arakawa. (1991). Automatic grasping and regrasping by space characterization for pick-and-place operations. In *Proceedings IROS '91 IEEE/RSJ International Workshop*. IEEE. Doi:10.1109/Iros.1991.174426

348

Thompson, E. A., Harmison, et al. (2006). Robot teleoperation featuring commercially available wireless network cards. *Journal of Network and Computer Applications*, *29*(1), 11–24. doi:10.1016/j.jnca.2004.11.001.

Thomsen, Fogel, & Krink. (2002). A clustal alignment improver using evolutionary algorithms. In *Proceedings of Evolutionary Computation, 2002*, (vol. 1, pp. 121–126). IEEE..

Ting & Macpherson. (2005). A limited set of muscle synergies for force control during a postural task. *Journal of Neurophysiology*, *93*, 609–613. PMID:15342720.

Todorov & Jordan. (2002). Optimal feedback control as a theory of motor coordination. *Nature Neuroscience*, *5*, 1226–1235. doi:10.1038/nn963 PMID:12404008.

Tomasello, M., Carpenter, M., Call, K., Behne, T., & Moll, H. (2005). Understanding and sharing intentions: The origins of cultural cognition. *The Behavioral and Brain Sciences*, *28*, 675–735. doi:10.1017/S0140525X05000129 PMID:16262930.

Torella, G., Gamma, F., & Palmesano, G. (2003). Neural networks for the study of gas turbine engines air system. In *Proceedings of the International Gas Turbine Congress 2003*. Tokyo, Japan: IGTC.

Triantafyllou, M. S., & Triantafyllou, G. S. (1995). An efficient swimming machine. *Scientific American*, *272*(3), 64–71. doi:10.1038/scientificamerican0395-64.

Triantafyllou, M., & Kumph, J. M. et al. (2000). *Maneuvering of a robotic pike*. Cambridge, MA: Massachusetts Institute of Technology.

Triton Logging. (2012). *Triton logging website*.

Tsuruta, K., Mikuriya, Y., & Ishikawa, Y. (1999). Micro sensor developments in Japan. *Sensor Review*, *19*(1), 7–42. doi:10.1108/02602289910255568.

Tzafestas, S. G. (1995). Neural networks in robot control. In Tzafestas, S. G., & Verbruggen, H. B. (Eds.), *Artificial Intelligence in Industrial Decision Making, Control, and Automation* (pp. 327–358). Dordrecht, The Netherlands: Springer. doi:10.1007/978-94-011-0305-3_11.

Tzafestas, S. G. (2009). *Human and nature minding automation*. Berlin: Springer.

Tzafestas, S. G., & Prokopiou, P. A. (1997). Compensation of teleoperator modeling uncertainties with sliding-mode controller. *Robotics and Computer-integrated Manufacturing*, *13*, 9–20. doi:10.1016/S0736-5845(96)00030-0.

Tzafestas, S. G., Prokopiou, P. A., & Tzafestas, C. S. (2001). A new partitioned robot neurocontroller: General analysis and application to teleoperator modeling uncertainties. *Machine Intelligence and Robot Control*, *3*, 7–26.

Usami, M., & Yamada, J. (2005). *Ubiquitous technology IC tag*. Tokyo, Japan: Ohmusha Publishing.

Uvarov. (1977). *Grasshoppers and locusts*.

van Brussel, H. (1989). The mechatronics approach to motion control. In *Proceedings International Conference on Motion Control: The Mechatronics Approach*. Antwerp, Belgium: IEEE.

van der Stappen, Wentink, & Overmars. (1999). Computing form-closure configurations. In *Proceedings of the IEEE International Conference on Robotics And Automation*. IEEE. Doi:10.1109/Robot.1999.770376

van der Stappen, Wentink, & Overmars. (2000). Computing immobilizing grasps of polygonal parts. *The International Journal of Robotics Research*, *19*(5), 467–479. doi:10.1177/02783640022066978.

van der Zee, L. F. (2009). *Design of a haptic controller for excavators*. Department of Electrical Engineering, University of Stellenbosch.

Van Pelt, J., Vajda, I., Wolters, P. S., Corner, M. A., & Ramakers, G. J. (2005). Dynamics and plasticity in developing neuronal networks in vitro. *Progress in Brain Research*, *147*, 173–188. doi:10.1016/S0079-6123(04)47013-7 PMID:15581705.

Vassilev, Job, & Miller. (2000). Towards the automatic design of more efficient digital circuits. In *Proceedings of the 2nd NASA/DoD workshop on Evolvable Hardware*. Washington, DC: IEEE Computer Society.

Vinayavekhin, Kudohf, & Ikeuchi. (2011). Towards an automatic robot regrasping movement based on human demonstration using tangle topology. In *Proceedings of the 2011 IEEE International Conference* (pp. 3332 – 3339). IEEE. Doi:10.1109/Icra.2011.5979648

Vogel, S. (1996). The thrust of flying and swimming. In *Life in moving fluids: The physical biology of flow*.

Volpe, R., Balaram, J., Ohm, T., & Ivlev, R. (1996). The rocky 7 mars rover prototype. In *Proceedings of Intelligent Robots and Systems '96 (Vol. 3*, pp. 1558–1564). IEEE.

Vukobratović, B. Surla, & Stokić. (1990). Biped loco-motion-dynamics, stability, control and application. New York: Springer-Verlag.

W3C_Member_Submission. (2004). *SWRL: A semantic web rule language combining OWL and RuleML*. Retrieved from http://www.w3.org/Submission/SWRL/

Wagenaar, D. A., Pine, J., & Potter, S. M. (2006). An extremely rich repertoire of bursting patterns during the development of cortical cultures. *BMC Neuroscience, 7*(11). doi: doi:10.1186/1471-2202-7-11 PMID:16464257.

Wakaumi, H., & Yokoyama, S. (1988). An exciting method for highly-sensitive magnetic sensors. In *Proceedings of the 12ᵗʰ Institute of Magnetic Engineers of Japan Conference*. IMEJ.

Wakaumi, H., Nakamura, K., Matsumura, T., & Yamauchi, F. (1989). *Automated wheelchair guided by magnetic ferrite marker*. Paper presented at the RESNA 12ᵗʰ Annual Conference. New Orleans, LA.

Wakaumi, H., Ajiki, H., Hankui, E., & Nagasawa, C. (2000). Magnetic grooved bar-code recognition system with slant MR sensor. *IEE Proceedings. Science Measurement and Technology, 147*, 131–136. doi:10.1049/ip-smt:20000369.

Wakaumi, H., Komaoka, T., & Hankui, E. (2000). Grooved bar-code recognition system with tape-automated-bonding head detection scanner. *IEEE Transactions on Magnetics, 36*, 366–370. doi:10.1109/20.822548.

Wakaumi, H., Nakamura, K., & Matsumura, T. (1989). A new automated wheelchair guided by magnetic ferrite marker lane. *NEC Research & Development, 95*, 62–68.

Wakaumi, H., Nakamura, K., & Matsumura, T. (1992). Development of an automated wheelchair guided by a magnetic ferrite marker lane. *Journal of Rehabilitation Research and Development, 29*, 27–34. doi:10.1682/JRRD.1992.01.0027 PMID:1740776.

Wallgrun, J. O., Frommberger, L., Wolter, D., Dylla, F., & Freksa, C. (2006). Qualitative spatial representation and reasoning in the SparQ-toolbox. In Barkowsky, T., Knauff, M., Ligozat, G., & Montello, D. R. (Eds.), *Spatial Cognition V* (pp. 39–58). Springer.

Wang, R. (2011). Bilateral control of teleoperation system with fuzzy singularly perturbed model. In *Proceedings International Conference on Intelligent Control and Information Processing (ICCIP 11)*, (pp. 941-945). ICCIP.

Wang, T., Liang, J., Shen, G., & Tan, G. (2005). *Stabilization based design and experimental research of a fish robot*. Paper presented at the IEEE/RSJ International Conference on Intelligent Robots and Systems (IROS). New York, NY.

Wang, L. X. (1994). *Adaptive fuzzy systems and control: Design and stability*. Englewood Cliffs, NJ: Prentice Hall.

Watanabe, K., Fukuda, T., & Tzafestas, S. G. (1992). An adaptive control for CARMA systems using linear neural networks. *International Journal of Control, 56*, 483–497. doi:10.1080/00207179208934324.

Watanabe, K., & Tzafestas, S. G. (1990). Learning algorithms for neural networks with the Kalman filters. *Journal of Intelligent & Robotic Systems, 3*, 305–319. doi:10.1007/BF00439421.

Wernli, R. L. (2000). *Low cost UUV's for military applications: Is the technology ready?* DTIC Document.

Westerberg, S., Manchester, et al. (2008). *Virtual environment teleoperation of a hydraulic forestry crane*.

Wikimedia Commons. (2012). Retrieved from http://commons.wikimedia.org

Wilke, Wang, & Ofria, Lenski, & Adami. (2001). Evolution of digital organisms at high mutation rates leads to survival of the flattest. *Nature, 412*, 331–333. doi:10.1038/35085569 PMID:11460163.

Williams, C. (2004). AUV systems research at the NRC-IOT. *An update*.

Willy, A., & Low, K. H. (2005). *Development and initial experiment of modular undulating fin for untethered biorobotic AUVs*. Paper presented at the IEEE International Conference on Robotics and Biomimetics (ROBIO). New York, NY.

Wolter, F., & Zakharyaschev, M. (2000). *Spatio-temporal representation and reasoning based on RCC-8.* Paper presented at the 7th Conference on Principles of Knowledge Representation and Reasoning (KR2000). Breckenridge, CO.

Wolters, P. S., Rutten, W. L., Ramakers, G. J., Van Pelt, J., & Corner, M. A. (2004). Long term stability and developmental changes in spontaneous network burst firing patterns in dissociated rat cerebral cortex cell cultures on multielectrode arrays. *Neuroscience Letters, 361*, 86–89. doi:10.1016/j.neulet.2003.12.062 PMID:15135900.

Xu, Y., & Au, S.-W. (2004). Stabilization and path following of a single wheel robot. *IEEE/ASME Transactions on Mechatronics, 9*(2), 407–419. doi:10.1109/TMECH.2004.828642.

Yadav, N., Khan, I. A., & Grover, S. (2010). Modelling and analysis of simple open-cycle gas turbine using graph networks. In *Proceedings of the International Conference of Electrical and Electronics Engineering*. IEEE.

Yamaguchi, T., & Nishimura, T. (2008). How to recommend preferable solutions of a user in interactive reinforcement learning? In *Proceedings of the International Conference on Instrumentation, Control and Information Technology (SICE2008)*, (pp. 2050-2055). Tokyo, Japan: The Society of Instrument and Control Engineers.

Yamaguchi, T., Nishimura, T., & Takadama, K. (2012). *Awareness based recommendation - Toward the cooperative learning in human agent interaction.* Paper presented at the International Conference on Humanized Systems 2012 (OS02_1003). Daejeon, Korea.

Yamaguchi, T., Nishimura, T., & Sato, K. (2011). How to recommend preferable solutions of a user in interactive reinforcement learning? In Mellouk, A. (Ed.), *Advances in Reinforcement Learning* (pp. 137–156). Rijeka, Croatia: InTech Open Access Publisher. doi:10.5772/13757.

Yamasaki, Nomura, & Sato. (2003). Possible functional roles of phase resetting during walking. *Biological Cybernetics, 88*, 468–496. PMID:12789495.

Yap, H. E., & Hashimoto, S. (2012). Attitude control of an airborne two-wheeled robot. In *Proceedings of Artificial Life and Robotics*. AROB.

Yee, S. K., & Milanovic, J. V. (2008). Overview and comparative analysis of gas turbine models for system stability studies. *IEEE Transactions on Power Systems, 23*(1). doi:10.1109/TPWRS.2007.907384.

Yi, S. Y., & Chung, M. J. (1998). Robustness of fuzzy logic for an uncertain dynamic system. *IEEE Transactions on Fuzzy Systems, 6*.

Yokokohji, Y., Imaida, et al. (2000). *Bilateral teleoperation: Towards fine manipulation with large time delay.*

Yoon, J., & Manurung. (2010). Development of an intuitive user interface for a hydraulic backhoe. *Automation in Construction, 19*(6), 779–790. doi:10.1016/j.autcon.2010.04.002.

Yoru, Y., Karakoc, T. H., & Hepbasli, A. (2009). Application of artificial neural network (ANN) method to exergetic analyses of gas turbines. In *Proceedings of the International Symposium on Heat Transfer in Gas Turbine Systems*. Antalya, Turkey: IEEE.

Yoshida, K., & Namerikawa, T. (2009). Stability and tracking properties in predictive control with adaptation for bilateral teleoperation. In *Proceedings 2009, American Control Conference*, (pp. 1323-1328). St. Louis, MO: ACC.

Yoshioka, T. (2004). *Radio frequency IC tags illustrated with pictures – The world of radio frequency identification expanding*. Tokyo, Japan: Ohmusha Publishing.

Yoshitake, S., Nagata, F., Otsuka, A., Watanabe, K., & Habib, M. K. (2012). Proposal and implementation of CAM system for industrial robot RV1A. In *Proceedings of the 17th International Symposium on Artificial Life and Robotics* (pp. 158–161). IEEE.

Youn, S.-J., & Oh, K.-W. (2007). Intention recognition using a graph representation. World Academy of Science, Engineering and Technology, 25.

Yu, J., & Wang, L. (2005). *Parameter optimization of simplified propulsive model for biomimetic robot fish.*

Yu, J., Hu, Y., Huo, J., & Wang, L. (2007). *An adjustable scotch yoke mechanism for robotic dolphin.* Paper presented at the IEEE International Conference on Robotics and Biomimetics (ROBIO). New York, NY.

Yu, J., Hu, Y., Fan, R., Wang, L., & Huo, J. (2007). Mechanical design and motion control of a biomimetic robotic dolphin. *Advanced Robotics*, *21*(3-4), 499–513. doi:10.1163/156855307780131974.

Zhang, & Gu. (2012). Grasp planning of 3D objects using genetic algorithm. In *Proceedings of the IEEE International Conference on Automation and Logistics*. IEEE. Doi:10.1109/Ical.2012.6308157

Zha, X. F., & Sriram, S. D. (2004). Feature-based component model for design of embedded system. In Gopalakrishnan, B. (Ed.), *Intelligent Systems in Design and Manufacturing, Proceedings of SPIE* (Vol. *5605*, pp. 226–237). Bellingham, WA: SPIE. doi:10.1117/12.571612.

Zhu, W.H., & Salcudean. (1999). Teleoperation with adaptive motion/force control. In *Proceedings of the IEEE International Conference on Robotics and Automation*, (pp. 231-237). Detroit, MI: IEEE.

Zhu, D., Gedeon, et al. (2011). Moving to the centre: A gaze-driven remote camera control for teleoperation. *Interacting with Computers*, (23): 85–95. doi:10.1016/j.intcom.2010.10.003.

Zhu, W. H., & Salcudean, S. E. (2000). Stability guaranteed teleoperation: An adaptive motion/force control approach. *IEEE Transactions on Automatic Control*, *45*, 1951–1959. doi:10.1109/9.887620.

Zhu, Y., Frey, H. C., & Asce, M. (2007). Simplified performance model of gas turbine combined-cycle systems. *Journal of Energy Engineering*. doi:10.1061/(ASCE)0733-9402(2007)133:2(82).

Zuo, & Sun. (1991). Force/torque closure grasp in the plane. In *Proceedings of the IEEE International Workshop on Intelligent Robots and Systems*. IEEE.

About the Contributors

Maki K. Habib received his Doctor of Engineering Sciences degree in intelligent and autonomous robots from the University of Tsukuba-Japan in 1990. He has wide international and cross-culture working experience in Japan, Switzerland, South Korea, Australia-Malaysia, Iraq, and he is currently a Professor at the Mechanical Engineering Department, School of Sciences and Engineering, American University in Cairo (AUC), Egypt. He has more than 27 year of experience teaching and researching in the area of Mechatronics and Robotics. He is also regional editor and associate editor of more than 10 reputable international journals. Professor Habib is also the director and the founder of the master programs in Robotics, Control, and Smart System (RCSS) at AUC. He edited six books, and he published more than 230 papers in internationally recognized books, journals, and conferences. His main area of research is focusing on evolution of Mechatronics, human adaptive and friendly mechatronics, autonomous navigation, service robots and humanitarian demining, telecooperation, distributed teleoperation and collaborative control, flexible automation, wireless sensor networks and ambient intelligence, biomimetic and biomedical robots.

J. Paulo Davim received his PhD degree in Mechanical Engineering from the University of Porto in 1997 and the Aggregation from the University of Coimbra in 2005. Currently, he is an Aggregate Professor at the Department of Mechanical Engineering of the University of Aveiro. He has more 25 years of teaching and research experience in manufacturing, materials, and mechanical engineering with special emphasis in machining and tribology. Recently, he has also interest in sustainable manufacturing and industrial engineering. He is the Editor-in-Chief of six international journals, guest editor of journals, books editor, book series editor, and scientific advisory for many international journals and conferences. Presently, he is an editorial board member of 20 international journals and acts as reviewer for than 70 prestigious ISI Web of Science journals. In addition, he has also published in his field of research as author and co-author more than 40 book chapters and 350 articles in journals and conferences (more 180 articles in ISI Web of Science, h-index 25+).

* * *

Shinya Aoi received the B.E., M.E., and Ph.D. degrees from the Department of Aeronautics and Astronautics, Kyoto University, Kyoto, Japan in 2001, 2003, and 2006, respectively. From 2003 to 2006, he was a Research Fellow of the Japan Society for the Promotion of Science (JSPS). From 2006 to 2007, he was a Center of Excellence (COE) Assistant Professor with Kyoto University. Since 2007, has been an Assistant Professor with the Department of Aeronautics and Astronautics, Graduate School of Engineering, Kyoto University. His research interests include dynamics and control of robotic systems, especially legged robots, and analysis and simulations of locomotion in humans and animals.

Hamid Asgari is a PhD Candidate in Department of Mechanical Engineering at University of Canterbury (UC), Christchurch, New Zealand. He received his M.Sc. in aerospace engineering from TMU and his B.Eng. in mechanical engineering from IUST. During more than 14 years of work experience in different industrial companies, he has had different positions and responsibilities such as chief engineer, R&D expert, maintenance planning engineer, and project coordinator. Mr. Asgari is a member of Iranian Society of Mechanical Engineers (ISME), American Society of Mechanical Engineers (ASME), and Iranian Society of Instrument and Control Engineers (ISICE).

Stephen Balakirsky is the project manager for the Knowledge Driven Planning and Modeling Project at the National Institute of Standards and Technology. His current research focuses on knowledge representations, robotic simulation, and robotic performance evaluation. Dr. Balakirsky is the principal architect for the open source Unified System for Automation and Robot Simulation (USARSim) project. He is also the chair of the industrial subgroup of the IEEE Robotics and Automation Society's (RAS) Ontologies for Robotics and Automation Working Group, the IEEE RAS Competition's Chair, and a past member of the executive committee of the RoboCup Federation. He received his doctor of engineering degree from the University of Bremen, and his Master and Bachelor of Science from the University of Maryland.

XiaoQi Chen is a Professor in Department of Mechanical Engineering at University of Canterbury. After obtaining his BE from South China University of Technology in 1984, he received China-UK Technical Co-Operation Award for his MSc study in Department of Materials Technology, Brunel University (1985 – 1986), and PhD study in Department of Electrical Engineering and Electronics, University of Liverpool (1986 – 1989). He was Senior Scientist at Singapore Institute of Manufacturing Technology (1992–2006) and a recipient of Singapore National Technology Award in 1999. His research interests include mechatronic systems, mobile robotics, assistive devices, and manufacturing automation.

Sebti Foufou received his PhD in computer science in 1997 from the University of Claude Bernard Lyon I, France, for a dissertation on parametric surfaces intersections. He worked as a research assistant at the University Lyon 1 from 1996 to 1998. He was with the department of computer science, University of Burgundy, France from 1998 to 2009. He also worked in 2005 and 2006, as guest researcher, at NIST, USA. He joined the department of computer science and engineering, Qatar University in Sept. 2009. His research topics of interest include 3D representations, geometric constraint solving, robotics, and PLM.

Stefanie Gutschmidt received the Preliminary Diploma (German equivalent to BS degree) in Mechanical Engineering from the Otto-von-Guericke University, Magdeburg, Germany, in 1997, and her MS degree from Rose Hulman Institute of Technology, Terre Haute, IN, USA, in 1999. She received her PhD degree in Applied Mechanics from Darmstadt University of Technology, Darmstadt, Germany in 2005. She worked as a postdoctoral research fellow at the Technion - Israel Institute of Technology, Haifa, Israel (2005-2008) and the University of Liege, Liege, Belgium (2008). She currently works as a Senior Lecturer in the Department of Mechanical Engineering at the University of Canterbury, Christchurch, New Zealand (since 2009). Her current research interests are in the following areas: nonlinear dynamics & vibrations, mechatronics, earthquake engineering, structural health monitoring using wireless smart sensors, energy harvesting based on structural vibration.

Christopher E. Hann received his BSC (hons) degree with first class honours in mathematics from the University of Canterbury in 1996, and the PhD degree in Mathematics from the same institute in 2001. He then worked as a teaching fellow in the mathematics department until 2003, and was in the Department of Mechanical Engineering as an FRST postdoctoral fellow from 2004-2006 and a Sir Charles Hercus Health Research Fellow and Senior Research Associate from 2007- mid 2010. He has worked in the Department of Electrical Engineering from mid 2010 – 2011 as a Lecturer and from 2012 as a Senior Lecturer and Rutherford Discovery Fellow. His research areas include Rocket systems and control, orbital mechanics, biomedical systems, system identification and control for a variety of industry applications and image processing and computer vision. Dr Hann has published over 250 refereed journal and conference papers and is an inventor on several patents.

Shuji Hashimoto received his B.S., M.S., and Dr.Eng. degrees in applied physics from Waseda University, Tokyo, Japan, in 1970, 1973, and 1977, respectively. He is currently a Professor in the Department of Applied Physics and the vice president of Waseda University, where he has also been the Director of the Humanoid Robotics Institute since 2000. He is the author of over 400 technical publications, proceedings papers, editorials, and books. His research interests are in human communication and KANSEI information processing, including image processing, music systems, neural computing, and humanoid robotics.

Zeid Kootbally is an Assistant Research Scientist with the Department of Mechanical Engineering at the University of Maryland, College Park. He received a Ph.D. in Computer Science from the University of Burgundy for a dissertation on "Moving Object Predictions in Dynamic Environments for Autonomous Ground Vehicles." Dr. Kootbally's current research is focused on simulation-based planning and control for kit building in manufacturing environments. The objective of this problem is to develop the measurement science and standards for planning and modeling by robots so that they are able to be more quickly re-tasked and are more flexible and adaptive.

Suguru N. Kudoh received his Master's degree in Biophysical Engineering (1995) and PhD from Osaka University, Japan (1998). He was a research fellow of Japan science and technology agency (JST) from 1997 to 1998, and a research scientist of National Institute of Advanced Industrial Science and Technology (AIST) from 1998 to 2009. Now he is an associate professor at Kwansei Gakuin University. The aim of his research is to elucidate relationship between dynamics of neuronal network and brain information processing. He also develops bio-robotics hybrid system in which a living neuronal network is connected to a robot body via control rules, corresponding genetically provided interfaces between a brain and a peripheral system.

Tüze Kuyucu has publications in evolvable hardware, swarm robotics, evolutionary computation, and biologically inspired computation. He used and developed bioinspired techniques for the design and optimization of digital circuits on real hardware, swarm control, fault tolerant wireless sensor networks, and modular robot control. His current research interests include mainly in the field of evolutionary computation, and particularly the study of emergent behavior, collective intelligence in robotics, genetic programming, evolvable hardware, swarms, and evolutionary optimization.

Hongbo Li received the Ph.D. degree from Tsinghua University, Beijing, China, in 2009. He is currently an assistant professor with the Department of Computer Science and Technology, Tsinghua University. His research interests include networked control systems and intelligent control.

Kai Liu received the Master degree from Tsinghua University, Beijing, China, in 2011. He is currently an Engineer with Beijing Institute of Control Engineering. His research interests include robotics and its application.

Andreas-Ioannis Mantelos received a B.Eng in Electronics and Computer Engineering (University of Portsmouth 2002) and M.Sc. in Mechatronics (University of Newcastle Upon-Tyne 2003). He is currently with the Intelligent Automation Systems Group, School of Electrical and Computer Engineering, National Technical University of Athens, Greece. He has a professional experience in software system design and implementation for automation applications. His research interests include Robotics, (especially Telerobotics, Web-based control systems, and software systems for mechatronic systems). He is a member of the Greek Technical Chamber.

Ellips Masehian is an assistant professor at the Faculty of Engineering, Tarbiat Modares University, Tehran, Iran. He received the B.S. and M.S. degrees in industrial engineering, both from Iran University of Science and Technology, Tehran, with honors, and a Ph.D. degree from Tarbiat Modares University. His research is focused on the application of heuristic, metaheuristic, and intelligent methods to combinatorial optimization, as well as single and multiple robot motion planning problems. Dr. Masehian has served in Editorial Boards of a number of journals as member, reviewer, and guest editor, and has been in Steering and Program Committees of many international conferences.

Sayyed Farideddin Masoomi received the BE degree in Mechanical Engineering - solids design from Islamic Azad University - Research and Science Branch, Tehran, Iran in 2006, and his MEngSt in Mechanical Engineering from the University of Auckland, New Zealand in 2009. He is currently pursuing the PhD degree and working as a tutor of control laboratory at University of Canterbury in New Zealand since 2010. His current research interests include autonomous mobile robots, biomimetic robots, dynamic modeling, and intelligent control systems.

Paul Milliken is a Scientist at Scion, the New Zealand Forest Research Institute. He received his BE in Mechanical Engineering from the University of Auckland in 1995 and his PhD in Production Technology from Massey University in 1999. He specialises in the development of mechanical and automated systems, computer programming, and mathematical modelling.

Bart Milne received his BE in Electrical & Electronic Engineering from the University of Canterbury, Christchurch, New Zealand in 2008, and his Graduate Diploma in Science in Mathematics from Victoria University, Wellington, New Zealand in 2010. In November 2011, he was awarded a three-year PhD scholarship to develop a teleoperated steep-slope forestry-harvesting machine in partnership with Scion Research and Future Forests Research in New Zealand. His current research interests include semi-autonomous mobile robots, dynamic modeling, and intelligent control systems.

Vu Trieu Minh is a professor at the Department of Mechatronics, Tallinn University of Technology, Tallinn, Estonia. He obtained his B.E of Mechanical from Hanoi University of Technology (HUT) in 1983, M. E of Industrial System Engineering and Ph.D. of Mechatronics from Asian Institute of Technology (AIT) in 1999 and 2004, respectively. He has previously worked in Vietnam, Thailand, Germany, Malaysia, Papua New Guinea, and Estonia. He has authored over thirty research papers and textbooks in the fields of model-based control algorithms, dynamical systems, advanced process control, and automotive engineering. He is a senior member of IEEE/CSS.

Seyed Javad Mousavi is a M.Sc. student of Information Technologies at the Faculty of Engineering, Tarbiat Modares University, Tehran, Iran. He received his B.S. in Physics from Sistan and Baluchestan University, Zahedan, Iran. His research interests are robot motion planning, regrasp planning, and simulation.

Fusaomi Nagata is currently Professor at Department of Mechanical Engineering, Tokyo University of Science, Yamaguchi, Japan. Prior to the university, he was a special researcher working on 3D design, machining and finishing using 3D CAD/CAM, NC machine tools and industrial robots in Fukuoka Industrial Technology Centre from 1988 to 2006. He obtained his D.E. degree in Intelligent Control Engineering from the Faculty of Engineering Systems and Technology, Saga University in 1999.

Takuma Nishimura graduated the Faculty of Advanced Engineering, Nara National College of Technology, Japan, and received his B.A. degree from The National Institution for Academic Degrees and University Evaluation, Japan, in 2010. He received his M.E. degree from Kyoto University, Japan in 2012. He joined NTT WEST from 2012. His research interests include museology, influential medium, recommender system, multiagent system, autonomous system, human-agent interaction, and reinforcement learning. He is a member of The Japanese Society for Artificial Intelligence.

Richard Parker is a Scientist at Scion, the New Zealand Forest Research Institute. He received his BSc in Zoology form the University of Canterbury in 1980 and his PhD from Massey University, New Zealand in 2010. He specialises in developing work practices to enhance productivity and safety in dangerous occupations such as forest harvesting and rural fire fighting. His research includes teleoperation and robotics with emphasis on removing workers from high-risk situations.

Anthony Pietromartire is a guest researcher at the National Institute of Standards and Technology (NIST), a federal laboratory of the United States Department of Commerce. He received a Master's Degree in Computer Science from the University of Burgundy specialized in Databases and Artificial Intelligence. Mr. Pietromartire's current research is focused on simulation-based planning and control for kit building in manufacturing environments. The objective of this problem is to develop the measurement science and standards for planning and modeling by robots so that they are able to be more quickly re-tasked and are more flexible and adaptive.

Balan Pillai, is a professor, holds degrees in engineering, namely Doctor of Science in Technology, from the Aalto University Schools of Technology and Science, and Master of Science (Business Economics) also from the Aalto University School of Economics. Currently he is affiliated as professor in Stanford University. Prof. Pillai has hands-on-experiences: over two-decades in paper and paper machine manufacturing, Security Printing industries, and over a decade of professorships in "hitecs" (that is Advanced Control Engineering and Cybertechnologies). Dr. Pillai is instrumental to several patents at home and abroad. He is a world-renowned speaker on high techs and has published over 140 scientific articles and academic books. Presently, his research area is focused on human-body sciences, including implants and biochips, which are aimed inserted into the human-body systems. Professor Pillai is also associating as a member at the IEEE DEST Committee researches in European Union, South Korea, Australia, and the US. He is also one of the Editorial Board Members at the *IMech, Part E, Institute of Mechanical Engineers*, journal published by SAGE Publication Company, Sussex, UK.

Raazesh Sainudiin received his PhD in Statistics from Cornell University, New York, USA. He was a Research Fellow at the Department of Statistics, University of Oxford, Oxford, UK. Dr. Sainudiin is a senior lecturer in Department of Mathematics and Statistics at University of Canterbury (UC), Christchurch, New Zealand. His research interests include statistical inference of stochastic processes embedded within stochastically evolving networks, such as statistical decision problems in population genetics, phylogenetics, ecological genetics, air-traffic management, and set-valued statistics.

Ken Saito was born in Japan in 1978. He received the B.S. degree in electronic engineering, the M.S. degree in electronic engineering, and the Ph.D. degree in engineering from Nihon University in 2001, 2004, and 2010, respectively. He was a Research Assistant at Department of Physics, College of Humanities & Sciences, Nihon University from 2007 to 2010, and Research Assistant at Department of Precision Machinery Engineering, College of Science and Technology, Nihon University from 2010 to 2011. He is now an Assistant Professor at Department of Precision Machinery Engineering, College of Science and Technology, Nihon University. His current research interests include hardware neural network, and micro-robot. Dr. Saito is a member of IEICE, IEEJ, INNS, and so on.

Vesa Salminen has been involved with several European and domestic research, education, and industrial implementation activities. His expertise areas are innovation leadership, spiral innovation process, hybrid innovation (radical, incremental, and life cycle), interdisciplinary design of mechatronics, industrial service and life cycle business and management of business transition. He has been affiliated as professor and research director at Lappeenranta University of Technology and research and education moderator at HAMK University of Applied Sciences. He also served the period 1999-2001 as senior research scientist at Massachusetts Institute of Technology, MIT, Boston. He is the managing director at a spinoff company Spiral Business Services Corp. that is providing consulting services for spiral innovation in network business environment. He has over 20 years industrial experience on various positions in several enterprises and in industrial organization Technology Industries of Finland.

Craig Schlenoff is the Group Leader of the Cognition and Collaboration Systems Group in the Intelligent Systems Division at the National Institute of Standards and Technology. His research interests include knowledge representation/ontologies, intention recognition, and performance evaluation techniques applied to autonomous systems and manufacturing. He previously served as the program manager for the Process Engineering Program at NIST and the Director of Ontologies at VerticalNet. He leads numerous million-dollar projects, dealing with performance evaluation of advanced military technologies. He received his Bachelors degree from the University of Maryland and his Masters degree from Rensselaer Polytechnic Institute, both in mechanical engineering.

Yoshifumi Sekine was born in Japan in 1944. He received the B.S. degree in electrical engineering and the M.S. degree in electrical engineering from Nihon University in 1966 and 1968, respectively. He is the doctor of engineering. He is now a Professor at Department of Electronic Engineering, College of Science and Technology, Nihon University. His current research interests include hardware neural network and micro-robot. Dr. Sekine is a member of IEICE, IEEJ, IEEE, and so on.

Mathieu Sellier is Senior Lecturer in Fluid Mechanics at the University of Canterbury, in Christchurch, New Zealand. His research interests span a wide range related to the mathematical modelling of fluid flows. Prior to his appointment at the University of Canterbury in 2006, he was a Postdoc at the Fraunhofer-ITWM in Kaiserslautern, Germany, developing numerical tools for the modelling and optimization of glass forming technologies. He got his PhD from the University of Leeds in 2006 for a thesis on the modelling of thin film flows and droplets on complex surfaces.

Katsunori Shimohara received the B.E. and M.E. degrees in Computer Science and Communication Engineering and the Doctor of Engineering degree from Kyushu University, Fukuoka, Japan, in 1976, 1978, and 2000, respectively. He was Director of the Network Informatics Laboratories and the Human Information Science Laboratories, Advanced Telecommunications Research Institute (ATR) International, Kyoto, Japan. He is currently a Professor at the Department of Information Systems Design, Faculty of Science and Engineering, ant the Graduate School of Science and Engineering, Doshisha University, Kyoto, Japan. He is also Director of the Research Center for Relationality-Oriented Systems Design, Doshisha University. His research interests include human communication mechanisms, evolutionary systems, human-system interactions, genome informatics, and socio-informatics. Dr. Shimohara is a member of IEEE, the Institute of Electronics, Information and Communication Engineers, the Japanese Society of Artificial Intelligence, the Human Interface Society, and the Society of Instrument and Control Engineers.

Zengqi Sun received the Ph.D. degree in Control Engineering in 1981 from the Chalmas University of Technology in Sweden. He is currently a Professor in the Department of Computer Science and Technology, Tsinghua University, China. His current research interests include intelligent control, robotics, fuzzy systems, neural networks, and evolution computing, etc.

Keiki Takadama received his M.E. degree from Kyoto University, Japan, in 1995, and got Doctor of Engineering Degree from the University of Tokyo, Japan, in 1998, respectively. He joined Advanced Telecommunications Research Institute (ATR) International from 1998 to 2002 as the visiting researcher and worked at Tokyo Institute of Technology from 2002 to 2006 as a lecturer. He moved to The University of Electro-Communications as associate professor in 2006 and is currently a professor from 2011. His research interests include multiagent system, autonomous system, human-agent interaction, reinforcement learning, learning classifier system, and emergent computation. He is a member of IEEE and a member of major AI- and informatics-related academic societies in Japan.

Minami Takato received the bachelor and master degrees from Nihon University in 2010 an 2012, respectively. Her research is the ceramic material and process applying for micro energy systems and micro-robot systems. She is currently candidate for the doctoral degree in engineering at Nihon University.

Ivan Tanev earned M.S. (with honors) and Ph.D. degrees from Saint-Petersburg State Electrotechnical University, Russia in 1987 and 1993, respectively, and Dr.Eng. degree from Muroran Institute of Technology, Japan, in 2001. He has been with the Bulgarian Space Research Institute (1987), Bulgarian Central Institute of Computer Engineering and Computer Technologies (1988-1989), Bulgarian National Electricity Company (1994-1997), Synthetic Planning Industry Co.Ltd., Japan (2001-2002), and ATR Human Information Science Laboratories (2002-2004), Japan. Since April 2004, he has been an associate professor at the Doshisha University, Japan. Dr. Tanev's research interests include evolutionary computations, multi-agent systems, and socioinformatics.

Spyros G. Tzafestas, an NTUA emeritus Professor and leader of the IAS research group since 2006, received B.Sc. in Physics, P.G. Diploma in Electronics (Athens University, 1963, 1964); D.I.C. and M.Sc. in Control (Imperial College, London University, 1966); Ph.D. (1969) and D.Sc. (1978) in Systems and Control (Southampton University). He is recipient of D.Sc. Eng. (Honoris Causa), Intl. Univ. Foundation (1987), Dr.-Ing. E.h., T.U. Munich (1997) and Docteur Honoris Causa, E.C. Lille (2003). From 1969 to 1973, he was a research leader at the Nuclear Research Center "Demokritos" (Athens), from 1973 to 1984, he was professor of systems and control engineering at Patras University, Greece, and from 1985 to 2006, he was professor of robotics and control at the National Technical University of Athens (NTUA), Greece. He has been Director of the NTUA Institute of Communication and Computer Systems (ICCS) for the period 1999-2009. He has published 30 research books and over 700 technical papers. He is the founding editor of the *Journal of Intelligent and Robotic Systems*, and editor of the Springer book series on Intelligent Systems, Control, and Automation. His research interests include control, intelligent systems, robotics, and automation. He is a Life Fellow of IEEE, a Fellow of IEE (now IET), and has received many awards for his achievements. His experience includes the organization of several conferences on intelligent systems, robotics, and control, and the leadership of several European R&D projects in IT, Control, Robotics, and Automation.

Fumio Uchikoba received the bachelor degree from Waseda University in 1983. He received the master degree from University of Electro-Communications in 1985. He entered the TDK Corporation in 1985. He was a visiting scientist at the Massachusetts Institute of Technology from 1988 to 1990. He joined department of precision machinery engineering, College of Science and Technology, Nihon University in 2003 as an associate professor. He is currently a professor in same university. He is the doctor of engineering.

Hiroo Wakaumi received his B.E. degree from Chiba University in 1973 and his Ph.D. degree from Tokyo Metropolitan University in 2000. Until 1992, he worked at NEC Corporation. Then he joined Tokyo Metropolitan College of Technology, whose name has been changed to Tokyo Metropolitan College of Industrial Technology in 2006, where he is currently a professor. His research interests include CCDs, display devices, high-voltage MOS ICs, magnetic sensing, optical sensing, and signal processing. He has published 44 papers in journals and international conferences, three books, and 80 general meeting papers including bulletins and magazines. He is now each senior member of IEEE and IEICE (The Institute of Electronics, Information, and Communication Engineers).

Keigo Watanabe is currently Professor at Department of Intelligent Mechanical Systems, Graduate School of Natural Science and Technology, Okayama University. Prior to the university, he was Professor of Department of Advanced Systems Control Engineering, Saga University, from 1998 to 2009. He obtained his D.E. degree in Aeronautical Engineering from Kyushu University in 1984.

Tomohiro Yamaguchi received his M.E. degree from Osaka University, Japan, in 1987. He joined Mitsubishi Electric Corporation in 1987 and moved to Matsushita Electric Industrial in 1988. He worked at Osaka University from 1991 to 1998 as a research associate and got Doctor of Engineering Degree from Osaka University in 1996. He moved to Nara National College of Technology as associate professor in 1998, and is currently a professor from 2007. His research interests include interactive recommender system, music information retrieval, multiagent reinforcement learning, autonomous learning agent, human-agent interaction, learning support system, human learning process, and mastery process. He is a member of The Japanese Society for Artificial Intelligence and The Society of Instrument and Control Engineers, Japan.

Huei Ee Yap received his B.Eng. degree in mechatronics engineering and information technology from the Australian National University, Canberra, Australia in 2006 and his M.Eng. degree in pure and applied physics from Waseda University, Tokyo, Japan, in 2009. He is currently working towards his PhD degree in pure and applied physics in the Graduate School of Advanced Science and Engineering, Waseda University, Tokyo, Japan. His research interests are in dynamic systems, non-linear system analysis, and control and embedded robotics.

Sho Yoshitake is a master course student at Graduate School of Science and Engineering, Tokyo University of Science, Yamaguchi, Japan. His current research interest is the robotic CAM system without using any robot language.

Index

CPSIA information can be obtained at www.ICGtesting.com
Printed in the USA
LVOW09*0219241013

358273LV00045BC/434/P

9 781466 642256